Practical Relativity

Practical Relativity

From First Principles to the Theory of Gravity

RICHARD N. HENRIKSEN

Queen's University, Kingston, Ontario

WILEY

A John Wiley and Sons, Ltd., Publication

This edition first published 2011

Registered office
John Wiley & Sons Ltd, The Atrium, Southern Gate, Chichester, West Sussex, PO19 8SQ, United Kingdom

For details of our global editorial offices, for customer services and for information about how to apply for permission to reuse the copyright material in this book please see our website at www.wiley.com.

Library of Congress Cataloguing-in-Publication Data
Henriksen, R. N.
Practical relativity : from first principles to the theory of gravity / Richard N. Henriksen.
p. cm.
Includes bibliographical references and index.
ISBN 978-0-470-74142-9 (cloth) – ISBN 978-0-470-74141-2 (pbk.) 1. Relativity (Physics) I. Title.
QC173.58.H36 2010
530.11 – dc22

2010023306

A catalogue record for this book is available from the British Library

ISBN: 978-0-470-74142-9 (H/B) 978-0-470-74141-2 (P/B)

Typeset in 10/12 Times by Laserwords Private Limited, Chennai, India
Printed and bound in Singapore by Markono Print Media Pte Ltd.

Cover images: Dice image from *Sights That Einstein Could Not Yet See – visualization of relativistic effects.* Ute Kraus, Hanns Ruder, Daniel Weiskopf, Corvin Zahn. *Latticework and Clocks* from Spacetime Physics, 2e by E.F Taylor and J. A. Wheeler © 1992 by Edwin Taylor and John Archibald Wheeler, used with permission.

To the memory of Gertrude and Norman, Emma and Halter;
and to Judith who is life's force incarnate.

Contents

Preface

This book is entitled *Practical Relativity*. Many of you (I hope that there will be many) will wonder why, once confronted with the dense forest of equations. I can only say that in places I have used words to explain conceptual points. However I think that what makes it 'practical' rests primarily on two other things. I have started at the beginning of the subject and gone nearly to the end. Moreover, insofar as I have been able to navigate between tedium and necessity, I have included all of the steps that lead to important results. This is, I think, a characteristic of 'practicality'. Both of these themes should be appreciated by serious students. Problems are included that elucidate the ideas, and these should be appreciated by professors. A solutions manual, containing answers to the problems, is available at www.wiley.com/go/henriksen.

My approach has been to regard fundamental principles 'eye to eye', so that any possible alternatives to the traditional arguments may become evident. Most derivations are from first principles. I have not cluttered the book with every possible application of the theory, but the grand classics are present. I believe, however, and I hope that you will agree, that the presentation of the necessary techniques has been comprehensive.

I have not used the latest mathematical treatments of vectors and tensors, as found for example in the Cartan calculus. My approach has been to remain as close as possible to familiar concepts of vectors, tensors and reference systems in the hope of capitalizing on received wisdom. I believe that this is another practical aspect.

There are many books on this subject, and in the course of my writing I have enjoyed reading many of them. They are referred to throughout the book. I do believe that the present book is not quite the same as the others, mainly due to the attempt to distinguish the positivist approach from the theoretical. While one measures, the other imagines although in the end the loop must be closed. I have also attempted to cast light on dark corners. I have enjoyed exploring the corners, and I hope that this book will also help you to explore and enjoy them.

"Why, I'd like nothing better than to achieve some bold adventure worthy of our trip."

Aristophanes, 450-385 BC

Acknowledgements

This book would not have been written without the opportunity to teach these subjects to many bright students, primarily at Queen's University. It could not have been written without the tradition of creative research that should be the pride of Queen's University. Colleagues set the standard, and I have been fortunate with these. The errors are of course mine. The inspiration is due to my wife Judith Irwin, who is a hard but loving taskmaster. Much credit is due to Wiley for overseeing the production process, as well as to Laserwords for laying out the pages elegantly.

RNH

Introduction

This book is written in six long chapters. The intention was to make each chapter a logical step on the way to relativistic electromagnetism and gravity, subjects that are the province of the last two chapters.

The first chapter starts from simple considerations of reference frames and vectors. The positivist attitude is emphasized. It is written entirely in the physical context of classical (Newtonian) space-time and mechanics. However, the notion of coordinate independence of the physical description leads inexorably to the apparatus of differential geometry. This is done deliberately, so that the reader will become accustomed to the formalism of relativity in an intuitive geometrical context.

The mystery of inertial frames is discussed at length in this chapter, with some connection to modern ideas. The discussion includes non-inertial frames, and the transformations between them. This leads to the introduction of time in the coordinate transformations and to a brief discussion of absolute time. Rotation matrices and angular velocity matrices are used to write Newton's second law in accelerated frames of reference. Contact is made with other, considerably less explicit, notation. Finally we emphasize that the necessity to define parallel transport of a vector already exists in Euclidian curvilinear coordinates. This is presented in a familiar (if awkward) notation, so that it is readily recognized later. This chapter assumes a familiarity with classical ideas at the level of advanced mechanics and neither is, nor was meant to be, gentle. It may be best to study it selectively.

The second chapter is devoted to the derivation of the Lorentz transformations in two distinct ways. The first method concentrates on the derivation of the Lorentz transformations as those transformations of space and time that leave the wave equation for light invariant. Considerable discussion is devoted to finding what the results would be, if other equations were taken as the source of the fundamental invariance. The essential step of allowing a time transformation, rather than insisting on absolute time, is shown to distinguish these transformations from earlier versions by Voigt and Poincaré.

After this derivation the question arises as to why such an invariance group should apply to all events in space-time. This question is answered by the operational or positivist derivation of the transformations first given by Einstein. We give a version that is based on light-clocks and the transformation of straight lines in a space-time diagram. We argue that such a linear transformation must be accompanied by a 'units transformation'. These scale factors are the usual, time dilation and Lorentz-Fitzgerald contraction. Putting these two concepts together yields the Lorentz transformations. Because of the maximal and invariant nature of the speed of light that is assumed 'a priori' in this approach, it is really a theory of ideal measurement. So long as what one can measure is reality, the implications of the transformations are 'real'.

The third chapter details many of the usual applications of the Lorentz transformations, together with some discussion that is perhaps rarer. Time dilation, the Döppler shift and the twin paradox start off the chapter. There are some astronomical applications. Time on a rotating disc is examined in the context of the Sagnac effect. The Lorentz-Fitzgerald contraction is discussed largely in terms of standard paradoxes, but once again the rotating disc is found to be instructive. Some gentle speculation is allowed here, since the topic has a history of errors. The velocity transformation is introduced and used to define the phenomenon of aberration of beams of relativistic particles. The limit is taken for photons and so the inherent transformation of angles appears, which is optical aberration.

Under the heading of geometrical optics we discuss such topics as Thomas precession and the appearance of moving objects. The derivations are not the most elegant possible! However, they do have the merit of revealing the essential unexpected phenomenon in the homogeneous Poincaré group. The astronomical phenomenon of 'light echos' is also introduced and then argued to be important using examples. A final topic in the chapter is dynamics with prescribed acceleration. This requires the transformation of particle acceleration between inertial observers. Hyperbolic motion is presented as an example of the horizon phenomenon.

In the fourth chapter we introduce Minkowski space-time and adapt all of the results of Chapter 1 regarding vectors and tensors to the four dimensions of space and time. At this stage we emphasize that it is not obligatory to conceive of space-time as a Minkowskian manifold, but that it is terribly convenient. We demonstrate this by rederiving the Lorentz transformations based on this principle in two ways. We assume first that metric moduli are invariant, and then that the metric itself should be invariant after synchronization. We also show that the four-vector treatment of velocity and acceleration allows previous results on their three-vector transformation properties to be readily obtained. These discussions serve principally to demonstrate the internal consistency of the Minkowskian metric space.

In the absence of real forces, we introduce the Lagrangian and Hamiltonian for a free particle and infer the momentum and energy. We show how one may impose constraints on the motion of a free particle to approximate relativistic forces. This is all done in generalized coordinates as well as Galilean coordinates. After deriving the action for a free particle, we observe that the Euler-Lagrange equations are equivalent to geodesic equations in the given metric. This leads to the equation of motion of a free particle that holds in any pure metric theory. Finally, the collisions of free particles are treated in terms of the conservation laws. The principles are extended to photons and applied to Compton and inverse-Compton scattering.

The fifth chapter is technically more challenging, but perhaps also more practical. The four-vector electromagnetic theory is presented in Galilean coordinates. Then in the traditional 'three plus one' split into space and time, Lagrangian and Hamiltonian methods for solving particle motion are introduced with examples. Many of the examples are important classics and some of them are solved in several different ways in order to elucidate the methods. The Lorentz equation of motion and the principle of relativity are used together to infer the transformation of electric and magnetic fields in an elementary way. One sees that these vectors are part of a larger object.

Next the three plus one split is abandoned, and Galilean four-vectors are used exclusively. This leads to the field tensor, electromagnetic field invariants and the tensor form of the field equations. The latter result requires, in part, a discussion of the action that holds for the matter coupled to the fields, when the vector potential is varied.

As a means of transiting to gravitational metrics, the Maxwell equations are generalized to metrics for which all components are in principle functions of space and time. Such dependence includes curvilinear coordinates and non-inertial coordinates, but the metric can also reflect a curved manifold. Finally, in this section, the Landau and Lifshitz approach based on a (locally) diagonalized metric is used to write the Maxwell equations in a recognizable form. This material is rather advanced and can be omitted without subsequent damage. It does, however, represent a useful exercise in the use of vectors, tensors and their duals. The final form of the equations can be used to discuss electromagnetism near rotating black holes and neutron stars.

Part II of the book deals with the implications of metrics that describe various gravitational fields. It is contained in one long Chapter 6. The chapter has a long prologue that is meant to introduce qualitatively the nature of the metric theory and its uses. The reader is free to pass on and let these speak for themselves.

The first major section explores the metric representation of a weak gravitational field. This is where the contact with Newtonian theory is established. The gravitational and cosmological frequency shifts are discussed in this section in order to form a complete set of such shifts, but their general nature is emphasized. Later the gravitational frequency shift is given a more general treatment. Simple tests are discussed briefly, with emphasis on the GPS system.

The next section deals with the general form of static and stationary metrics. The Lagrange equations are used to find the Christoffel symbols when the metric is spherically symmetric. The Hamilton-Jacobi and Eikonal equations are introduced for general metrics. These are used to discuss the energy of a particle and the proper frequency of a photon.

The third section presents the metrics for two of the best known and important strong gravitational fields, due to Schwarzschild and Kerr. Each metric has its own subsection. The nature of the Schwarzschild horizon is clarified by introducing freely-falling (inertial) observers, following LeMaître. We see that this field of inertial observers is completely determined by the metric in Schwarzschild form. The classic calculations of orbital precession and light deflection are given in detail in two independent approaches.

The Kerr metric is less manageable, but we find the meaning of its horizon and ergosphere. Frame dragging and energy extraction are discussed, as is the upper limit to the specific angular momentum.

In the last two sections we give first the conservation laws of matter as the true divergence of the energy (density)-momentum (flux) tensor. This is then used in the discussion of the matter sources of the gravitational metrics.

Following Gauss and Riemann, the curvature is identified as the distinction between what is merely the Minkowski metric in generalized coordinates and the gravitational metrics. This leads to defining the Riemann, Ricci and Einstein tensors. The Riemann curvature tensor is shown to be equivalent to the non-commutation of the second-order true derivatives. The Einstein equations are given and the Bianchi identities are shown to be essential to the conservation laws of matter. Finally, a brief discussion of the modification necessary to include the cosmological constant (or vacuum energy density from another point of view) is given.

No detailed calculations using the Einstein equations are presented. These are left to other texts, although in principle the reader has the techniques with which to launch himself into the heart of this grand subject.

Part I

The World Without Gravity

www.wiley.com/go/henriksen

1
Non-Relativity for Relativists

Dura lex, sed lex (The law is hard but it is the law)

1.1 Vectors and Reference Frames

In this section we discuss our fundamental concepts as drawn from experience. This ends in frustration since experience is approximate, most things are known relative to other things, and our concepts often seem to be defined in terms of themselves. Thus 'fundamental' argument resembles the circular snake devouring its tail (the *Ouroboros*). However we must make a beginning, and so we confront our first definition and its algebraic implications.

What is an inertial reference frame? I prefer to parse this question into two principal questions. By 'reference frame' we mean some well-defined system of assigning a measured time and a measured position to an 'event'. For the moment an 'event' is point-like, as for example the time at which a particle or the centre of mass of an extended body takes a particular spatial position. The reference frame also implies an 'observer' who records the measurements. The resulting numbers are the 'coordinates' of the event in this reference frame. By 'inertial' we mean a reference frame in which Newton's first law of motion[1] applies to sufficiently isolated bodies. This axiom requires not only that the coordinates of a body be determinable from moment to moment, but also that fixed spatial directions be defined. Neither one of these definitions is particularly exact or obvious and yet they are fundamental to our subject. Thus we continue their exploration in the next two sections.

[1] To paraphrase: 'A body either remains at rest or continues with constant speed in a straight line, unless acted upon by external forces'. Newton's second law also holds only in an inertial frame for 'real' forces.

Practical Relativity: From First Principles to the Theory of Gravity Richard N. Henriksen

1.1.1 Reference Frames

Although this is not strictly necessary, location is normally specified relative to a set of objects that have no relative motion between them. Some fixed point within this set of objects is chosen as the reference point or 'origin' from which all distances are measured. On small enough scales that we can reach continuously, the measurement is made by placing a standard length along a straight line between the points of interest. We call this standard length a 'ruler' or a 'unit' and we assume that we can determine a 'straight line'. On larger scales, various more subtle methods are required.

Our most familiar example is the Earth itself. On small scales we have no difficulty in establishing a rigid frame of reference by assuming Euclidean geometry. That is, we assume that the Earth is 'flat' so that trigonometry and an accurate ruler suffice to measure distance. When lasers are used we are assuming that even the near space above the surface of the Earth is Euclidean and that light follows the straight lines. On larger scales the Earth is found to be a sphere, so that its surface does not obey Euclidean geometry. Position has to be assigned by latitude and longitude, which requires the use of a combination of accurate clocks and astronomical observations in the measurements. Distance is computed between points using the rules of spherical trigonometry, rather than the Euclidean rule of Pythagoras[2] (see e.g. Figure 1.1).

The Earth is not exactly a rigid sphere, but a global reference frame precise enough to detect this fact became generally available only with the advent of the Global Positioning System (GPS) of satellites. This remarkable development, based on multiple one-way

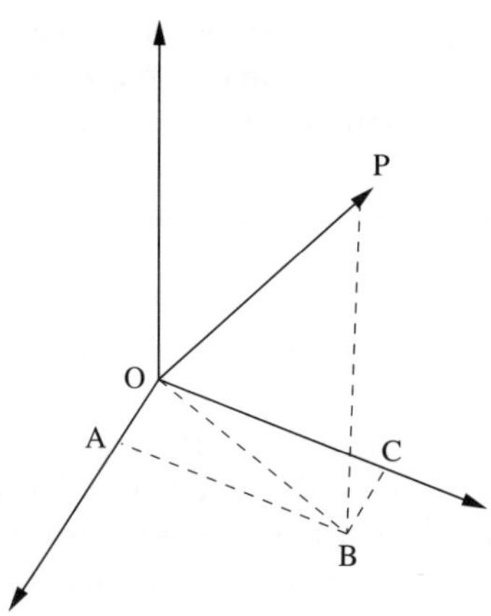

$$OB^2 = AB^2 + OA^2$$
$$OP^2 = OB^2 + BP^2$$
$$OP^2 = AB^2 + OA^2 + BP^2$$

Figure 1.1 *The three perpendicular axes emanating from O are reference directions. Each axis is rigid and the projections of OP on these axes furnish the Cartesian weights or components. The theorem of Pythagoras gives the distance OP in terms of these*

[2] It is not clear how many cultures discovered this result, but in European cultures it is generally recognized by this name. It was apparently known in ancient Mesopotamia [1].

radar ranging, has allowed us to measure the ebb and flow of oceans and continents in a non-rigid, spheroidal global frame. However, it assumes principles that we have yet to examine, and that will be the subject of much of this book.

Thus the procedure to define a 'rigid' frame of spatial reference always involves assumptions about the nature of the world around us, and it is these that we must carefully examine subsequently. Moreover such a reference frame is always an idealization. Errors are involved in determining practical spatial coordinates on every scale, so that our knowledge of distance is always approximate. Moreover the degree of idealization increases with spatial extent of the reference frame, as it becomes progressively more difficult to maintain rigidity.

In parallel with spatial position, we have managed recently to establish a global measure of time that allows us to say whether or not events occurred simultaneously. This means that a single number can be assigned to a global point-like event (e.g. the onset of an earthquake in China or sunrise at Stonehenge on Midsummer's Day). The number is assigned by each of a network of synchronized atomic clocks distributed over the reference frame of the Earth. The sequence of such numbers defines 'coordinate time' for the Terrestrial Reference Frame. The difference between such numbers that encompass the beginning and end of an extended event (such as a lifetime) may be called a 'duration' for brevity. In practice, only durations of finite length are meaningful since no measurement can be made with infinite precision, but we normally assume that they can be arbitrarily small in principle. Figure 1.2 shows an ideal rigid reference frame with synchronized clocks at each spatial point.

The creation of a terrestrial coordinate time has been accomplished through the global synchronization of atomic clocks (within limits) rather than by astronomical measures such as day count and Sun angle. The latter does not establish a global reference time as any 'jet-lagged' traveller knows well! Once again this global clock synchronization involves principles and corrections that we have yet to discuss, but which will be one of our principal preoccupations.

Our direct experience of time tends, however, to be local rather than global. It is an event that includes oneself whose duration is measured by our clock, our schedule, our heart beat or our ageing process. Such local time is proper to us and is generally referred to as 'proper time'. The 'origin' of either coordinate or proper time is just as arbitrary as is the spatial origin, and may be chosen for convenience.

There are many reasons, however, why proper time does not run at the same rate as coordinate time. These reasons are physical as well as psychological. One physical reason is that our bodies age according to a thermodynamic time measured by increasing entropy, and the rate is different for different individuals. Another is the differing set of inertial frames that an individual occupies relative to the terrestrial reference frame. This unexpected dependence we shall explore in subsequent sections. The psychology of time is not within the competence of this author, but 'apparent' proper time is notoriously variable!

The complications involved in defining reference frames have been elegantly revealed by our exploration of the solar system. The planets do not form a rigid system of reference. A global reference frame on Mars moves relative to a global reference frame on the Earth, so that a rigid reference frame encompassing the two is not possible. One solution is to construct an imaginary rigid frame whose origin is at the centre of the Sun. The three independent directions required to encompass all space in the Cartesian fashion are not fixed in the Sun, which is not rigid either, but rather with respect to very distant objects in the Universe (such as quasars) that appear fixed to us. Coordinates determined along these directions are useful to determine the momentary position of the centres of mass of the planets. Ultimately, however, we are forced to have recourse to systems proper to each planet, such as latitude and longitude for the Earth, and these are neither fixed nor constantly oriented with respect to the Cartesian reference axes.

Time measurements in the solar system have also revealed difficulties with a pan-planet coordinate time. For example, assuming nothing faster than our electromagnetic signals, Martian events happen later for an Earth observer than they do for a Martian observer such as a Martian Rover Vehicle (and vice versa for Earth events observed on Mars, such as the initiation of a command signal on the Earth). Electromagnetic signals propagate in a vacuum with the speed of 'light', which is almost universally labelled as c and which has the approximate numerical value 0.2998 metres per nanosecond (we know it to much greater accuracy). Thus although we can experience a Martian duration delayed by the travel time of our signals (and slightly distorted due to the motion of Mars relative to the Earth), we cannot share proper times. Moreover there can be no electromagnetic connection between the Earth and Mars during this travel time.

There is, then, since at present c is the fastest signal we know, a causality gap wherein nothing on Earth can affect Mars and vice versa. This a-causal gap varies roughly from 4 minutes to a little less than 12 minutes depending on the relative positions of Earth and Mars. We have met such an effect previously when astronauts were on the Moon, but the gap was only of the order of two seconds. Our intercontinental calls by way of satellites in synchronous orbit have an a-causal gap roughly equal to a third of a second, which is barely noticeable in conversation.

One might think that by using atomic clocks synchronized on Earth and Mars we could agree on simultaneous events after the fact at least, and so establish a pan-planet coordinate time, which would be 'absolute' in the solar system. However, we shall see that even the most perfect atomic clocks cannot remain synchronized in the presence of relative velocity between reference frames, provided that signals of only finite speed are available to us.

The sort of reference frame that we can construct at the centre of the Sun is composed of an inferred origin plus geometric straight lines and it has no proper 'observer'. Time and space in this frame are constructed from events measured by atomic clocks and observers located elsewhere, after correction to the solar origin. These corrections are an example of a general mapping from local coordinates to 'generalized' coordinates at the centre of the Sun. Such mappings will be discussed in greater detail below. Although useful as fictitious standards and widely used in the theory of gravity, these virtual reference frames are distinct from a tangible reference frame that is inhabited by 'observers' capable of measuring the coordinates of events directly.

The conclusion to this discussion so far may be summarized algebraically by stating that a reference frame allows an observer to assign coordinates to point-like events according to

$$\{x^a\} \equiv \begin{pmatrix} t \\ q^1 \\ q^2 \\ q^3 \end{pmatrix}. \tag{1.1}$$

The notation on the left indicates a set of four quantities as a takes on the successive values $\{0, 1, 2, 3\}$, equal to the set of quantities in the column four-vector on the right (in order beginning at the top). Thus $x^0 = t$, $x^1 = q^1$ and so on. If there is any danger of confusing the raised indices with powers in a given context, we will enclose them in brackets. For brevity we write the column vector usually as x^a.

The quantity t is simply the coordinate time for the reference system and the set $\{q^i\}$ where $i = 1, 2, 3$ give the spatial position. Generally curly brackets are meant to indicate a set, but more usually they are simply understood. These may be the familiar Cartesian set $\{x, y, z\}$ (see Figure 1.3) or they might be spherical polar coordinates $\{r, \theta, \phi\}$ (θ is co-latitude, ϕ is longitude and r is the distance from the origin); or in fact any other set of three numbers that defines a spatial position. As such they are 'generalized coordinates'.

We shall use this convention whereby letters early in the alphabet (before h) shall take on four values 0, 1, 2, 3 as above for x^a, while those later in the alphabet will run from 1 to 3, as above for q^i. All four quantities in x^a may be taken as pure numbers, each

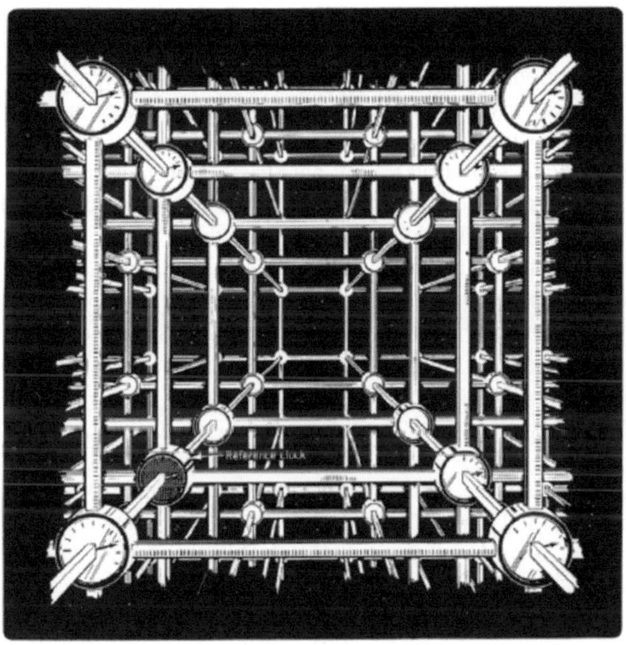

Figure 1.2 *After a rigid spatial frame of reference is established locally by measurement and synchronization, it might appear as shown in this cartoon. Each ruler indicates a unit of distance and any point on the grid is located with three numbers giving the three independent spatial steps relative to the reference point. The fourth number is the coordinate time, which is the same over the grid. The reference point is shown as having the reference clock with which all of the other clocks are synchronized. Extended to infinity, the grid is the instantaneous world of the reference observer O and friends. It is their inertial frame of reference. Source: Reproduced with permission from Taylor & Wheeler,* Spacetime Physics *(1966) W.H. Freeman & Company (See Plate 1.)*

giving the value of the corresponding quantity in terms of standard units when length or time is involved, or giving the radian measure for angles.

By space we mean primarily the relative position of events, and especially the distance between them.[3] We can locate a particular point or position by a three vector, called a position vector, that may be written as the column vector x^i (i.e. a 3×1 matrix)

$$\mathbf{r} \equiv \begin{pmatrix} x \\ y \\ z \end{pmatrix} \tag{1.2}$$

in Cartesian coordinates. As we have discussed previously, such coordinates are relative to an origin and to a choice of three orthogonal fixed directions. A notation that emphasizes this is

$$\mathbf{r} = x\widehat{\mathbf{e}}_x + y\widehat{\mathbf{e}}_y + z\widehat{\mathbf{e}}_z. \tag{1.3}$$

It should be emphasized that when an entire vector is distinguished by an index, the index does not refer to a component of the vector, but is rather the name of that vector. Such indices are frequently placed in brackets to indicate this distinction, but we shall try to avoid this notational complication except when absolutely necessary.

The vectors $\widehat{\mathbf{e}}_i$ have only directional information since they each have a standard unit magnitude, but together they define three orthogonal directions, which are the Cartesian axes as labelled by their subscripts. They are strictly constant vectors, since each indicates the same direction at every point in space if it is transported parallel to itself. This 'parallel transport' is essentially defined by keeping these vectors pointing at the distant objects used to define the Cartesian axes, which are sufficiently distant that there is no parallax (apparent motion) during the displacement.

The three unit vectors in Equation (1.3) are the Cartesian coordinate 'base vectors' since they clearly possess the property (here we denote $\{x, y, z\}$ by i or j as each takes on the respective values $\{1, 2, 3\}$)

$$\widehat{\mathbf{e}}_i \cdot \widehat{\mathbf{e}}_j = \delta_{ij}, \tag{1.4}$$

where the 'Kronecker delta' δ_{ij} is the $(ij)^{th}$ element of a diagonal 3×3 matrix that has ones on the diagonal and zeros elsewhere. It is the component form of the 'identity matrix' (sometimes called the 'unit matrix'), which we write as

$$\underline{\underline{1}} = \begin{pmatrix} 1, 0, 0 \\ 0, 1, 0 \\ 0, 0, 1 \end{pmatrix}. \tag{1.5}$$

The double underline notation generally signifies a 3×3 matrix. This matrix operating respectively with any vector or any 3×3 matrix returns the vector or the matrix unchanged.

[3] For the Ancients, pure space is the 'void' or vacuum and has no particular physical properties. It is the theatre for events concerning objects. Newton invented 'absolute space' to define an absolute standard of rest. The quantum ground state is almost a return to this.

Property (1.4) allows any other vector **A** with arbitrary magnitude and direction to be written as a weighted sum over these base vectors according to

$$\mathbf{A} = A_x\widehat{\mathbf{e}}_x + A_y\widehat{\mathbf{e}}_y + A_z\widehat{\mathbf{e}}_z, \tag{1.6}$$

where evidently by the property (1.4) the 'weights' are found from

$$A_i \equiv \widehat{\mathbf{e}}_i \cdot \mathbf{A}. \tag{1.7}$$

These 'weights' are in fact the Cartesian 'components' of **A**, and they give the magnitude of the arbitrary vector as

$$A \equiv |\mathbf{A}| \equiv \sqrt{\mathbf{A}\cdot\mathbf{A}} = \sqrt{A_x^2 + A_y^2 + A_z^2}, \tag{1.8}$$

and its direction by the set of 'direction cosines' A_i/A. The coordinates of a spatial point (x, y, z) are the weights of the position vector in the Cartesian reference system.

The 'displacement vector' between two widely separated positions is given by

$$\Delta\mathbf{r} = (\Delta x)\widehat{\mathbf{e}}_x + (\Delta y)\widehat{\mathbf{e}}_y + (\Delta z)\widehat{\mathbf{e}}_z, \tag{1.9}$$

so that $(\Delta x, \Delta y, \Delta z)$ are the Cartesian components or weights of the displacement vector. The magnitude of this vector gives the distance $\Delta\ell = |\Delta\mathbf{r}|$ between the two points according to the theorem of Pythagoras. This is essentially why the magnitude and the vector 'dot product' are defined as they are. They contain the theorem of Pythagoras for the measure of distance between two points along the straight line joining them, in Euclidean space. Normally the theorem is defined in two dimensions, but the extension to three dimensions is immediate by projection (e.g. Figure 1.1).

It is more useful from the point of view of using displacements along curves to consider infinitesimal steps. These may be added up finally to give the total finite distance ℓ along the curve, by considering each of $\{x, y, z\}$ to be a parametric function of the arc length ℓ. That is we work with the differential form of Pythagoras as

$$d\ell^2 \equiv d\mathbf{r}\cdot d\mathbf{r} \equiv d\mathbf{r}^2 = dx^2 + dy^2 + dz^2, \tag{1.10}$$

where the last expression assumes Cartesian coordinates.

We can conveniently write the infinitesimal displacement vector $d\mathbf{r}$ in the form

$$d\mathbf{r} = \frac{\partial\mathbf{r}}{\partial x}dx + \frac{\partial\mathbf{r}}{\partial y}dy + \frac{\partial\mathbf{r}}{\partial z}dz, \tag{1.11}$$

since in these very special Cartesian coordinates the partial derivatives are simply equal to the corresponding base vectors (e.g. Equation (1.3)). Moreover the base vectors are normalized and orthogonal in Cartesian coordinates, so we have directly from Equation (1.11) $d\mathbf{r}^2 = d\ell^2$ consistently with Equation (1.10).

In general the prescription for the measure of distance between two close points in terms of the respective coordinates of the two points is called a 'metric'. In Euclidean space the differential form of the Pythagorean metric is always fundamental, and the different forms it may take depend only on the choice of coordinates.

Whenever, as for the Earth, the shape of a boundary surface of an object is not planar, then even if we continue to believe that the 'space' between objects is Euclidean, we may want to use coordinates more suited to the curved boundary than to the space. This implies that we introduce 'curvilinear' coordinates $\{q^i\}$, such that a constant value of one of these coordinates coincides with the curved surface. This would be the constant radius for a spherical Earth. Such coordinates are an example of generalized coordinates q^i.

In order for generalized coordinates to be acceptable for a physical observer, there should be a smooth 'one to one' mapping between these and the Cartesian coordinates. This mapping takes the form[4]

$$x = x(\{q^i\}), y = y(\{q^i\}), z = z(\{q^i\}), \tag{1.12}$$

since every point of space has one and only one Cartesian position. Thus the position vector of a point can be assumed to have the functional form $\mathbf{r}(\{q^i\})$. In a Cartesian reference frame this is explicitly

$$\mathbf{r} = x(\{q^i\})\widehat{\mathbf{e}}_x + y(\{q^i\})\widehat{\mathbf{e}}_y + z(\{q^i\})\widehat{\mathbf{e}}_z. \tag{1.13}$$

Normally the motion of a point is included in the functions $q^i(t)$, as is motion of a point particle in Newtonian dynamics. Occasionally t may occur explicitly in the transformations (1.13), but this is almost always due to a relative motion between the two reference frames, and will be ignored until later in this chapter.

Thus $d\mathbf{r}$ could be calculated by partial differentiation of Equation (1.13) with respect to the $\{q^i\}$, since the Cartesian base vectors are constant vectors. But the resulting components would be in the Cartesian reference system rather than in the generalized reference system.

However, generalized coordinates come with their own preferred directions in space. These directions are the natural ones to choose for the base vector directions. When this choice is made, we speak of a 'coordinate basis'. Our previous special choice $\{\widehat{\mathbf{e}}_x, \widehat{\mathbf{e}}_y, \widehat{\mathbf{e}}_z\}$ constitutes a Cartesian coordinate basis. This basis is 'normalized' and 'orthogonal' since the base vectors are both normalized to unity and are mutually orthogonal, which state we will term 'orthonormal' henceforward for brevity.

The simplest way to define generalized base vectors is by analogy with Equation (1.11), where we replace the Cartesian coordinates by the generalized coordinates. Consequently

$$d\mathbf{r} = \frac{\partial \mathbf{r}}{\partial q^i} dq^i. \tag{1.14}$$

In this last equation we introduce the Einstein summation convention in order to shorten our descriptions. This regards a product of terms in which there are repeated alphabetic indices as being summed from 1 to 3 if they are from the back part of the alphabet (say

[4] We use the physics convention where the name of a function is not changed with its functional form under a mapping, and for the moment the functional dependence is indicated by the set notation to include all three variables.

past the letter h as above), while they are summed from 0 to 3 if they are from the front part of the alphabet. Hence the expression (1.14) contains three terms on the right.

Each of the partial derivatives in Equation (1.14) is a vector in the increasing direction of the corresponding q^i coordinate. They are independent if the members of the set $\{q^i\}$ are independent, since then they define three non-coplanar lines in space. Hence the three partial derivatives of the position vector define three directions of a coordinate basis. These vectors are evidently written formally as

$$\mathbf{e}_i = \frac{\partial \mathbf{r}}{\partial q^i}, \tag{1.15}$$

and may be calculated explicitly when the form (1.13) is known (see Problem 1.2 part (a) for a simple case). This procedure constructs a 'coordinate basis'. One can choose base vectors whose directions are not directly related to the directions of the coordinate axes, but this will not concern us until later.

Any vector may also be written as a weighted sum of these base vectors, where once again the weights are the vector's components. However, such base vectors possess in general some very grave complications relative to the Cartesian base vectors. They are in general not constant from point to point, not mutually orthogonal (i.e. they do not have the property of Equation (1.4)) and they are not normalized! This means in particular that it is not easy even to identify the same vector at two different points using generalized coordinates, since the weights will vary even for a constant vector due to the varying base vectors. The fundamental definition of the same vector at two separate points is that it has the same magnitude and direction; that is, it is 'parallel transported' between the two points.

To illustrate parallel transport algebraically, we return to the Cartesian reference system where this operation is accomplished simply by holding weights (components) constant under the displacement. Because of the general nature of the mapping between Cartesian coordinates and generalized coordinates, the generalized coordinates vary in a complicated fashion when a vector is parallel transported (see the example discussion at the end of this chapter). Thus consider a relatively simple choice of generalized coordinates such as the orthonormal spherical polars r, θ, ϕ (see e.g. Figure 1.3). Since spherical coordinates are mutually orthogonal, we may adopt an orthonormal coordinate basis $\{\widehat{\mathbf{e}}_r, \widehat{\mathbf{e}}_\theta, \widehat{\mathbf{e}}_\phi\}$. The weights of a vector in this basis are calculated just as for Cartesian weights by Equation (1.7). We will call such weights 'physical components' since the physical dimensions of the vector are preserved in the weight. Other weight representations to be discussed below do not necessarily possess this property. We therefore distinguish the physical or orthonormal component by placing the coordinate label in brackets, that is $A_{(q)}$.

To continue with our discussion of constant vectors, we can find geometrically using Figure 1.3 the mapping for a vector $\mathbf{A}$ directed along the x axis from its Cartesian component A_x to physical components in spherical polars as,

$$\begin{pmatrix} A_{(r)} \\ A_{(\theta)} \\ A_{(\phi)} \end{pmatrix} \equiv \begin{pmatrix} A_x \; cos\phi \sin\theta \\ A_x \; cos\phi \cos\theta \\ -A_x \; sin\phi \end{pmatrix}. \tag{1.16}$$

So holding A_x constant by parallel transport to another point of space implies varying spherical polar coordinates as $\{\theta, \phi\}$ vary. This argument is continued in more detail in Problem 1.1, part(c).

Problems

1.1 **(a)** Using the Euclidean geometry of Figure 1.3, show that

$$\begin{pmatrix} \widehat{\mathbf{e}}_r \\ \widehat{\mathbf{e}}_\theta \\ \widehat{\mathbf{e}}_\phi \end{pmatrix} \equiv \begin{pmatrix} \cos\phi\sin\theta & \sin\phi\sin\theta & \cos\theta \\ \cos\phi\cos\theta & \sin\phi\cos\theta & -\sin\theta \\ -\sin\phi & \cos\phi & 0 \end{pmatrix} \begin{pmatrix} \widehat{\mathbf{e}}_x \\ \widehat{\mathbf{e}}_y \\ \widehat{\mathbf{e}}_z \end{pmatrix} \tag{1.17}$$

(b) If $\underline{\underline{\mathbf{S}}}$ is the 3×3 matrix in part (a), show that the transpose $\widetilde{\underline{\underline{\mathbf{S}}}}$ (that is, the matrix with the rows and columns interchanged) is in fact the inverse of $\underline{\underline{\mathbf{S}}}$ so that $\widetilde{\underline{\underline{\mathbf{S}}}}\,\underline{\underline{\mathbf{S}}} = \underline{\underline{\mathbf{1}}}$, the unit matrix. Consequently show that

$$\begin{pmatrix} \widehat{\mathbf{e}}_x \\ \widehat{\mathbf{e}}_y \\ \widehat{\mathbf{e}}_z \end{pmatrix} \equiv \begin{pmatrix} \cos\phi\sin\theta & \cos\phi\cos\theta & -\sin\phi \\ \sin\phi\sin\theta & \sin\phi\cos\theta & \cos\phi \\ \cos\theta & -\sin\theta & 0 \end{pmatrix} \begin{pmatrix} \widehat{\mathbf{e}}_r \\ \widehat{\mathbf{e}}_\theta \\ \widehat{\mathbf{e}}_\phi \end{pmatrix} \tag{1.18}$$

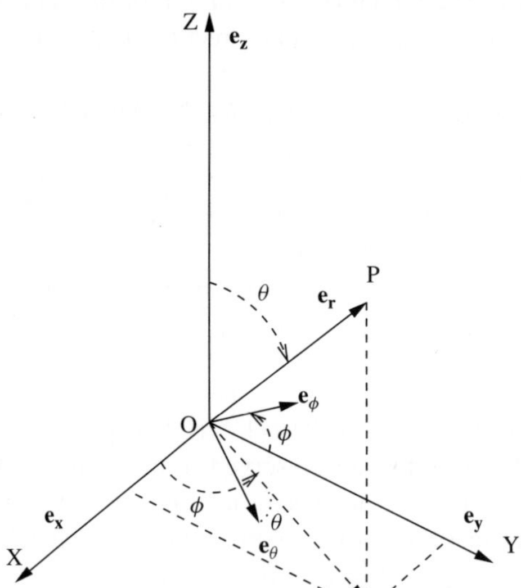

Figure 1.3 *The figure shows the base vectors for both Cartesian and spherical polar coordinates. Although the Cartesian base vectors or reference directions are the same at all points of space, the spherical polar base vectors vary dramatically with the direction* (θ, ϕ) *on the sphere. Note that* $\mathbf{e}_\theta$ *is perpendicular to the plane of* $\mathbf{e}_r$*,* $\mathbf{e}_\phi$ *and so points below the X-Y plane. A point O is generally held fixed as an 'origin', while P is any other point of space*

(c) Show that for an arbitrary vector **A** (which includes the position vector) the spherical polar weights or components and the Cartesian weights or components are expressed, each set in terms of the other, by the same matrices as in parts (a) and (b) respectively.

1.2 **(a)** From Problem 1.1 part (a), you can write the Cartesian components of a position vector $\mathbf{r} = r\widehat{\mathbf{e}}_r$. Thus use Equation (1.14) to write the orthogonal but un-normalized base vectors in spherical polar coordinates in terms of the Cartesian base vectors as

$$\begin{pmatrix} \mathbf{e}_r \\ \mathbf{e}_\theta \\ \mathbf{e}_\phi \end{pmatrix} \equiv \begin{pmatrix} \partial\mathbf{r}/\partial r \\ \partial\mathbf{r}/\partial\theta \\ \partial\mathbf{r}/\partial\phi \end{pmatrix} = \begin{pmatrix} \cos\phi\sin\theta & \sin\theta\sin\phi & \cos\theta \\ r\cos\phi\cos\theta & r\sin\phi\cos\theta & -r\sin\theta \\ -r\sin\phi\sin\theta & r\cos\phi\sin\theta & 0 \end{pmatrix} \begin{pmatrix} \widehat{\mathbf{e}}_x \\ \widehat{\mathbf{e}}_y \\ \widehat{\mathbf{e}}_z \end{pmatrix} \tag{1.19}$$

Note that again by Problem 1.1 part (a), we can write this last result more simply as

$$\begin{pmatrix} \mathbf{e}_r \\ \mathbf{e}_\theta \\ \mathbf{e}_\phi \end{pmatrix} = \begin{pmatrix} \widehat{\mathbf{e}}_r \\ r\widehat{\mathbf{e}}_\theta \\ r\sin\theta\,\widehat{\mathbf{e}}_\phi \end{pmatrix}. \tag{1.20}$$

(b) Show that had we used the physical displacement component $d\ell_{(\alpha)} = \sqrt{g_{\alpha\alpha}}dq^\alpha$ in the base vector definitions of Equation (1.14), then we would have obtained the orthonormal coordinate base vectors for spherical polar coordinates, as we found them in Problem 1.1, part (a). There is no sum over Greek indices, and see below for the definition $g_{\alpha\alpha} \equiv \mathbf{e}_\alpha \cdot \mathbf{e}_\alpha$.

The required weights at each point for an arbitrary vector are not so easy to find as for the Cartesian basis, because the generalized base vectors do not satisfy Equation (1.4). However, it happens that we can always find another set of three vectors $\{\mathbf{e}^i\}$ that obey a slightly more subtle Equation (1.4) in the form

$$\mathbf{e}^i \cdot \mathbf{e}_j = \delta^i_j. \tag{1.21}$$

These vectors are called reciprocal base vectors for fairly obvious reasons. The symbol δ^i_j is again the Kronecker delta, but the up and down indices remind us of the necessity of a set of reciprocal base vectors. The reciprocal base vector may be thought of as a row vector (a 1×3 matrix, rather than a 3×1 matrix for a column vector) that is matrix multiplied into the column base vector to effect a 'dot', that is a scalar, product.

One practical definition of these reciprocal base vectors is (we use Greek letters to indicate that no sums are intended when they are repeated in a product)

$$\mathbf{e}^\alpha \equiv \frac{\mathbf{e}_\beta \wedge \mathbf{e}_\gamma}{\mathbf{e}_\alpha \cdot (\mathbf{e}_\beta \wedge \mathbf{e}_\gamma)}. \tag{1.22}$$

Here the ordered set $\{\alpha, \beta, \gamma\}$ take on successively the three cyclic (i.e. even) permutations of the set $\{1, 2, 3\}$, in order to define the three reciprocal vectors. The denominator is invariant under these permutations. The usual vector 'cross product' is indicated by the symbol $\wedge$ rather than a multiplication symbol, and is sometimes called the 'wedge' product. This definition yields immediately (since $\mathbf{e}_\beta \wedge \mathbf{e}_\gamma \parallel \mathbf{e}_\alpha$)

$$\mathbf{e}_\epsilon \cdot \mathbf{e}^\alpha = \delta_\epsilon^\alpha, \tag{1.23}$$

as required.

An alternative definition of a reciprocal basis that is useful for some purposes is

$$\mathbf{e}^j \equiv \nabla q^j. \tag{1.24}$$

By remembering the definition (1.15), and by using the Cartesian representation of both $\mathbf{r}$ (1.13) and of the gradient operator ∇ $(\widehat{\mathbf{e}}_i \partial_i)$, we see that

$$\mathbf{e}^j \cdot \mathbf{e}_i = \frac{\partial q^j}{\partial x}\frac{\partial x}{\partial q^i} + \frac{\partial q^j}{\partial y}\frac{\partial y}{\partial q^i} + \frac{\partial q^j}{\partial z}\frac{\partial z}{\partial q^i} \equiv \frac{\partial q^j}{\partial q^i} = \delta^j_{\ i}. \tag{1.25}$$

It is readily shown (see Problem 1.3) that two reciprocal sets that satisfy the condition of Equation (1.21) are identical.

Problems

1.3 Show that if two sets of reciprocal base vectors $\mathbf{e}^i$ and $\mathbf{n}^i$ exist such that for each set Equation (1.21) is satisfied, then their difference is orthogonal to every base vector in the space, and so must be zero.

1.4 Use the definition (1.24) to find the reciprocal vectors in spherical polar coordinates to be $\{\mathbf{e}^r, \mathbf{e}^\theta, \mathbf{e}^\phi\} = \{\widehat{\mathbf{e}}_r, \widehat{\mathbf{e}}_\theta/r, \widehat{\mathbf{e}}_\phi/r\sin\theta\}$.

Verify that definition (1.22) yields the same result.

1.5 Consider the primed oblique coordinate axes as shown in Figure 1.4.

The inverse mapping is

$$\begin{pmatrix} x \\ y \\ z \end{pmatrix} = \begin{pmatrix} x' + y'\cos\beta \\ y'\sin\beta \\ z'\cos\alpha \end{pmatrix} \tag{1.26}$$

Use Equation (1.15) to find the base vectors, and Equation (1.24) to find the reciprocal base vectors for the primed reference axes. Verify that the condition (1.21) is obtained.

The component or weight of an arbitrary vector along each axis of a non-orthogonal basis now follows from the desired representation $\mathbf{A} = A^i\mathbf{e}_i$ and the property (1.21) as

$$A^j \equiv \mathbf{e}^j \cdot \mathbf{A}, \tag{1.27}$$

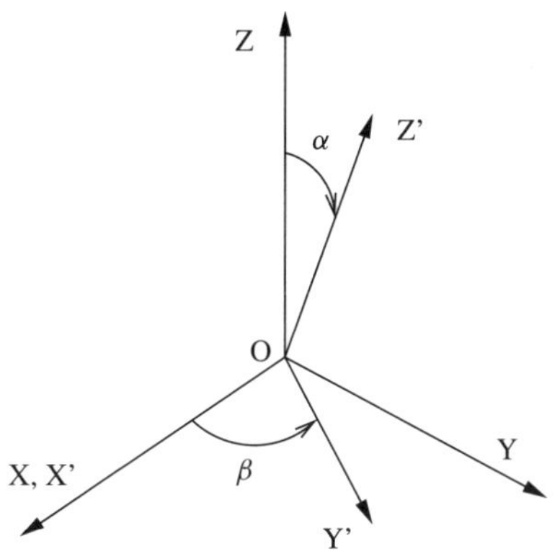

Figure 1.4 *An example of oblique reference axes, but rectilinear as in the Cartesian case*

for j any one of $1, 2, 3$. The upper index indicates once again the necessity of using the reciprocal vector. Similarly the representation $\mathbf{r} = \mathbf{e}_i q^i$ shows the generalized coordinates to be the weights for the position vector, just as are $\{x, y, z\}$ in the Cartesian system.

We may express the differential theorem of Pythagoras in generalized coordinates using Equation (1.14) and remembering the definition of the base vectors as

$$d\ell^2 \equiv d\mathbf{r}^2 = \mathbf{e}_i \cdot \mathbf{e}_j dq^i dq^j \equiv g_{ij} dq^i dq^j. \tag{1.28}$$

We have used here the Einstein summation convention to express a double sum and we have introduced the important definition

$$g_{ij} \equiv \mathbf{e}_i \cdot \mathbf{e}_j. \tag{1.29}$$

This two index quantity comprises a 3×3 matrix as i and j take the three possible values, which we shall refer to either as the 'metric matrix', because of its appearance in the Pythagorean distance measure, or occasionally as the 'metric tensor' because of its behaviour under general mappings between coordinates, which behaviour defines tensors as we shall see subsequently.

We may write the metric matrix explicitly from the Equations (1.15) and (1.13) as

$$g_{jk} = \frac{\partial x^i}{\partial q^j}\frac{\partial x^i}{\partial q^k}, \tag{1.30}$$

where the sum over i from 1 to 3 gives three terms in Cartesian x,y,z for fixed j and k.

Problem

1.6 Use both definitions (1.29) and (1.30) together with the results of Problem 1.1 to write the theorem of Pythagoras in spherical polar coordinates as

$$d\ell^2 = dr^2 + r^2 d\theta^2 + r \sin^2\theta d\phi^2. \tag{1.31}$$

A further irritating complication in the case of non-orthogonal generalized coordinates is that one can use the metric matrix to form another description of an arbitrary vector **A**, in addition to the weights or components calculated from Equation (1.27). Consider the quantities

$$\underline{\underline{\mathbf{g}}}\mathbf{A} \equiv g_{ij}A^j \equiv A_i. \tag{1.32}$$

By placing $\underline{\underline{\mathbf{g}}}$ in front of a column vector we imply matrix multiplication as indicated by the summation convention in the second statement of this equation, and the quantity A_i is defined by the final statement. To see what this means geometrically we may use Equation (1.29) for the metric matrix to write

$$A_i \equiv \mathbf{e}_i \cdot \mathbf{e}_j A^j, \tag{1.33}$$

which we see is equivalent to

$$A_i \equiv \mathbf{e}_i \cdot \mathbf{A} \tag{1.34}$$

because of the base vector representation of A. Such components form a set of weights for the arbitrary vector **A** that are just as good as those found by scalar product with the reciprocal vectors in Equation (1.27). However, although not identical, both sets of components are unique to the vector concerned in a given coordinate system.

Consequently in summary, in a completely general reference frame we have a choice of representation for an arbitrary vector. Either

$$\mathbf{A} = \mathbf{e}_i A^i, \tag{1.35}$$

as above in terms of the base vectors, or

$$\mathbf{A} = \mathbf{e}^i A_i \tag{1.36}$$

in terms of the reciprocal base vectors. The components of each type follow easily from the property (1.21). The $\{A^i\}$ are referred to as contravariant components (because they transform to other coordinates in a manner opposite to the base vectors) and the $\{A_i\}$ are called covariant components (because they transform as do the base vectors). Example (1.1) discusses these transformations explicitly. The two sets are not independent since they are related through the metric matrix by Equation (1.32) in a given reference system of coordinates. The contravariant components may be thought of as forming column vectors while the covariant coordinates form row vectors. When g_{ij} is applied to the position vector we obtain the covariant coordinates $\{q_i\}$ rather than the contravariant coordinates $\{q^i\}$.

Example 1.1

We may find the transformation of the base vectors under a change of generalized coordinates from q^i to q'^i from the definition as

$$\mathbf{e}'_i \equiv \frac{\partial \mathbf{r}}{\partial q'^i} = \frac{\partial q^k}{\partial q'^i}\frac{\partial \mathbf{r}}{\partial q^k} \equiv \frac{\partial q^k}{\partial q'^i}\mathbf{e}_k \tag{1.37}$$

On the other hand from the alternative definition (1.24) of the reciprocal base vectors

$$\mathbf{e}'^i \equiv \nabla q'^i = \widehat{\mathbf{e}}_j \frac{\partial q'^i}{\partial x^j} = \Sigma_j \frac{\partial q'^i}{\partial q^k} \widehat{\mathbf{e}}_j \frac{\partial q^k}{\partial x^j} = \frac{\partial q'^i}{\partial q^k} \mathbf{e}^k. \tag{1.38}$$

We see then that the base vectors and the reciprocal base vectors transform inversely. Similarly, for an arbitrary vector $\mathbf{A}$, we see that for the covariant components

$$A'_i = \mathbf{e}'_i \cdot \mathbf{A} = \frac{\partial q^k}{\partial q'^i} \mathbf{e}_k \cdot \mathbf{A} = \frac{\partial q^k}{\partial q'^i} A_k, \tag{1.39}$$

as for the base vectors. Making the same argument with the reciprocal base vectors we find that the contravariant components of $\mathbf{A}$ transform like the reciprocal base vectors.

Equation (1.30) shows the metric matrix to be real and symmetric (its transpose is equal to itself). As such it possesses an inverse $\underline{\underline{\mathbf{g}}}^{-1}$ that seems at first to be another complicating quantity. However, it really serves to give us an operation of 'raising indices' that corresponds to the lowering operation of Equation (1.32). We see that (the ij component of $\underline{\underline{\mathbf{g}}}^{-1}$ is denoted simply g^{ij})

$$g^{ij} A_j = g^{ij} g_{jk} A^k = \delta^i_k A^k \equiv A^i, \tag{1.40}$$

since by definition the inverse of a matrix by itself yields the identity matrix. Similarly we can 'lower an index' according to the relation (1.32). When the coordinates are orthogonal, the diagonal components of g^{ij} are simply the inverse of the quantities g_{ij}.

This operation of 'raising' or 'lowering' indices is an algorithmic description of the procedure to obtain the contravariant component from the covariant and conversely. This is the whole story for the lowering and raising of indices of vectors. When required, we shall see that for tensors the raising or lowering operation can be applied to each index of the tensor.

Both sets of components are necessary for vectors because under a change in generalized coordinates, the vector dot product of two vectors $\mathbf{A}$ and $\mathbf{B}$ is only a scalar invariant (as numbers should be) if it is defined as

$$\mathbf{A} \cdot \mathbf{B} \equiv A^i B_i \equiv A_i B^i. \tag{1.41}$$

This is because the inverse transformation rules of the contravariant and covariant components cancel in such a summed product. Our definition of lowering and raising indices also yields the important result for the scalar product in generalized coordinates in the form

$$\mathbf{A} \cdot \mathbf{B} \equiv g_{ij} A^i B^j \equiv g^{ij} A_i B_j. \tag{1.42}$$

At this point it is useful to regard the base vectors and the reciprocal base vectors in a slightly different fashion. Consider the 3 matrices $e_{(i)j}$ and $e^{(i)j}$ formed from the components of the three base vectors and the reciprocal base vectors respectively. Here we use the bracket index to distinguish the base vector 'label' from the component or axis index. The lower index j on the base vector tells us to write each base vector along a row, while the upper index j tells us to write each reciprocal base vector in a column. That is, we have taken the covariant components of the base vectors and the contravariant components of the reciprocal base vectors in some non-coordinate basis. In the coordinate basis the base vector matrix is diagonal.

Now recalling the function of the reciprocal base vectors we have

$$e^{(i)k} e_{(j)k} = \delta^i_j. \tag{1.43}$$

Mutiplying this by $e_{(i)\ell}$ and letting the summation convention apply also to the bracketed base labels we obtain

$$(e_{(i)\ell} e^{(i)j}) e_{(j)k} = e_{(j)\ell} \tag{1.44}$$

whence we conclude that the bracketed term on the left satisfies

$$e_{(i)\ell} e^{(i)j} = \delta^j_\ell. \tag{1.45}$$

That is, the sum over the base vector label yields the Kronecker delta in the component indices. This is somewhat surprising but it may be verified by direct matrix multiplication (by reversing the order of the factors in the sum on the left of (1.45)). One must recall the definitions of the base vectors (1.15) and of the reciprocal base vectors (1.24), and use Cartesian components for each vector. In Example (1.2) we use this representation to find the result for the inverse metric analogous to (1.29) as

$$g^{\ell m} = \mathbf{e}^{(\ell)} \cdot \mathbf{e}^{(m)}. \tag{1.46}$$

Example 1.2

Let us recall that by the existence of the inverse metric matrix

$$g^{\ell i} g_{ij} = \delta^\ell_j. \tag{1.47}$$

Moreover we know from Equation (1.29) and the definition of the dot product that

$$g_{ij} \equiv e^k_{(i)} e_{(j)k} \tag{1.48}$$

so

$$g^{\ell i} e^k_{(i)} e_{(j)k} = \delta^{\ell.}_j \tag{1.49}$$

We can now multiply this last equation by $e^{(j)m}$ and use property (1.45) to obtain $g^{(\ell)i} e^m_{(i)} = e^{(\ell)m}$. Multiplying this last result by $e^{(j)}_m$ and using (1.43) on the left gives finally $g^{\ell j} = \mathbf{e}^{(\ell)} \cdot \mathbf{e}^{(j)}$ as in Equation (1.46).

Once again we emphasize that in the special case where the base vectors $\{\mathbf{e}_i\}$ are mutually orthogonal and normalized (orthonormal in short), then the denominator in Equation (1.22) is unity and $\widehat{\mathbf{e}}^\alpha = \widehat{\mathbf{e}}_\alpha$. In this case components of an arbitrary vector follow just as in Equation (1.7), and they are unique. When using such base vectors, which include Cartesian base vectors, we needn't distinguish components and reciprocal components by lower and upper indices. Whenever possible we indicate the unique component along the q^i direction for an arbitrary vector by $A_{(q)}$.

We can refer to these unique components as 'physical' weights or components because due to the orthonormality of the base vectors they possess the true physical dimensions of the vector. When, as above, we refer to the 'usual' cross product we mean to employ physical components so that $\mathbf{C} = \mathbf{A} \wedge \mathbf{B}$ means

$$C_{(\alpha)} = \epsilon_{\alpha jk} A_{(j)} B_{(k)} \tag{1.50}$$

where the epsilon symbol is $+1$ if $\{\alpha, j, k\}$ is a cyclic (even) permutation of $\{1, 2, 3\}$ and -1 if it is an acyclic (odd) permutation. It is zero if any two indices are the same.

If we use base vectors that are orthogonal but not normalized, we have to be more careful in defining 'physical' components. With such a choice, the metric matrix remains in a particularly simple form. For by Equation (1.29) we see that it will still be diagonal, if not simply δ_{ij}. The up and down indices remain because $\mathbf{e}^\alpha$ and $\mathbf{e}_\alpha$ are not identical in general, although they are parallel. In fact Equation (1.22) shows that for this special case (recall that there is no sum over repeated Greek indices and the vertical bars indicate the magnitude of a vector)

$$\mathbf{e}^\alpha = \frac{1}{|\mathbf{e}_\alpha|}\widehat{\mathbf{e}}_\alpha \equiv \frac{1}{\sqrt{g_{\alpha\alpha}}}\widehat{\mathbf{e}}_\alpha, \tag{1.51}$$

where the last part of the identity follows from the definition (1.29) for an orthogonal basis. Consequently we can write the physical component of an arbitrary vector $\mathbf{A}$ in two ways, namely

$$A_{(\alpha)} \equiv \frac{\mathbf{e}_\alpha}{|\mathbf{e}_\alpha|} \cdot \mathbf{A} \equiv \frac{A_\alpha}{|\mathbf{e}_\alpha|} \equiv \frac{A_\alpha}{\sqrt{g_{\alpha\alpha}}} \tag{1.52}$$

in terms of the covariant component, and

$$A_{(\alpha)} \equiv \frac{\mathbf{e}^\alpha}{|\mathbf{e}^\alpha|} \cdot \mathbf{A} \equiv \frac{A^\alpha}{|\mathbf{e}^\alpha|} = \sqrt{g_{\alpha\alpha}} A^\alpha \tag{1.53}$$

in terms of the contravariant component. The last step in the latter equation follows from $\mathbf{e}^\alpha \cdot \mathbf{e}_\alpha = 1$, for remembering that the base and reciprocal base vectors are parallel we find from this relation

$$|\mathbf{e}^\alpha| = \frac{1}{|\mathbf{e}_\alpha|}. \tag{1.54}$$

For a diagonal metric matrix, the inverse matrix is also diagonal with matrix elements $g^{\alpha\alpha} = 1/g_{\alpha\alpha}$. Thus by the preceding relation $\sqrt{g^{\alpha\alpha}} = |\mathbf{e}^\alpha|$ since $\sqrt{g_{\alpha\alpha}} = |\mathbf{e}_\alpha|$. In fact it is the case for quite general reference frames (see Example 1.2) that

$$g^{\alpha\beta} \equiv \mathbf{e}^\alpha \cdot \mathbf{e}^\beta, \tag{1.55}$$

in parallel with Equation (1.29). For orthogonal base vectors this relation yields the diagonal elements, as above.

A simpler proof than that given above follows from our definitions of contravariant and covariant components of a general vector **A** as

$$A^\alpha \equiv \mathbf{e}^\alpha \cdot \mathbf{A} = \mathbf{e}^\alpha \cdot \mathbf{e}^\beta A_\beta \equiv g^{\alpha\beta} A_\beta. \tag{1.56}$$

For general **A** this proves (1.55).

To calculate the wedge or cross product as a contravariant component we use Equations (1.50) and (1.52), (1.53) together to write (we must sum over greeks explicitly)

$$C^\alpha = \Sigma_{\beta\gamma} \left(\epsilon_{\alpha\beta\gamma} \frac{A_\beta B_\gamma}{\sqrt{g_{\alpha\alpha}}\sqrt{g_{\beta\beta}}\sqrt{g_{\gamma\gamma}}} \right). \tag{1.57}$$

Since no two of $\{\alpha, \beta, \gamma\}$ can be the same, the denominator on the right is simply $\sqrt{g}$ for an orthogonal reference system. If C^α is indeed to be contravariant, then $\epsilon_{\alpha\beta\gamma}/\sqrt{g}$ should be a three index contravariant 'tensor' under transformation between orthogonal reference systems. We therefore write it as

$$e^{\alpha\beta\gamma} \equiv \epsilon_{\alpha\beta\gamma}/\sqrt{g}. \tag{1.58}$$

Consequently we may write the cross product in orthogonal reference systems as

$$C^i = e^{ijk} A_j B_k \tag{1.59}$$

or as

$$C_i = e_{ijk} A^j B^k \tag{1.60}$$

where it is easy to show that

$$e_{ijk} \equiv \sqrt{g}\epsilon_{ijk}. \tag{1.61}$$

Fortunately in view of the density of the preceding discussion one can choose generalized coordinates in many cases to be orthogonal, although their base vectors are neither constant in space nor normalized in general.[5] Examples of these choices are the familiar 'curvilinear coordinates' such as spherical, cylindrical or spheroidal coordinates as discussed in the Problems.

[5] The full generality summarized above for non-orthogonal base directions is really only necessary in the modern theory of gravity, or whenever there is some compelling reason to adopt non-orthonormal components.

Although each of the q^i will increase in the direction of the corresponding curvilinear axis, they do not in general measure length along these axes. In fact we see from Equation (1.28), together with g_{ij} being diagonal for orthogonal axes, that it is rather (recall that there is no summation over Greek indices)

$$d\ell^{\alpha} \equiv \sqrt{g_{\alpha\alpha}}dq^{\alpha} \tag{1.62}$$

that measures length along such axes. We recognize this as the physical component of the displacement vector by Equation (1.53). When general curvilinear coordinates are used with an orthonormal base, it is this physical component of the coordinates that is implied.

We can understand the relation between Cartesian and curvilinear coordinates in a more intuitive fashion. Suppose that at a given point in space we have three orthogonal directions that do not coincide with the three orthogonal Cartesian directions. These directions may vary from point to point in general, but at each point they differ from the Cartesian directions *only by a rigid rotation* provided that the same 'handedness'[6] and the physical displacements of Equation (1.62) are used (see Figure 1.6 for an example).

A rigid rotation in space of the coordinate axes may be described by the operation of an 'orthogonal' 3×3 matrix. To see why this is so and what the orthogonality property is, consider an arbitrary column vector **A** expressed in generalized orthonormal coordinates as (q is summed and we bracket it to emphasize the orthonormal basis and physical component)

$$\mathbf{A} = \widehat{\mathbf{e}}_{(q)}A_{(q)}. \tag{1.64}$$

The x, y, z Cartesian components are now found as

$$A_{(i)} = \widehat{\mathbf{e}}_i \cdot \widehat{\mathbf{e}}_{(q)}A_{(q)}, \tag{1.65}$$

where, as i takes on values from 1 to 3, we obtain respectively the x, y, z components. The index q is summed from 1 to 3 over the curvilinear coordinates.

We can write this last expression in explicit matrix form as (see e.g. Problem 1.1, part (c))

$$\begin{pmatrix} A_x \\ A_y \\ A_z \end{pmatrix} \equiv \begin{pmatrix} \widehat{\mathbf{e}}_x \cdot \widehat{\mathbf{e}}_{(1)} & \widehat{\mathbf{e}}_x \cdot \widehat{\mathbf{e}}_{(2)} & \widehat{\mathbf{e}}_x \cdot \widehat{\mathbf{e}}_{(3)} \\ \widehat{\mathbf{e}}_y \cdot \widehat{\mathbf{e}}_{(1)} & \widehat{\mathbf{e}}_y \cdot \widehat{\mathbf{e}}_{(2)} & \widehat{\mathbf{e}}_y \cdot \widehat{\mathbf{e}}_{(3)} \\ \widehat{\mathbf{e}}_z \cdot \widehat{\mathbf{e}}_{(1)} & \widehat{\mathbf{e}}_z \cdot \widehat{\mathbf{e}}_{(2)} & \widehat{\mathbf{e}}_z \cdot \widehat{\mathbf{e}}_{(3)} \end{pmatrix} \begin{pmatrix} A_{(1)} \\ A_{(2)} \\ A_{(3)} \end{pmatrix}, \tag{1.66}$$

where the dot products of the orthonormal base vectors are the direction cosines of each of the Cartesian directions relative to the local curvilinear axes. The values of these

[6] This handedness may be expressed as

$$\widehat{\mathbf{e}}_{\alpha} \cdot (\widehat{\mathbf{e}}_{\beta} \wedge \widehat{\mathbf{e}}_{\gamma}) = \pm 1, \tag{1.63}$$

where α, β, γ are an even permutation of $\{1, 2, 3\}$ as before and $\widehat{\mathbf{e}}$ indicates normalization. A right-handed system has this expression equal to $+1$.

cosines may be calculated geometrically in any specific case (see e.g. Problem 1.7). Symbolically we write this last relation in matrix notation as

$$\mathbf{A}_{(c)} = \underline{\underline{\mathbf{S}}}\mathbf{A}_{(q)}, \tag{1.67}$$

where the subscript (c) means Cartesian components and the subscript (q) means orthonormal curvilinear. The matrix $\underline{\underline{\mathbf{S}}}$ is the 'rotation matrix', although it is not yet in its simplest form.

However, the 'curvilinear components' might be (in an important special case) simply relative to another Cartesian reference frame that is rotated relative to the first. In that case, both sets of components would be constant for a constant vector. There is no real distinction between such frames for a constant relative orientation, but when one frame is inertial (see below) and the other is rotating relative to it rather than merely rotated, there is a very strong distinction.

A rigid rotation does not in fact require nine parameters to be specified and the direction cosines are all inter-related by orthonormality. Only three independent parameters are required to specify a rigid rotation and these are usually taken as the 'Euler angles', simply related to the usual polar angles. These afford a much more convenient form for $\underline{\underline{\mathbf{S}}}$ (see Problem 1.12).

Problems

1.7 From Figure 1.5, show by careful resolution that the base vectors in the system $\{\zeta, \kappa, Z'\}$ are related to the Cartesian base vectors by the matrix operation

$$\begin{pmatrix} \widehat{\mathbf{e}}_\zeta \\ \widehat{\mathbf{e}}_\kappa \\ \widehat{\mathbf{e}}_{Z'} \end{pmatrix} \equiv \begin{pmatrix} \cos\phi & \sin\phi & 0 \\ -\sin\phi\cos\theta & \cos\phi\cos\theta & \sin\theta \\ \sin\phi\sin\theta & -\cos\phi\sin\theta & \cos\theta \end{pmatrix} \begin{pmatrix} \widehat{\mathbf{e}}_x \\ \widehat{\mathbf{e}}_y \\ \widehat{\mathbf{e}}_z \end{pmatrix} \tag{1.68}$$

1.8 Verify the limiting cases for which $\theta = 0$, and $\phi = 0$ respectively in Problem 1.7. These matrices are often labelled $\underline{\underline{\mathbf{S}}}_\phi$ and $\underline{\underline{\mathbf{S}}}_\theta$ respectively. Each of these is defined in its own set of axes and so $\underline{\underline{\mathbf{S}}}_\theta$ may also be obtained by the permutation $\{3 \rightarrow 1 \rightarrow 2\}$ on the rows and columns of $\underline{\underline{\mathbf{S}}}_\phi$. Verify this and so deduce also $\underline{\underline{\mathbf{S}}}_\psi$ in its own axes for the final Euler rotation of Figure 1.5.

1.9 Show that the matrix found in Problem 1.7 can be written as $\underline{\underline{\mathbf{S}}}_\theta\,\underline{\underline{\mathbf{S}}}_\phi$.

1.10 Find the complete Euler angle matrix that gives the primed base vectors in terms of the Cartesian base vectors by calculating $\underline{\underline{\mathbf{S}}} \equiv \underline{\underline{\mathbf{S}}}_\psi\,\underline{\underline{\mathbf{S}}}_\theta\,\underline{\underline{\mathbf{S}}}_\phi$.

1.11 Show that for each of the individual Euler angle rotation matrices of the preceding question $\widetilde{\underline{\underline{\mathbf{S}}}_\alpha}\,\underline{\underline{\mathbf{S}}}_\alpha = \underline{\underline{\mathbf{1}}}$ where $\widetilde{\underline{\underline{\mathbf{S}}}_\alpha}$ is the transpose of $\underline{\underline{\mathbf{S}}}_\alpha$ and $\underline{\underline{\mathbf{1}}}$ is the unit 3×3 matrix. Here α is any of ϕ, θ or ψ. Note that this property of 'orthogonality' also applies to $\underline{\underline{\mathbf{S}}}$ by using the transpose of a matrix product.

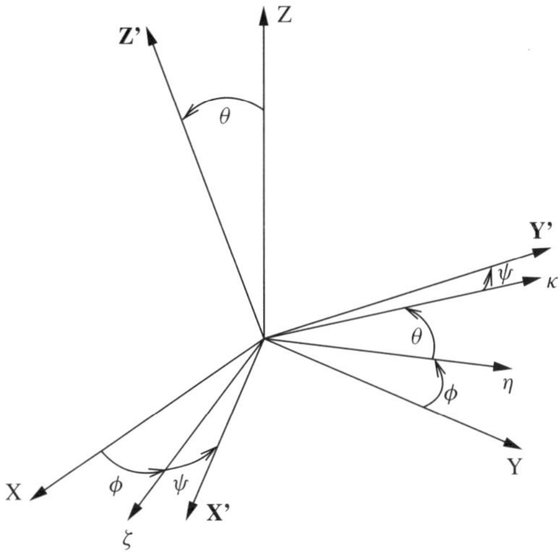

Figure 1.5 *The figure shows a Cartesian reference frame* $\{X, Y, Z\}$*, together with axes* $\{\zeta, \eta, Z\}$ *that are rotated through the angle* ϕ *about the Z axis. The direction labelled* ζ *is frequently referred to as a line of 'nodes' since the plane of* X', Y' *intersects the X, Y plane along this line. The line marked* η *is rotated about the line of nodes by the angle* θ *to coincide with the direction* κ*. This moves Z to* Z' *and* ζ, κ, Z' *are the 'line of node axes'. Finally the axes* ζ *and* κ *are rotated about* Z' *through an angle* ψ *to coincide with the axes* X' *and* Y' *respectively. By varying these angles in the ranges* $0 \leq \phi \leq 2\pi$*,* $0 \leq \theta \leq \pi$*,* $0 \leq \psi \leq 2\pi$*, any orientation of the primed axes can be obtained relative to the Cartesian axes. They are one choice of 'Euler' angles*

1.12 Prove that for any arbitrary vector **A**, the transformation from unprimed Cartesian axes to primed Cartesian axes is given by $\mathbf{A}' = \underline{\underline{\mathbf{S}}}\mathbf{A}$. Note that the inverse transformation is given easily by the orthogonality property.

If we wish to express the transformation of coordinates under rotation then we must apply the above analysis to a position vector, for which $\mathbf{A}_{(c)} \equiv \{x^i\}$ and $\mathbf{A}_{(q)} \equiv \{\ell^i\}$, and where the length or 'physical' coordinate $d\ell^i$ was defined in Equation (1.62). This definition may be integrated whenever the $g_{\alpha\alpha}$ do not depend on q^α, as assumed here. The more general case may be included simply by replacing x^i by dx^i and ℓ^i by $d\ell^i$ in the rest of this section.

Using again the summation convention, the matrix multiplication to transform position vectors becomes explicitly

$$\{x^i\} = S^i{}_j \ell^j. \tag{1.69}$$

The index i designates the rows and the offset lower index j designates the columns in the matrix. Each row i forms a 'row vector' as j takes on its three values, and it is to be multiplied element by element with the column vector $\{\ell^j\}$. Indeed, when we defined the dot product in Equation (1.41) in terms of upper and lower indices, we were multiplying a column vector $\{A^i\}$ by a row vector $\{B_i\}$. We use $\{x^i\}$ for the Cartesian components to indicate column vector, even though they are equivalent to the physical components $\{x, y, z\}$.

Now we may write the squared magnitude of the position vector as $x^i x^i$ in Cartesian components and $\ell^i \ell^i$ in general orthogonal coordinates, since we need not distinguish row and column vectors in such coordinates. Hence by Equation (1.69)

$$x^i x^i = S^i{}_j \ell^j S^i{}_k \ell^k \equiv S^i{}_j S^i{}_k \ell^j \ell^k, \tag{1.70}$$

where the summation convention applies to all three repeated indices, and we have rearranged a numerical product corresponding to one element of the summation. However, under a rotation of axes the magnitude of the position vector has not changed so that $x^i x^i = \ell^j \ell^j$. Consequently to describe a rigid rotation of axes the matrix $\underline{\underline{\mathbf{S}}}$ must satisfy

$$S^i{}_j S^i{}_k = \delta_{jk}. \tag{1.71}$$

This is the property of orthogonality. The argument depends upon the existence of coordinate invariant 'length' or distance, as given by the theorem of Pythagoras. We will make much use of coordinate invariants in the next chapter.

If we interchange the rows and columns of the first matrix in Equation (1.71) we have a simpler way of writing it in matrix notation as

$$\underline{\underline{\widetilde{\mathbf{S}}}}\,\underline{\underline{\mathbf{S}}} = \underline{\underline{\mathbf{1}}}, \tag{1.72}$$

since $S^i_j \equiv \widetilde{S}^j_i$. Here $\underline{\underline{\widetilde{\mathbf{S}}}}$ indicates the transpose rotation matrix and $\underline{\underline{\mathbf{1}}}$ is the unit matrix, with ones on the diagonal and zeros elsewhere (that is $(\underline{\underline{\mathbf{1}}})_{ij} = \delta_{ij}$). This shows that orthogonality means that the inverse matrix to an orthogonal matrix is its transpose. It also shows by taking the determinant of the expression that

$$\det \underline{\underline{\mathbf{S}}} = \pm 1. \tag{1.73}$$

The value -1 only occurs if one of the Cartesian axes is reflected in a plane mirror normal to it so that it changes direction and hence 'handedness' under the 'rotation'. The handedness also changes if all of the coordinate directions are similarly reflected.

We may note for consistency that Equation (1.30), used with the ℓ replacing the q, together with Equation (1.69) yields

$$g_{jk} \equiv S^i{}_j S^i{}_k \equiv \delta_{jk}, \tag{1.74}$$

by the orthogonality of rigid rotation.

This completes our discussion of the geometry of space, that is, of the relative position of objects as measured by directions and distance in some reference frame. These relations are normally expressed as vectors, and we have been at pains to present these intuitively but generally. Fortunately we do not need the full discussion immediately. We notice that throughout this discussion of vectors, time has been ignored. That is, we do not use vectors to measure the relative position of events in space and time. This is fundamentally because we do not have the equivalent of the theorem of Pythagoras for the relative distance of point-like 'events', and we do not consider time as a direction. The majority of the next chapter is dedicated to establishing a metric similar to the Pythagorean metric for a limited kind of space-time.

In the next section we shall need to consider relatively moving reference frames. In such cases time does enter into the transformation of coordinates between such frames. However, it remains for the moment Newtonian, that is not subject to transformation, being universal.

1.1.2 Inertial Reference Frames

Inertial reference frames are defined as those in which Newton's first law applies to a sufficiently isolated body. Moreover Newton's second law defines a 'real' force **F** acting on a (non-isolated) body in such frames of reference according to

$$\mathbf{F} \equiv m\mathbf{a}, \tag{1.75}$$

where **a** is the acceleration of an object measured in an inertial frame, and m is the 'inertial mass'. The acceleration is determined once the coordinates in the inertial reference frame are chosen, and are measured at appropriate time intervals along the object's trajectory. The apparent acceleration of a body relative to a non-inertial reference frame is different from that produced by the real forces, and these extra accelerations may be considered as arising from 'fictional' forces, occasionally referred to as 'inertial' forces. Of course if, as we speculate below after Mach, the inertial frames are produced by some influence of the mean Universe, then in some sense these inertial forces are also 'real'. Einstein's theory of gravity places such forces on the same foundation as gravitational forces. Classically 'real' forces are normally assumed not to vary with the reference frame; however, the Lorentz force on a charge moving in a magnetic field is a disturbing exception with which we shall have to deal later.

Returning to Equation (1.75), with sufficient accuracy in the measurements we can calculate the second time derivative of the Cartesian components in Equation (1.3) to obtain

$$\mathbf{a} = \ddot{x}\,\widehat{\mathbf{e}}_x + \ddot{y}\,\widehat{\mathbf{e}}_y + \ddot{z}\,\widehat{\mathbf{e}}_z, \tag{1.76}$$

where $\ddot{x}$ is a convenient notation for d^2x/dt^2.

The inertial mass may be found in units of a standard body (at present the international prototype kilogram kept in Sèvres near Paris at the International Bureau of Weights and Measures) by subjecting them both to the same external influence (such as the action of an extended standard spring) and measuring the ratio of the resultant accelerations in an inertial reference frame. In practice, since the gravitational force on a body is proportional to its inertial mass and independent of its atomic or molecular composition, we merely have to compare the 'weight'[7] of a body to the weight of the standard kilogram

[7] That is, the gravitational force on the body according to the second law at the surface of the Earth.

Figure 1.6 *This beautiful picture of the disc of our galaxy shows it to be an arc in the sky. In ancient times it might have been regarded as the projection of a straight line on the stellar sphere of the sky. In fact it is real. We live on a rotating disc of stars and dust and gas. The interesting point is that the terrestrial coordinate system is obviously rotated with respect to the natural system of the disc. Globally our terrestrial coordinates are spherical polar with the axis defined by the rotation axis. Galaxy coordinates would be oriented relative to the rotation axis of the galaxy. Astronomers must convert frequently between the two systems, using methods outlined in the text.* Source: *Reproduced with permission from cielosdelteide.com. Copyright 2010 Daniel López (See Plate 2.)*

in order to measure its inertial mass. Strictly this has to be at the same point on the Earth's surface.

This proportionality of the classical 'real' gravitational force to inertial mass is summarized in the famous 'inverse square law' of Newton

$$\mathbf{F}_{12} = G\frac{m_1 m_2}{r_{12}^2}\widehat{\mathbf{r}}_{12}, \tag{1.77}$$

which gives the real force on body 1 due to body 2 in terms of the product of the inertial masses $m_1 m_2$, the inverse square of the distance between them r_{12}^2 and the unit vector drawn from body 1 to body 2, $\widehat{\mathbf{r}}_{12}$. Newton's universal constant is denoted more or less universally by G. This law shows, together with Equation (1.75), that all bodies acted on gravitationally by a particular body (body 2 here, which is usually the Earth in our experience) will experience the same acceleration towards that body.

This latter behaviour, together with the lack of dependence on any other property such as composition, appears mysterious when gravity is compared to any of the other three fundamental physical forces, namely, the electromagnetic and the strong and weak nuclear forces.[8] We may consider in particular the familiar electromagnetic force. In

[8] A modern theoretical triumph has been to unite the electromagnetic force (united by Maxwell) with the weak nuclear force into the electro-weak interaction (Weinberg-Salam); however we await the Higgs field for complete experimental confirmation.

that case the acceleration imposed by the electric force on bodies at rest with the same inertial mass is proportional to the electric 'charge'.[9] The electromagnetic force on moving charges is also proportional to the charge.

Both Galileo [2] and Newton [3] demonstrated this universal acceleration of the gravitational force. They showed that the periods of pendula made from different quantities of different materials were the same, when moving under gravity at the surface of the Earth. A more precise determination came from the torsion balance experiments of Eötvös. However, it was finally Einstein who recognized this fact as having deep significance for the relationship of inertial forces to gravity, and finally banished the distinction.

We delay most of this discussion to later chapters, but we should observe here that we have so far given only an operational definition of an inertial frame of reference, without considering *why* or *what* it is in some physical sense. Experimentally we recognize many reference frames that are *not* inertial according to the first law, namely reference frames that are accelerated relative to a given inertial reference. This acceleration may be rotational or linear or both. The Earth itself, which we normally adopt as our inertial reference, is not strictly inertial since it rotates about its axis (e.g. Figure 1.8) and revolves about the Sun. Moreover all bodies on its surface are certainly non-isolated, being subject to the gravitational force exerted by the Earth and to the reaction of its solid surface. Fortunately the extent to which horizontal motion can be made inertial by reducing friction, and the degree to which the Earth reference frame is itself inertial, are sufficient to allow the empirical discovery of the first law.

In an effort to improve on the surface of the Earth as an inertial frame, we might correct all dynamic measurements to the reference frame at the centre of the Sun. But then we would inevitably encounter the acceleration of the Sun in the galaxy (e.g. Figure 1.6) and of the galaxy in the local group of galaxies, and so on in an increasing hierarchy until we are confronted with the mean Universe.

Newton himself clearly perceived this problem and short-circuited the process by assuming the existence of 'absolute space', with respect to which inertial frames were unaccelerated [3]. However, there is no independent evidence for such a thing, as was argued forcefully by Mach [4] (see also [5]), and it was discarded by Einstein. The modern understanding of the Universe is much more in accord with Mach's view that it is the mean Universe that physically determines inertial (freely falling) frames and inertial mass. However, we still do not have a proven physical mechanism that creates the inertial mass of a body.

The modern cosmological understanding of absolute 'rest' stems from the discovery of the cosmic microwave background (CMB) [6]. This discovery allows us to define a mean rest frame of the Universe as one whose velocity relative to us renders the temperature of this radiation isotropic to about[10] one part in 10^5. This may furnish the archetypal inertial frame.

It is also the case that there is an apparent temperature associated with linear aceleration relative to the quantum vacuum [7]. Although the temperature is exceedingly small for

[9] Electric charge must also be defined in terms of a standard procedure, such as the quantity required to release a mole of a monovalent ion in electrolysis. The real force between charges is summarized in Coulomb's law if at rest, and by the Lorentz force if in motion.

[10] There are random fluctuations in the temperature of about this order associated with the birth of small-scale structure in the Universe.

most practical accelerations, it would in principle allow a local inertial frame to be found to any measurable accuracy by minimizing this temperature. Unfortunately, the quantum vacuum is not yet incorporated into standard cosmology.

To understand the part of Mach's principle which treats inertial mass, we note that the fictional forces that appear under acceleration relative to an inertial frame are proportional to the inertial mass of a body, just as is the 'real' force of gravity. In fact the inertial mass of a body only appears when the body is compelled to depart from an inertial state. If the inertial state is determined relative to the mean Universe, it seems natural to suppose that there is some physical influence due to the mean Universe by which a body is given its inertial mass. This was Mach's view. We are left to wonder why acceleration with respect to this frame of reference produces 'fictional', that is 'inertial forces'. Familiar examples of these are the centrifugal or Coriolis forces that appear under rotation. What physical mechanism allows the mean Universe to establish the preferred frames of reference in which inertial forces do not appear? In the absence of an answer, Mach is not yet satisfied.

Such a mechanism is at present unproven, but not unimagined. The hypothetical Higgs field whose quanta are thought necessary to give mass to the gauge particles of particle physics [8] is thought to pervade all space. Thus it could be that this field defines the fundamental inertial frame[11] together with its mass-creating property. Before these remarks become public, the Large Hadron Collider (LHC) operated by CERN (originally 'Conseil Europèen pour La Recherche Nuclèaire') may have found the Higgs Boson as evidence for the Higgs field and as confirmation of the electro-weak unification.

In practice the hierarchical observational regression to the mean Universe in search of the archetypal inertial frame is carried out by astronomy as the search for the local 'standard of rest'. 'Local' implies good enough for the desired precision of dynamic measurements on the spatial scale of interest. The surface of the Earth is quite noticeably not inertial and the centre of the Earth is not much better. The centre of the Sun may serve as a solar system standard of rest, but not if one wishes to understand the velocity dispersion of stars in our neighbourhood or the structure of our galaxy (the 'Milky Way'). It is now quite common to refer velocities to the reference frame determined by the CMB as discussed above.

Assuming that we can adopt an inertial frame sufficient for the purposes at hand (all measurement is ultimately imperfect), we soon discover the first principle of relativity due essentially to Galileo, namely that *a reference frame moving with an arbitrary constant velocity v relative to a given inertial frame is also inertial*.

This Galilean principle means that Newton's laws (including the third law) have the same significance in this infinite set of reference frames. However the coordinates of events are not invariant between such frames. The principle excludes rotation so that if we restrict ourselves to orthogonal base vectors we can imagine relatively translating, rigid Cartesian inertial systems. Let the origin of one system be O, say a local standard of rest, and that of the another be O'. Let the relative velocity between them be the constant vector $\dot{\mathbf{R}} \equiv \mathbf{u}$. Then from Figure 1.7, which is representational of Euclidean space with the dimension normal to the paper expressed only by perspective[12] together

[11] It will have to be inserted into the dynamic Universe rather than the Euclidean one.

[12] Our figures are not always representational, in which case they are 'diagrams', as when we attempt to represent space-time.

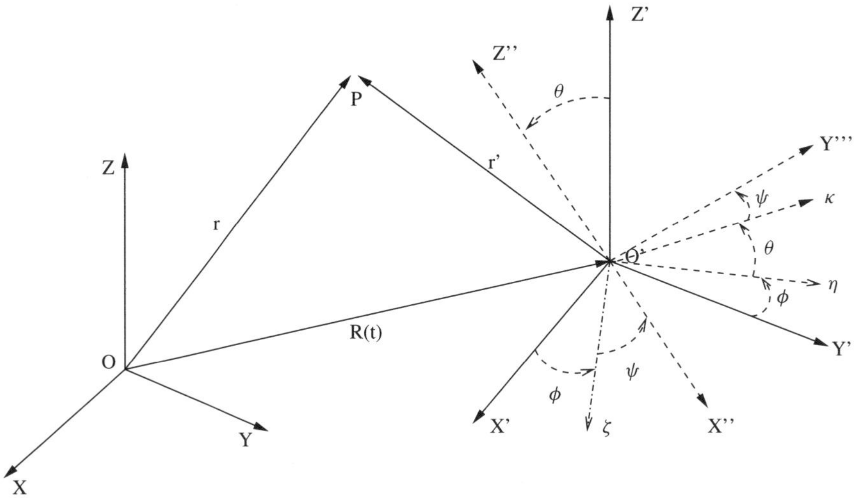

Figure 1.7 *This shows a Cartesian reference frame {X, Y, Z}, relative to which a linearly accelerated Cartesian reference frame {X′, Y′, Z′} is located by the vector* **R**(t). *This latter frame is not inertial since points in the inertial frame are accelerated with respect to it unless* $\dot{\mathbf{R}}$ *is constant, but the reference directions are inertial since they remain parallel to inertial directions. Relative to these accelerated axes the figure shows a third set of axes {X″, Y″, Z″} which are rotating relative to inertial directions as described by the Euler angles* ϕ(t), θ(t), ψ(t). *These double-primed axes are doubly non-inertial axes. The 'line of node axes' are those rotated only by* ϕ *relative to the primed axes and are labelled* (ζ, η, Z'). *The intermediate set rotated by both* ϕ *and* θ *are labelled* (ζ, κ, Z'')

with simple vector addition, we have that the two position vectors ($\mathbf{r}$ relative to O and $\mathbf{r}'$ relative to O') are related by

$$\mathbf{r} = \mathbf{r}' + \mathbf{u}t. \tag{1.78}$$

We have made use of our arbitrary time origin to set $t = 0$ at the instant where O and O' coincide, and we ignore for the moment the rotating double-primed axes and possible time dependence of $\mathbf{u}$. Since both unprimed and primed reference systems are Cartesian, we have the representation of Equation (1.3) for each of $\mathbf{r}$ and $\mathbf{r}'$. The arbitrary vector $\mathbf{u}$ would then necessarily be represented in terms of its Cartesian weights, that is components, to express fully Equation (1.78). If one reference frame is rotated with respect to the other as are the double primed axes, then $\mathbf{r}'$ would have the form $\mathbf{r}''$ in this frame. We would apply a one-time rotation to bring one set of axes parallel to the other according to $\mathbf{r}' = \tilde{\underline{\underline{\mathbf{S}}}}\mathbf{r}''$. The representation of Equation (1.78) would then be in the mutually parallel axes. We know that the orthogonality property of a rotation $\underline{\underline{\mathbf{S}}}$ ensures that $\mathbf{r}'^2 = \mathbf{r}''^2$ so that distance is preserved under the rotation.

Suppose now that we wish to compare the distances between two closely spaced points at rest in the local standard of rest. Then we find $d\mathbf{r}^2$ from Equation (1.78) as

$$d\mathbf{r}^2 = d\mathbf{r}'^2 + 2\mathbf{r}' \cdot \mathbf{u}dt + \mathbf{u}^2 dt^2. \tag{1.79}$$

But there is no reason to prefer O to O', so that these two inertial observers should agree on this objective distance, that is $d\mathbf{r}^2 = d\mathbf{r}'^2$. Fortunately all is well if O and O' have synchronized clocks that allow the measurements to be carried out simultaneously, for then $dt = 0$ and the distance is agreed between any two inertial observers, and hence by all.

This assumption of a universally agreed time, kept presumably by synchronized atomic clocks, normally passes without comment. We have already discussed the mechanism for achieving this above and we have hinted at some difficulties both practical and in principle (for example, the impossibility of sharing proper time with a Mars observer). However, it seems a very natural assumption based on experience, and Newton famously adopted it explicitly, no doubt after long reflection on the meaning of t in his second law. It is worth quoting Newton on this point (as translated in [3]) since no-one has expressed this assumption better.

> Absolute, true, and mathematical time, of itself, and from its own nature flows equably without regard to anything external, and by another name is called duration.

Such a time coordinate, which we refer to as 'Newtonian coordinate time', one hopes to measure for events by arbitrarily accurate clocks. We see why Newton insists on this unique time if we calculate the acceleration used in his second law in terms of a transformed coordinate time $T(t)$, where t is Newtonian coordinate time. We find by two applications of the chain law that

$$\frac{d^2\mathbf{r}}{dt^2} = \frac{d^2T}{dt^2}\frac{d\mathbf{r}}{dT} + \left(\frac{dT}{dt}\right)^2 \frac{d^2\mathbf{r}}{dT^2}. \tag{1.80}$$

If the second law is to have the same form in the new time, then the transformation $T(t)$ must be linear so that the first term vanishes and the second term is the new acceleration multiplied by a constant. This arbitrary multiplicative constant equal to $(dT/dt)^2$ amounts to a change in units. Thus to within a unit of measurement, Newton's mathematical time must be unique in a given inertial frame and independent of any choice of reference frame.

However, Newton goes on to indicate technical difficulties in obtaining the measure of this time [3]:

> Relative, apparent, and common time, is some sensible and external (whether accurate or inequable) measure of duration by the means of motion, which is commonly used *instead of true time*; such as an hour, a day, a month, a year.

In other words absolute time exists, but we only approximate its measure in practice. Most of our introduction to relativity in the next chapter will involve adopting the positivist view: namely that there is no reason to assume the existence of something that we can not necessarily measure. In this respect, absolute time and absolute space are similar, and we shall have to dispense with both of them.

Figure 1.8 *The Earth rotates! Well at least these star trails circling the pole star 'Polaris', indicate that either the distant stars rotate around us or that we rotate. However, measurements of 'inertial forces' on Earth show that they are present, according to the assumption of terrestrial rotation. The stars indicate an inertial frame on average that astronomers refer to as a 'local standard of rest'. On the scale of the visible stars it becomes apparent that this inertial reference is not exact and one must regress to larger scales. The regression continues until the mean Universe itself is reached.* Source: *Reproduced with permission from Dr F.-J. (Josch) Hambsch, http://www.astronomie.be/hambsch/namibia06/startrails1.htm (See Plate 3.)*

Equation (1.78), by differentiation with respect to absolute time, also gives us the Galilean transformation of apparent velocities of moving points between inertial observers as

$$\mathbf{v} = \mathbf{v}' + \mathbf{u}, \tag{1.81}$$

where $\mathbf{v} \equiv d\mathbf{r}/dt$ and $\mathbf{v}' \equiv d\mathbf{r}'/dt$. For the moment, then, a prime indicates a quantity defined for the frame of O'. Below it will sometimes mean $\mathbf{r}$ with components relative to the axes of O'. This ambiguity is difficult to avoid in vector equations except when the 'resolution' of vectors along relatively rotated axes is done explicitly by matrix operators.

The acceleration is clearly invariant under the transformation (1.78) between inertial observers, as it must be if real forces defined by the second law of Newton are to be invariant under the change of inertial observer.

In order to emphasize the peculiar special nature of inertial frames, it is useful to consider transformations to non-inertial frames. It is convenient in the discussion of rotation to consider relatively rotating Cartesian reference frames. Moreover we think of the rotation matrix $\underline{\underline{\mathbf{S}}}$ in an *active* sense of Equation (1.69). That is, instead of rotating the reference frame, we consider the vector to be rotated in the opposite sense. In that case Equation (1.69) gives the relation between the components of the new and old vectors in the *same* Cartesian reference frame. The form of $\underline{\underline{\mathbf{S}}}$ does not depend on whether the vector is rotated in one sense or the reference frame is rotated in the opposite sense (see e.g. Goldstein p. 136). Thus the base vectors together with any vector fixed relative to them are rotated into new positions in inertial space.

As before let O be the origin of an inertial Cartesian reference frame. We let the Cartesian frame with origin O' be non-rotating with respect to O, but now it is non-inertial due to an arbitrary motion $\mathbf{R}(t)$ that may include acceleration. Then, as was already used in Equation (1.78), Figure 1.7 shows that

$$\mathbf{r} = \mathbf{r}' + \mathbf{R}(\mathbf{t}), \tag{1.82}$$

for any point located both with respect to O and O' when $\mathbf{R}(\mathbf{t})$ locates O' relative to O in Newtonian time. However, let us now consider a set of points that form a three-dimensional rigid physical structure or body. We have seen that all Cartesian reference frames are such rigid structures. Thus we can very well imagine a Cartesian reference frame based on three perpendicular axes drawn through points in the rigid body in question. All points in the body will by virtue of its structure have fixed coordinates relative to these 'body axes'. We suppose now that these body axes together with the body are rotating in some general way relative to inertial axes. The position vector of a point relative to body axes will be denoted $\mathbf{r}''$. We may now write that $\mathbf{r}'' = \underline{\underline{\mathbf{S}}}\mathbf{r}'$, for the rotation of $\mathbf{r}'$ into $\mathbf{r}''$, using the active nature of $\underline{\underline{\mathbf{S}}}$. Note that $\mathbf{r}'$ is the position vector of a point in the rigid body relative to (resolved along) the primed axes of Figure 1.7.

We recall here that for continuing rotation rather than once rotated, the angles that occur in the rotation matrix will be functions of time. For a rigidly rotating body, however, $\mathbf{r}''$ is not a function of time. Hence by differentiating the expression $\mathbf{r}'' = \underline{\underline{\mathbf{S}}}\mathbf{r}'$ we obtain

$$\mathbf{0} = \underline{\underline{\dot{\mathbf{S}}}}\mathbf{r}' + \underline{\underline{\mathbf{S}}}\dot{\mathbf{r}}', \tag{1.83}$$

where the time derivative of a matrix is simply the matrix of the time derivatives. Multiplying the latter equation on the left by $\underline{\underline{\widetilde{\mathbf{S}}}}$, using the orthogonality property of $\underline{\underline{\mathbf{S}}}$, and rearranging, we have an expression for the velocity of a point in the rotating body expressed in non-rotating axes as

$$\dot{\mathbf{r}}' = -\underline{\underline{\widetilde{\mathbf{S}}}}\,\underline{\underline{\dot{\mathbf{S}}}}\mathbf{r}'. \tag{1.84}$$

The matrix product that occurs in Equation (1.84) gives the velocity of a fixed point in a rigid body relative to inertial axes, by matrix multiplication with its inertial position vector. We refer to it as the angular velocity 'matrix' or 'operator' and give it the name $\underline{\underline{\mathbf{\Omega}}}'$ so that

$$\underline{\underline{\mathbf{\Omega}}}' \equiv \underline{\underline{\widetilde{\mathbf{S}}}}\,\underline{\underline{\dot{\mathbf{S}}}}. \tag{1.85}$$

Hence the rotational velocity of a point in a rotating body becomes, as expressed in components along the singly primed axes,

$$\dot{\mathbf{r}}' = -\underline{\underline{\mathbf{\Omega}}}'\mathbf{r}'. \tag{1.86}$$

Notice that unlike $\underline{\underline{\mathbf{S}}}$, $\underline{\underline{\mathbf{\Omega}}}'$ is always an active operator that transforms $\mathbf{r}'$ into $\dot{\mathbf{r}}'$.

Problem

1.13 **(a)** Show that the angular velocity matrices associated with each of the Euler angles (see Figure 1.7) may be written in their respective frames of reference as

$$\underline{\underline{\boldsymbol{\Omega}}}_{\phi} = \dot{\phi}\underline{\underline{\boldsymbol{\sigma}}}_{z}, \tag{1.87}$$

$$\underline{\underline{\boldsymbol{\Omega}}}_{\psi} = \dot{\psi}\underline{\underline{\boldsymbol{\sigma}}}_{z}, \tag{1.88}$$

$$\underline{\underline{\boldsymbol{\Omega}}}_{\theta} = \dot{\theta}\underline{\underline{\boldsymbol{\sigma}}}_{x}, \tag{1.89}$$

where

$$\underline{\underline{\boldsymbol{\sigma}}}_{z} \equiv \begin{pmatrix} 0 & 1 & 0 \\ -1 & 0 & 0 \\ 0 & 0 & 0 \end{pmatrix}, \tag{1.90}$$

and

$$\underline{\underline{\boldsymbol{\sigma}}}_{x} \equiv \begin{pmatrix} 0 & 0 & 0 \\ 0 & 0 & 1 \\ 0 & -1 & 0 \end{pmatrix}. \tag{1.91}$$

(b) Verify that these matrices produce the same rotational velocity when operating on a position vector in their axes, as do the corresponding vector rotational velocities, $\boldsymbol{\omega}_{\phi} = \dot{\psi}\widehat{\mathbf{e}}_{z}$, $\boldsymbol{\omega}_{\theta} = \dot{\theta}\widehat{\mathbf{e}}_{\zeta}$, $\boldsymbol{\omega}_{\psi} = \dot{\psi}\widehat{\mathbf{e}}_{z''}$.

Both the velocity and the position vector are described relative to inertial axes, and so the angular velocity matrix as defined above is also relative to these axes. We label it with a prime to remind us of our translating frame with inertial axes, but it would clearly have the same form for the inertial observer O, who uses parallel axes. To avoid awkwardness in description, we often say that a vector that is expressed in terms of components along certain axes is 'resolved along' these axes. Operationally this is effected by the operation of the rotation matrix or of its inverse, when each is used in the passive sense. We may also say that the angular velocity matrix is 'resolved' along inertial axes. This is effected operationally by a 'passive similarity transformation' as is defined immediately below.

If we wish to express Equation (1.84) in rotating axes, it suffices to act on both sides of the equation with the rotation matrix $\underline{\underline{\mathbf{S}}}$ according to Equation (1.67). We are using now the passive interpretation of the rotation matrix. This gives the velocity of a point in the rigid body that is *defined in inertial axes but resolved along rotating axes* as

$$\underline{\underline{\mathbf{S}}}\dot{\mathbf{r}}' = -\underline{\underline{\mathbf{S}}}\,\underline{\underline{\boldsymbol{\Omega}}}'\,\widetilde{\underline{\underline{\mathbf{S}}}}\mathbf{r}'', \tag{1.92}$$

where we have used the orthogonality property to substitute the inverse active mapping $\mathbf{r}' = \widetilde{\underline{\underline{\mathbf{S}}}}\mathbf{r}''$. This equation now expresses the angular velocity of a point in a rotating body wholly in rotating (double primed) axes, so that the matrix operator appearing here must be the angular velocity matrix expressed in these rotating axes $\underline{\underline{\boldsymbol{\Omega}}}''$. Hence we have the important formula for transforming the angular velocity matrix between sets of relatively rotating Cartesian reference frames as

$$\underline{\underline{\boldsymbol{\Omega}}}'' = \underline{\underline{\mathbf{S}}}\,\underline{\underline{\boldsymbol{\Omega}}}'\,\underline{\underline{\widetilde{\mathbf{S}}}}. \tag{1.93}$$

Such a matrix transformation is referred to as a 'similarity transformation'. The angular velocity operators are similar, merely resolved along different reference directions.

Problem

1.14 Show that the total angular velocity (see below) of a body may be written as

$$\boldsymbol{\omega} = \boldsymbol{\omega}_\phi + \boldsymbol{\omega}_\theta + \boldsymbol{\omega}_\psi, \tag{1.94}$$

where each vector is represented by its components in ('resolved along') the same set of axes. Recall that $\underline{\underline{\mathbf{S}}} = \underline{\underline{\mathbf{S}}}_\psi\,\underline{\underline{\mathbf{S}}}_\theta\,\underline{\underline{\mathbf{S}}}_\phi$ in Equation (1.85) and that the Leibnitz rule of differentiation applies to matrices. You will have to recognize angular velocity matrices transformed to the inertial (primed) axes from the inverse of Equation (1.93) and the appropriate rotation to inertial axes.

Example 1.3

By the result of Problem 1.14 we can write the vector angular velocities in the axes in which they are defined as the column vectors

$$\boldsymbol{\omega}_\phi = \begin{pmatrix} 0 \\ 0 \\ \dot{\phi} \end{pmatrix}, \tag{1.95}$$

$$\boldsymbol{\omega}_\theta = \begin{pmatrix} \dot{\theta} \\ 0 \\ 0 \end{pmatrix}, \tag{1.96}$$

$$\boldsymbol{\omega}_\psi = \begin{pmatrix} 0 \\ 0 \\ \dot{\psi} \end{pmatrix}. \tag{1.97}$$

This means that we can now resolve the vector sum

$$\boldsymbol{\omega} = \boldsymbol{\omega}_\phi + \boldsymbol{\omega}_\theta + \boldsymbol{\omega}_\psi \tag{1.98}$$

along inertial or rotated axes as desired.

Thus to obtain $\boldsymbol{\omega}_\phi$ in rotating axes we must calculate $\underline{\underline{\mathbf{S}}}_\psi\,\underline{\underline{\mathbf{S}}}_\theta\,\boldsymbol{\omega}_\phi$, which gives

$$\dot{\phi}\sin\theta\sin\psi\,\widehat{\mathbf{e}}_1 + \dot{\phi}\sin\theta\cos\psi\,\widehat{\mathbf{e}}_2 + \dot{\phi}\cos\theta\,\widehat{\mathbf{e}}_3,$$

along rotated axes as shown in Figure 1.5. To obtain $\boldsymbol{\omega}_\theta$ in rotated axes we calculate $\underline{\underline{\mathbf{S}}}_\psi \boldsymbol{\omega}_\theta$ which yields

$$\dot{\theta} \cos\psi\, \widehat{\mathbf{e}}_1 - \dot{\theta} \sin\psi\, \widehat{\mathbf{e}}_2.$$

Since $\boldsymbol{\omega}_\psi$ is already in rotated axes, the vector sum yields

$$\begin{aligned} \boldsymbol{\omega} = \;& (\dot{\theta}\cos\psi + \dot{\phi}\sin\theta\sin\psi)\widehat{\mathbf{e}}_1 + \\ & (-\dot{\theta}\sin\psi + \dot{\phi}\sin\theta\cos\psi)\widehat{\mathbf{e}}_2 + \\ & (\dot{\psi} + \dot{\phi}\cos\theta)\widehat{\mathbf{e}}_3. \end{aligned} \tag{1.99}$$

Problem

1.15 Follow a similar procedure to that of the previous example in order to resolve the vector angular velocity along inertial axes. Since $\boldsymbol{\omega}_\phi$ is already in inertial axes you will only have to calculate $\widetilde{\underline{\underline{\mathbf{S}}}}_\phi \boldsymbol{\omega}_\theta$ and $\widetilde{\underline{\underline{\mathbf{S}}}}_\phi \widetilde{\underline{\underline{\mathbf{S}}}}_\theta \widetilde{\underline{\underline{\mathbf{S}}}}_\psi \boldsymbol{\omega}_\psi$ where the inverse matrices are essential (why?). You should find

$$\begin{aligned} \boldsymbol{\omega} = \;& (\dot{\phi} + \dot{\psi}\cos\theta)\widehat{\mathbf{e}}_Z \\ & (\dot{\theta}\sin\phi - \dot{\psi}\sin\theta\cos\phi)\widehat{\mathbf{e}}_Y \\ & (\dot{\theta}\cos\phi + \dot{\psi}\sin\theta\sin\phi)\widehat{\mathbf{e}}_X. \end{aligned} \tag{1.100}$$

If we write the similarity transformation of Equation (1.93) in index notation we have our first example of the transformation of a multi-index quantity or tensor. For, respecting the row vector times column vector convention, the natural way to write the components of $\underline{\underline{\boldsymbol{\Omega}}}'$ is from Equation (1.85)

$$\Omega'^i{}_k = \tilde{S}^i{}_j \dot{S}^j{}_k, \tag{1.101}$$

where the sum over j correctly multiplies the ith row vector in $\widetilde{\underline{\underline{\mathbf{S}}}}$ into the kth column vector in $\dot{\underline{\underline{\mathbf{S}}}}$. The offset in the indices is used to clarify further which is the column index (upper, first) and which is the row index (lower, right). With a similar understanding of the notation, Equation (1.93) becomes

$$\Omega''^\ell{}_m = S^\ell{}_i \Omega'^i{}_k \tilde{S}^k{}_m. \tag{1.102}$$

If we remember Equation (1.69) and take x^j there to be the rotated Cartesian x''^j here, and the coordinates ℓ^j to be the inertial Cartesian x'^j, then differentiating we obtain

$$S^i{}_j = \frac{\partial x''^i}{\partial x'^j}. \tag{1.103}$$

Moreover, because of the uniqueness of the matrix inverse we can write it as

$$\tilde{S}^k{}_m = \frac{\partial x'^k}{\partial x''^m}, \tag{1.104}$$

since then $\tilde{S}^k_m S^m_j = \delta^i_j$ as required. Consequently Equation (1.102) can also be written in a form similar to the transformation of tensors under general coordinate transformations (although here we discover it between rotated Cartesian coordinates) as

$$\Omega''^{\ell}{}_m = \frac{\partial x''^{\ell}}{\partial x'^i}\frac{\partial x'^k}{\partial x''^m}\Omega'^i{}_k. \tag{1.105}$$

Note that the transformation of the row or contravariant index i is inverse to the transformation of the column or covariant index k.

Avoiding coordinate notation for the moment, we return to Equation (1.93) for $\underline{\underline{\mathbf{\Omega}}}''$. Sustituting the definition of $\underline{\underline{\mathbf{\Omega}}}'$ this gives $\underline{\underline{\mathbf{\Omega}}}''$ in terms of the rotation matrix as

$$\underline{\underline{\mathbf{\Omega}}}'' = \underline{\underline{\dot{\mathbf{S}}}}\underline{\underline{\tilde{\mathbf{S}}}}. \tag{1.106}$$

Although the rotation matrix can be quite complex, the angular velocity matrix is actually rather simple. By differentiating the orthogonality condition in the form of Equation (1.72) one obtains

$$\underline{\underline{\dot{\tilde{\mathbf{S}}}}}\underline{\underline{\mathbf{S}}} + \underline{\underline{\tilde{\mathbf{S}}}}\underline{\underline{\dot{\mathbf{S}}}} = \underline{\underline{\mathbf{0}}}. \tag{1.107}$$

Since by the respective definitions the transpose and the derivative operations commute, this last equation says that the transpose of the second term (which is the transpose of $\underline{\underline{\mathbf{\Omega}}}'$) equals the negative of the transpose of the first term (which is $\underline{\underline{\mathbf{\Omega}}}'$). Hence *the angular velocity matrix is an antisymmetric matrix*. As such it has only three independent off-diagonal elements that we group together into a column vector $\{\omega'^k\}$. The location of these off-diagonal elements in $\underline{\underline{\mathbf{\Omega}}}'$ is assigned according to

$$\Omega'^{\alpha}{}_{\beta} = \epsilon^{\alpha}{}_{\beta k}\omega'^k. \tag{1.108}$$

Here the epsilon symbol is simply $+1$ if $\{\alpha, \beta, k\}$ are an even (cyclic) permutation of $\{1, 2, 3\}$, -1 if an odd (acyclic) permutation, and zero if any two of the indices are the same. We do this in order to write the rotational velocity of Equation (1.86) in a more intuitive form. Thus in component notation, Equation (1.86) now becomes

$$\dot{x}'^i = -\Omega'^i{}_j x'^j = -\epsilon^i{}_{jk}\omega'^k x'^j \equiv \epsilon^i{}_{kj}\omega'^k x'^j, \tag{1.109}$$

or in vector notation

$$\dot{\mathbf{r}}' = \boldsymbol{\omega}' \wedge \mathbf{r}'. \tag{1.110}$$

Since this is true for any point in the rigid body, we call $\boldsymbol{\omega}'$ the 'angular velocity' of the body and normally treat it as column vector. An arbitrary vector $\mathbf{A}$ whose components are fixed in rotating axes (so it is a rotating vector) will also satisfy this latter equation.

We note in passing the useful mixed matrix-vector relation that follows from the preceding argument in the form

$$\underline{\underline{\mathbf{\Omega}}}'\mathbf{r}' = -\boldsymbol{\omega}' \wedge \mathbf{r}'. \tag{1.111}$$

If rather than differentiate Equation (1.72) we differentiate the equivalent $\underline{\underline{\mathbf{S}}}\,\underline{\underline{\widetilde{\mathbf{S}}}} = \underline{\underline{\mathbf{1}}}$ with respect to time, we find that $\underline{\underline{\mathbf{\Omega}}}''$ is also antisymmetric and may therefore be written in terms of quantities $\{\omega''^k\}$ just as was done above for $\underline{\underline{\mathbf{\Omega}}}'$. These quantities are related to the $\{\omega'^k\}$ implicitly through the transformation (1.93). This can be done in any set of axes so that, in particular, Equation (1.111) can be written in whatever reference frame is in use.

In fact it may be shown using Equation (1.93) that the three elements of the angular velocity matrix transform as a column vector under any orthogonal transformation except one that involves a change of handedness (see footnote 6), for which $\det \underline{\underline{\mathbf{S}}} = -1$. In that case the $\boldsymbol{\omega}$ vector component normal to a reflecting mirror does not change sign under reflection (which changes the handedness), since both it and the corresponding Cartesian axis change sign (the $\boldsymbol{\omega}$ vector may be taken to lie along one of these axes). Such a quantity is called an 'axial vector' to distinguish it from normal vectors whose components do change sign under a change of handedness. Any cross product of normal vectors behaves this way, as is readily seen by reversing the direction of all of the axes of the reference frame. Then the components of the normal vectors all change sign, but the cross product does not. Since this distinction does not occur except under reflection, which is not a rigid rotation, we normally ignore this exception and treat angular velocities and cross products in general as true vectors.

Problems

1.16 Show that under the simple rotation $\underline{\underline{\mathbf{S}}}_{\phi}$, the angular velocity in the rotating axes $\underline{\underline{\mathbf{\Omega}}}''(\phi)$ is equal to the angular velocity in the inertial axes in the opposite sense, $\underline{\underline{\mathbf{\Omega}}}'(-\phi)$.

1.17 **(a)** Consider the combined rotation $\underline{\underline{\mathbf{S}}} = \underline{\underline{\mathbf{S}}}_{\theta}\,\underline{\underline{\mathbf{S}}}_{\phi}$. Show that $\underline{\underline{\mathbf{\Omega}}}' - \underline{\underline{\mathbf{\Omega}}}'_{\phi} + \underline{\underline{\widetilde{\mathbf{S}}}}_{\phi}\,\underline{\underline{\widetilde{\mathbf{S}}}}_{\theta}\,\underline{\underline{\dot{\mathbf{S}}}}_{\theta}\,\underline{\underline{\mathbf{S}}}_{\phi}$. Note that the last term is the theta rotation in the line of node axes (refer to Figure 1.7) transformed to inertial axes.

(b) Show that the last term in the previous expression is explicitly

$$\dot{\theta}\begin{pmatrix} 0 & 0 & -\sin\phi \\ 0 & 0 & \cos\phi \\ \sin\phi & \cos\phi & 0 \end{pmatrix} \tag{1.112}$$

Hence show that the negative of this operation on the position vector $\mathbf{r}'$ yields the same result as $\dot{\theta}\mathbf{e}_{\zeta} \wedge \mathbf{r}'$(see Figure 1.7, $\mathbf{e}_{\zeta}$ is along the line of nodes).

1.18 **(a)** For the same combined rotation in ϕ and θ as in the previous problem, show that in the rotated axes

$$\underline{\underline{\mathbf{\Omega}}}'' = \underline{\underline{\mathbf{\Omega}}}''_{\theta} + \underline{\underline{\mathbf{S}}}_{\theta}\,\underline{\underline{\dot{\mathbf{S}}}}_{\phi}\,\underline{\underline{\widetilde{\mathbf{S}}}}_{\phi}\,\underline{\underline{\widetilde{\mathbf{S}}}}_{\theta}. \tag{1.113}$$

(b) Calculate the last term in part (a) of this problem to be explicitly

$$\dot{\phi}\begin{pmatrix} 0 & \cos\theta & \sin\theta \\ -\cos\theta & 0 & 0 \\ \sin\theta & 0 & 0 \end{pmatrix} \tag{1.114}$$

Hence confirm that the negative of this operation on the position vector $\mathbf{r}$ in the line of node axes yields the same result as $\dot{\phi}\mathbf{e_z} \wedge \mathbf{r}$ (see Figure 1.7).

Let us turn to consider the acceleration of a point relative to rotating rigid axes. We cease now to regard all points as fixed relative to these axes. The rotating axes continue to furnish a rigid non-inertial reference frame, but the position of a given point relative to these axes may now be a function of time $\mathbf{r}''(t)$. We know that the operation of rotating the position vector of a point in the translating inertial axes $\mathbf{r}'$ is effected actively by $\underline{\underline{\mathbf{S}}}(t)$ according to $\mathbf{r}'' = \underline{\underline{\mathbf{S}}}\mathbf{r}'$. Differentiating this with respect to time gives

$$\dot{\mathbf{r}}'' = \dot{\underline{\underline{\mathbf{S}}}}\,\underline{\underline{\widetilde{\mathbf{S}}}}\mathbf{r}'' + \underline{\underline{\mathbf{S}}}\dot{\mathbf{r}}', \tag{1.115}$$

where we have used the inverse (active) transformation in the first term on the right. We may also resolve this equation along the translating inertial axes by multiplying by $\underline{\underline{\widetilde{\mathbf{S}}}}$ on the left to obtain after slight rearrangement

$$\dot{\mathbf{r}}' = \underline{\underline{\widetilde{\mathbf{S}}}}\dot{\mathbf{r}}'' - \underline{\underline{\mathbf{\Omega}}}'\mathbf{r}'. \tag{1.116}$$

We have used $\mathbf{r}' = \underline{\underline{\widetilde{\mathbf{S}}}}\mathbf{r}''$ and the definition of $\underline{\underline{\mathbf{\Omega}}}'$. This illustrates again the operational method of resolving vectors along the axes of relatively rotating frames, but we wish to continue our discussion with Equation (1.115).

We may recognize $\underline{\underline{\mathbf{\Omega}}}''$ in the first term on the right of Equation (1.115) so that using the definition of angular velocity and solving for the second term on the right we have

$$\underline{\underline{\mathbf{S}}}\dot{\mathbf{r}}' = \dot{\mathbf{r}}'' + \boldsymbol{\omega}'' \wedge \mathbf{r}''. \tag{1.117}$$

The term on the left is the operational form of resolving the velocity in translating inertial axes along the rotating axes.

In a more intuitive form[13] Equation (1.117) becomes

$$(\mathbf{v}')_{O''} = \mathbf{v}'' + \boldsymbol{\omega}'' \wedge \mathbf{r}''. \tag{1.118}$$

In any vector equation the various vectors involved must be resolved along the same set of axes. In this latter equation we see this explicitly for the rotating axes just as in Equation (1.117), since $()_{O''}$ also means resolved along rotating axes. In Equation (1.116) it is resolved operationally along translating inertial axes.

[13] There is a certain confusion in using $\mathbf{v}'$ and $\mathbf{v}''$ as follows, since $\mathbf{v}''$ is not the velocity in translating inertial axes resolved along rotating axes. It is rather the velocity defined relative to rotating axes.

However, it is the nature of vectors that they may be resolved formally along any set of axes, and so the operational form in Equation (1.116) is often written simply as

$$\mathbf{v}' = \mathbf{v}'' + \boldsymbol{\omega}' \wedge \mathbf{r}', \tag{1.119}$$

where the resolution along any set of axes common to all of the vectors is implicit. Unfortunately this is a case where $\mathbf{v}''$ means a velocity calculated with respect to rotating axes. If written explicitly, the double primed vector $\mathbf{v}''$ in the formula would be enclosed in brackets with subscript O'. Although this representation (1.119) is convenient, the precise statement is always in terms of the operational Equations (1.116) or (1.117).

By repeating all of these steps for an arbitrary vector $\mathbf{A}'$ in place of the position vector, we can write the equivalent of Equation (1.116) along translating inertial axes as,

$$\frac{d\mathbf{A}'}{dt} = \left(\frac{\partial \mathbf{A}''}{\partial t}\right)_{O'} + \boldsymbol{\omega}' \wedge \mathbf{A}', \tag{1.120}$$

and $\left(\frac{\partial \mathbf{A}''}{\partial t}\right)$ is defined in the double-primed axes. The partial derivative refers to holding the base vectors fixed during this operation. Because of the arbitrariness of $\mathbf{A}$, this latter equation is sometimes [9] written as an operator equation between time derivatives in the form

$$\frac{d()_{O'}}{dt} = \frac{\partial ()_{O''}}{\partial t} + \boldsymbol{\omega}' \wedge (). \tag{1.121}$$

The subscripts indicate the reference frame in which the vectors to be differentiated with respect to time are defined, and the partial derivative is taken holding the base vectors in rotating axes fixed.

We return to our discussion of the second time derivative of position by writing Equation (1.115) in the form

$$\dot{\mathbf{r}}'' = \underline{\underline{\boldsymbol{\Omega}}}''\mathbf{r}'' + \underline{\underline{\mathbf{S}}}\dot{\mathbf{r}}'. \tag{1.122}$$

The first operation on the right is an active transformation of the position vector into an angular velocity, while the second operation is a passive resolution of the velocity along rotating axes.

We differentiate this expression once more with respect to time to obtain

$$\ddot{\mathbf{r}}'' = \dot{\underline{\underline{\boldsymbol{\Omega}}}}''\mathbf{r}'' + \underline{\underline{\boldsymbol{\Omega}}}''\dot{\mathbf{r}}'' + \dot{\underline{\underline{\mathbf{S}}}}\dot{\mathbf{r}}' + \underline{\underline{\mathbf{S}}}\ddot{\mathbf{r}}'. \tag{1.123}$$

Now we wish to substitute from Equation (1.116) for $\dot{\mathbf{r}}'$ in the second last term, but in that expression we write $\underline{\underline{\boldsymbol{\Omega}}}' \equiv \tilde{\underline{\underline{\mathbf{S}}}}\dot{\underline{\underline{\mathbf{S}}}}$ as its definition in terms of the rotation matrix and we also write $\mathbf{r}'$ as $\tilde{\underline{\underline{\mathbf{S}}}}\mathbf{r}''$. We may then substitute the resulting expression for $\dot{\mathbf{r}}'$ in our last equation, rearrange the various terms and use the definition of $\underline{\underline{\boldsymbol{\Omega}}}'' \equiv \dot{\underline{\underline{\mathbf{S}}}}\tilde{\underline{\underline{\mathbf{S}}}}$ where appropriate to find

$$\ddot{\mathbf{r}}'' = 2\underline{\underline{\boldsymbol{\Omega}}}''\dot{\mathbf{r}}'' - \underline{\underline{\boldsymbol{\Omega}}}''^2\mathbf{r}'' - \dot{\underline{\underline{\boldsymbol{\Omega}}}}''\mathbf{r}'' + \underline{\underline{\mathbf{S}}}\ddot{\mathbf{r}}'. \tag{1.124}$$

This expression gives a precise form for the acceleration measured in a generally rotating Cartesian reference frame. When multiplied by an arbitrary mass, the first term on the right is the Coriolis force while the second term is the centrifugal force. The third term is sometimes called the Poincaré force and it is only present when there is rotational acceleration. We recognize the last term as the resolution of the real acceleration in the inertial frame (the real forces per unit mass) resolved along the rotating axes. *None of the other 'forces' are real*. They arise solely because we have used a reference frame that is not inertial, and they have the property of the gravitational force in that they produce always the same acceleration regardless of mass or composition. In the non-inertial frame these accelerations are real enough in that their effects are readily observed. Once again we see that inertial frames embody a deep mystery regarding the physical influence that prefers them.

We can use Equation (1.111) to write Equation (1.124) in a more familiar vector form as

$$\ddot{\mathbf{r}}'' = -2\boldsymbol{\omega}'' \wedge \dot{\mathbf{r}}'' - \boldsymbol{\omega}' \wedge \boldsymbol{\omega}' \wedge \mathbf{r}'' - \dot{\boldsymbol{\omega}}'' \wedge \mathbf{r}'' + (\ddot{\mathbf{r}}')_{O''}. \tag{1.125}$$

If finally we recall Equation (1.82) and differentiate it twice we may write

$$\ddot{\mathbf{r}}' = \ddot{\mathbf{r}} - \ddot{\mathbf{R}}. \tag{1.126}$$

Consequently Equation (1.125) may be used to relate the acceleration in rotating accelerated axes to that in the inertial frame as

$$\ddot{\mathbf{r}}'' = -2\boldsymbol{\omega}'' \wedge \dot{\mathbf{r}}'' - \boldsymbol{\omega}' \wedge \boldsymbol{\omega}' \wedge \mathbf{r}'' - \dot{\boldsymbol{\omega}}'' \wedge \mathbf{r}'' + (\ddot{\mathbf{r}} - \ddot{\mathbf{R}})_{O''}. \tag{1.127}$$

In this expression $\ddot{\mathbf{r}}$ is known in terms of the real forces in the inertial frame, divided by the mass of the object in question.

All of this discussion has served mainly to introduce concepts using the classical description (since Newton) of the world, that will help us to better understand the modern relativistic view. However, the classical view has been in its own way 'relativistic' ever since Galileo recognized the invariance between inertial frames. We have emphasized the peculiar nature of these frames by finding the apparent forces akin to gravity that appear when inertial frames are abandoned. The picture has long served us well and continues to do so in most parts of reality and indeed in applied physics, so the formulaic results of this chapter must be our starting point for the next stage of our adventure. They are 'practical relativity' in their own right.

As a final demonstration of the usefulness of the methods used, we conclude this chapter with formal examples that give a practical solution for identifying a constant vector in physical curvilinear coordinates. With such an identification, the true change in a vector with position may also be calculated wholly in terms of these coordinates.

Example 1.4

An arbitrary vector $\mathbf{A}$ may be expressed in terms of its Cartesian components $\mathbf{A}_{(c)}$ and in terms of physical components relative to a rotated curvilinear system as $\mathbf{A}_{(q)}$. Our technique replaces an explicit variation of the base vectors with an matrix operation, so the base vectors should be treated as constants under differentiation while the matrix elements vary. On rotating back passively from curvilinear to Cartesian components we have

$$\mathbf{A}_{(c)} = \widetilde{\underline{\underline{\mathbf{S}}}}\mathbf{A}_{(q)}. \tag{1.128}$$

Under parallel displacement (which maintains a constant vector) $d\mathbf{A}_{(c)} = 0$, from which we obtain $0 = d\underline{\underline{\widetilde{\mathbf{S}}}}\mathbf{A}_{(q)} + \underline{\underline{\widetilde{\mathbf{S}}}}\delta\mathbf{A}_{(q)}$ and hence

$$\delta\mathbf{A}_{(q)} = -\left(\underline{\underline{\mathbf{S}}}\, d\underline{\underline{\widetilde{\mathbf{S}}}}\mathbf{A}\right)_{(q)}. \tag{1.129}$$

This gives the apparent change in the curvilinear physical coordinates for the parallel displacement of a vector, which we distinguish by the δ operator rather than the Leibnitz operator d. Note that the row of the 3×3 matrix operating on $\mathbf{A}$ in this expression is designated by q. This gives the q component of $\delta\mathbf{A}$.

Consequently the *true* changes with position of the curvilinear physical components of an arbitrary vector are given by

$$\nabla\mathbf{A}_{(q)} \equiv d\mathbf{A}_{(q)} - \delta\mathbf{A}_{(q)}. \tag{1.130}$$

A 'true derivative' with respect to each physical displacement along a coordinate axis $d\ell_{(j)}$ follows as $\nabla\mathbf{A}_{(q)}/d\ell_{(j)}$. In a more convenient notation we have

$$\nabla_{(j)}\mathbf{A}_{(q)} \equiv \frac{\partial\mathbf{A}_{(q)}}{\partial\ell_{(j)}} + \left(\underline{\underline{\mathbf{S}}}\,\frac{\partial\underline{\underline{\widetilde{\mathbf{S}}}}}{\partial\ell_{(j)}}\mathbf{A}\right)_{(q)}. \tag{1.131}$$

Together with the examples, this argument illustrates the subtleties of differentiating vectors in curvilinear coordinates. This uses only our familiar Euclidian space and classical mechanics. Much later we shall need this concept in a general space. Although the concept is the same, the notation will be made more elegant (consider for example using Equation (1.69) to write Equation (1.131) in index notation).

Example 1.5

In this example we calculate the changes in the physical components of a vector (using cylindrical-polar coordinates $\{r, \phi, z\}$) under rotation through an angle ϕ about the z axis using the formula (1.131).

Under such a rotation we know

$$\underline{\underline{\mathbf{S}}} \equiv \begin{pmatrix} \cos\phi & \sin\phi & 0 \\ -\sin\phi & \cos\phi & 0 \\ 0 & 0 & 1 \end{pmatrix} \tag{1.132}$$

whence we find

$$\frac{\partial\underline{\underline{\widetilde{\mathbf{S}}}}}{\partial\phi} = \begin{pmatrix} -\sin\phi & -\cos\phi & 0 \\ \cos\phi & -\sin\phi & 0 \\ 0 & 0 & 0 \end{pmatrix}. \tag{1.133}$$

This allows us to find $\partial\underline{\underline{\mathbf{S}}}/\partial\ell_{(\phi)}$ by multiplying the last equation by $1/r$, and we observe that operating with $\partial_{(r)}$ or $\partial_{(z)}$ on $\underline{\underline{\widetilde{\mathbf{S}}}}$ gives zero. Consequently the second term in Equation (1.131) is zero for the r and z derivatives, and so the true derivative along these axes reduces to the usual derivative. For the derivative in the ϕ direction, the second term by direct calculation of the indicated matrix product becomes

$$\frac{1}{r}\begin{pmatrix} 0 & -1, & 0 \\ 1 & 0 & 0 \\ 0 & 0 & 0 \end{pmatrix}\begin{pmatrix} A_{(r)} \\ A_{(\phi)} \\ A_{(z)} \end{pmatrix}. \tag{1.134}$$

Thus from formula (1.131) we obtain, by remembering to hold the unit vectors constant in the first term and considering that each row in the matrix corresponds in order to r, ϕ, z,

$$\nabla_{(\phi)} A_{(r)} = \frac{1}{r}\frac{\partial A_{(r)}}{\partial \phi} - \frac{A_{(\phi)}}{r}, \tag{1.135}$$

$$\nabla_{(\phi)} A_{(\phi)} = \frac{1}{r}\frac{\partial A_{(\phi)}}{\partial \phi} + \frac{A_{(r)}}{r}, \tag{1.136}$$

and all other true derivatives equal the usual derivatives.

Example 1.6

We can use the true change of a vector to calculate its true time rate of change. Thus for the position vector r in cylindrical polar coordinates we have the true velocity as

$$\mathbf{v} \equiv \frac{\nabla \mathbf{r}}{dt}, \tag{1.137}$$

which explicitly, by Equation (1.130) and the term (1.134) when multiplied by $\dot{\phi}$, becomes

$$\mathbf{v} = \frac{d\mathbf{r}}{dt} + \begin{pmatrix} 0 & -\dot{\phi} & 0 \\ \dot{\phi} & 0 & 0 \\ 0 & 0 & 0 \end{pmatrix}\begin{pmatrix} r \\ 0 \\ 0 \end{pmatrix}, \tag{1.138}$$

that is $\mathbf{v} = \dot{r}\widehat{\mathbf{e}}_r + \dot{\phi} r \widehat{\mathbf{e}}_\phi$. The same procedure may be applied to the velocity in order to obtain the true acceleration when the trajectory of a particle is prescribed in cylindrical polar coordinates. We find

$$\mathbf{a} \equiv \frac{\nabla \mathbf{v}}{dt} = \frac{d\mathbf{v}}{dt} - \dot{\phi}\underline{\underline{\sigma}}_z \begin{pmatrix} \dot{r} \\ r\dot{\phi} \\ 0 \end{pmatrix}, \tag{1.139}$$

where $\dot{\phi}\underline{\underline{\sigma}}_z$ is the matrix in the expression for $\mathbf{v}$ and $\underline{\underline{\sigma}}_z$ appeared in Problem 1.13(a). This gives

$$\mathbf{a} = (\ddot{r} - (\dot{\phi})^2 r)\widehat{\mathbf{e}}_r + 2\dot{\phi}\dot{r}\widehat{\mathbf{e}}_\phi + \ddot{\phi} r \widehat{\mathbf{e}}_\phi. \tag{1.140}$$

It is left as an exercise to show that $\dot{\ell}_{(j)}\nabla_{(j)}\mathbf{v}$ will give the same result if partial derivatives are correctly summed in the end to give total derivatives.

References

1. Ferguson, K. (2008) *The Music of Pythagoras*, Walker & Co., New York, pp. 80–81.
2. Galileo, G. (2002) Dialogues Concerning Two Sciences, Paragraph 129, in *On the Shoulders of Giants* (ed. S. Hawking), Running Press, Philadelphia, p. 462.
3. Newton, I. (2002) *Principia*, Definition II, in *On the Shoulders of Giants* (ed. S. Hawking), Running Press, Philadelphia, p. 738.
4. Mach, E. (1933) *Die Mechanik* (9th edn), Brockhaus, Wiesbaden, pp. 226–7.
5. Wheeler, J. A. (1962) In *Gravitation and Relativity* (eds H.-Y. Chiu and W. F. Hoffman), Benjamin, New York.
6. COBE four year results (1996) Astrophysical Journal, **464**, L1, and references therein: Wilkinson Microwave Anisotropy Probe (WMAP), five year results, 2008.
7. Unruh, W. G. (1976) *Phys. Rev. D*, **14**, 870U.
8. Higgs, P. W. (1964) *Physics Letters*, **12**, 132.
9. Goldstein, H. (1980) *Classical Mechanics* (2nd edn), Addison-Wesley, Reading, MA. p. 176.

2

Invariance of Physical Law Under Change of Inertial Frame of Reference

Nothing comes from nothing.

Attributed to Epicurus by Lucretius in De Rerum Natura.

2.1 Prologue

We have seen in Chapter 1 that inertial frames of reference are of special significance for classical Newtonian dynamics. They are defined as those systems of reference in which Newton's first law holds in the absence of 'real' forces, which quantities are defined in turn in these same frames by the second law. Under the Galilean transformation of coordinates (Equation (1.78) together with $t' = t$) between inertial frames, these laws remain unchanged. We say that they are invariant under a change of inertial frame of reference, or that they are independent of the relative velocity between observers at rest in separate inertial frames (briefly 'inertial observers').

This invariance of the laws does *not* imply that separate inertial observers possess the same description of the world. As the train glides at a constant rate out of the station for the despatcher, the station glides in the opposite direction at a constant rate for the passenger. This variation of physical description relative to inertial observer is properly described as 'relativity'. It appears even though the dynamic laws for each observer are the same, save perhaps for initial conditions.

It is after all the train that changes inertial frames (taking the surface of the Earth to be inertial) initially, and this change has measurable effects that may be detected by any observer. Even after the train has settled into constant motion relative to the station,

Practical Relativity: From First Principles to the Theory of Gravity Richard N. Henriksen

a pendulum set into motion by the acceleration may still be swinging. This motion is periodic in the train frame but not in the reference frame of the station. There is then 'symmetry breaking' between inertial observers. We will have to allow for such physical consequences ultimately; but they arise during the change of inertial frame and so do not affect our arguments concerning given, different, inertial observers.

So all is well between Newtonian dynamics and Galilean inertial invariance; the laws of dynamics are invariant under the Galilean transformation. However, a long struggle by physicists to understand electricity and magnetism ended in 1873 with James Clerk Maxwell's unification of electricity and magnetism published in his *Treatise on Electricity and Magnetism*. This work summarized and unified experimental results and physical insights, as accumulated by many international workers, in a set of equations that are known universally as Maxwell's equations (occasionally Clerk-Maxwell's equations, to recognize his origins more closely).

It is difficult to exaggerate the triumph of this physical theory. It provides an explanation for light and a prediction for its propagation speed. The equations have withstood every experimental test inside and outside of matter. They have led to uncountable practical applications including the still-burgeoning field of wireless communication. Moreover, this theory of the electromagnetic field has guided the development of quantized gauge field theory and, as we shall argue, is largely responsible for discarding the Galilean transformation between inertial coordinates in favour of the Lorentz transformation. For in fact these equations, despite all of their triumphs, are not invariant under the Galilean transformation.

Should they be? The first philosophical reaction to this apparent preference for reference frame was that these 'light' waves described motion relative to a Universal reference frame of 'rest', similar to the absolute space of Newton. However the substance at rest in this frame had to be dynamic in order to support a mechanical interpretation of light waves, and it was generally referred to as the 'aether'. This aether frame was an obvious 'primordial' inertial frame relative to which all other inertial frames move with a constant velocity.

Gradually it was realized that terrestrial laboratories in which Maxwell's equations were discovered were only located in approximate inertial frames. There were accelerations relative to the aether, due to the rotation and revolution of the Earth. The resulting change in velocity of earth-based measurements relative to the aether was expected to lead to measurable changes in the propagation speed of the light waves. Such changes were never detected despite the best efforts of experimenters such as A.A. Michelson and E.W. Morley.

These workers did not interpret their null result to imply the absence of an aether frame, but their hypothesis of an 'aether boundary layer' on the Earth was eventually abandoned due in part to the astronomical phenomenon of aberration detected by J. Bradley in 1729. Bradley explained the apparent small elliptical motion of a distant star in terms of the annual revolution of the Earth relative to the aether frame (which we would now call an archetypal inertial reference). In this frame of reference the light travelled in a fixed direction towards us at speed c. But if the light ultimately 'jumped' to an aether frame moving with the Earth, then one would expect no annual change in the apparent location of the star since the relative velocity between the Earth and the aether boundary layer is always zero. We shall discuss this effect again in subsequent chapters.

The difficulty preoccupied theoretical physicists such as H.A. Lorentz, and inspired *ad hoc* solutions. One example was the proposal that lengths contracted parallel to their motion when moving relative to the aether frame (the famous 'contraction' due to Lorentz and G.F. Fitzgerald). Given the peculiar properties of the aether and the forces produced by acceleration with respect to it, this was a plausible proposal according to the mechanical standards of explanation of the day. However, the seeds of a more abstract mathematical understanding were germinating.

Lorentz found the transformations of coordinates and fields that did leave Maxwell's equations invariant. However, he was dismayed to find that this required a time transformation as well as a transformation of spatial coordinates, since according to Newton there could be only one absolute time. Henri Poincaré studied the mathematical properties of these transformations, rederived them elegantly, showed that they formed a group[1], and enunciated clearly the idea that perhaps physical law was universally invariant under these Lorentz transformations. Since dynamic law was known to be invariant under change of inertial reference frame, Poincaré was led to the idea that the Lorentz transformations were the correct transformations between inertial frames. Indeed he calculated the transformation of forces under these transformations. The postulate of the universality of physical law between inertial frames of reference came to be known as the *principle of relativity*.

However, it seems that Poincaré never explicitly considered the implications of the time transformation or its operational reality (see additional discussion below). Although he knew that the equation of motion of a light front was invariant under the Lorentz group, he does not seem to have realized the operational implications of an invariant speed of light. These missing elements in the mathematical structure erected by Lorentz and Poincaré were furnished in brilliant fashion by Albert Einstein, working contemporaneously with Poincaré. Poincaré did describe the Lorentz group as a rotation in space-time, which anticipates the bold invention of H. Minkowski of space-time as a 'metric space'. The latter concept means roughly that there is a measure for distances in our space-time diagrams below. We delay introducing this concept in detail until later in this book.

In the next section we will deduce and illustrate these various conclusions. Although our emphasis will be on practical applications, it should not be overlooked that the inertial frames of reference remain 'special' in this theory. Their preferred character implies still the existence of an aether. This would be a standard frame of reference, relative to which inertial frames move at constant velocity. The only advance is that the standard frame might be any inertial frame so that there is no definition of absolute 'rest'. But its origin remains a mystery in this 'special theory of relativity', restricted to inertial frames.

There is a classical theory that assumes a preferred frame at rest relative to the cosmological background radiation (CMB) and parameterizes the resulting corrections to the Lorentz transformations so that they may be limited quantitatively by experiment. The theory (RMS) is due to Robertson [1]; and then Mansouri and Sexl [2] and will appear again in the next section. Unfortunately it does not provide a mechanism for the existence of inertial frames.

---

[1] A group is a collection of elements that is closed under the group operation and which possess an inverse and an identity relative to this operation. The elements here would be the collection of inertial frames and the operation is the Lorentz transformation. The Galilean transformations also are a group operation on the set of inertial observers.

In summary, the Lorentz group of transformations replaces the Galilean group of transformations, but not the philosophy of inertial frames. It seems that there are two ways in which to improve on this situation. In one, due to Ernst Mach, one seeks some influence of the mean Universe that creates the inertial frames and indeed inertia. In the other, due to Albert Einstein, one attempts to remove the arbitrary nature of inertial frames by subjecting the intrinsic structure of space-time to local physical law. To date this has been perfected only for gravity, which is described as the non-flatness of space-time metrics. Electromagnetism and the sub-atomic forces have not been similarly incorporated into the structure of space-time.

2.2 The Theory of Light or Electromagnetic Waves

2.2.1 Wave Propagation Speed

Maxwell's equations in free 'space' (which is generally interpreted as the absence of matter in an inertial frame of reference) reduce to a wave equation with a constant propagation speed c, in the form

$$\frac{1}{c^2}\partial_t^2\Phi - \nabla^2\Phi = 0. \tag{2.1}$$

Here Φ, the wave field, may be regarded as the electric scalar potential and is thus independent of coordinates. Experimentally the time and position that appear in this equation refer to the measured properties of an electromagnetic wave for a *nearly* inertial observer. However, following Poincaré we interpret this equation to apply strictly in an inertial frame, ideally any inertial frame rather than one to be identified with the aether. The sources of the field are on a boundary of the matter-free region, and are prescribed in time.

The constancy of c with frequency and position assumes a non-dispersive and homogeneous inertial vacuum, however short the wave period. This inertial vacuum may be termed the aether in the sense discussed in the prologue, although any inertial frame of reference may be so regarded.

The nature of exactly what part of the free electromagnetic field is constrained to move with the speed c bears some examination. It is not a phase velocity if the field is composed of a superposition of fields from extended sources. Consider for example a plane wave incident on a plane reflecting boundary. If the normal to the plane is in the y direction and the normal to the plane wave makes an angle α with the negative y direction in the $x-y$ plane, then the reflected plane wave will make an angle α with the positive y direction as shown in Figure 2.1.

Within the limits of the idealization, the superposed wave field is found to have a phase velocity $v_{ph} = c/\sin\alpha$ in the x direction along the reflecting plane. Under the infinite plane assumptions, however, this pattern is absolutely uniform in x at each height y and so no signal, in the sense of pattern modulation, is actually propagating at this speed. The pattern is pre-determined by the initial infinite plane and plane wave conditions.

The previous example is an electromagnetic analogy of the mechanical model that consists in closing very long scissors. The point of contact at a distance ℓ from the

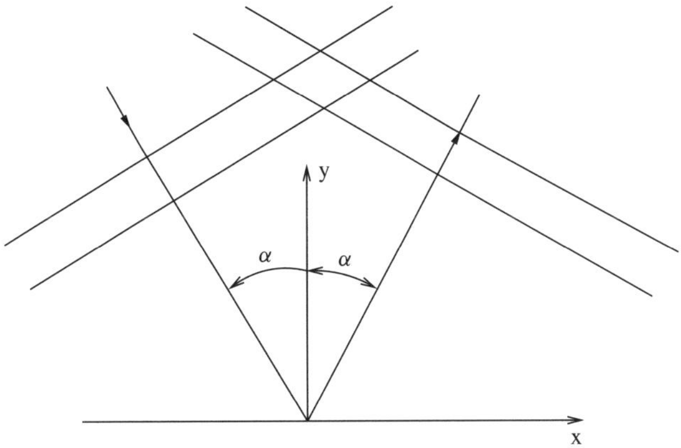

Figure 2.1 The sketch indicates a plane wave incident on a plane reflecting surface and obeying Snell's law at the angle α. Both planes are strictly infinite to avoid edge effects. The z dimension in the figure is suppressed

pivot of the scissors moves with speed $v_{ph} = \dot{\theta}\ell/\sin\theta$ when the opening angle is θ and the angular speed is $\dot{\theta}$. This arbitrarily fast phase speed has also been pre-determined initially by the extended construction of the scissors. Modulation of the contact signal is not possible at super-luminal speed, since commands to open or close are limited to the elastic shear wave speed along the scissor arms.

The physical propagation is best discussed in terms of a wave 'front', a surface that defines the boundary between a wave field created by distant sources and the inertial space that is not yet affected by these sources. Our concept of causality requires this to be possible. A simple example is a spherical wave front emanating from a point source at the origin. It is convenient to introduce the function $\chi \equiv r\Phi$, since then Equation (2.1) becomes the one-dimensional wave equation for χ in t and r, namely

$$\frac{1}{c^2}\partial_t^2\chi - \partial_r^2\chi = 0, \tag{2.2}$$

where r is the radial distance from the source. The solution of this equation is readily visualized. A new point source announces itself to the world behind a spherical bubble that it radiates initially. At a greater radius than that of this first wave at any time, there is no knowledge of the new source (dispersive media are not being discussed here). There is then a discontinuity of the wave quantities on this spherical surface, which property defines a 'spherical wave front'.

It is not difficult to see that a superposition of point sources at different locations, each of which begins radiating at the same time, can create at a time t a wave front which has a complicated shape in space.

This is illustrated in Figure 2.2 for two point sources in space-time.

From Figure 2.2 we see that we can write $R(x, y) = ct = \sqrt{x^2 + y^2}$ on the large sphere part of the composite wave front, where the coordinates of a general point P on the composite front are x, y. On the smaller wave front section $R = c(t - t_o)\sqrt{1 + \frac{d^2}{c^2(t-t_o)^2} - \frac{2d}{c(t-t_o)}\cos\phi}$ and $|\cos\phi| = \sqrt{1 - y^2/x^2}$. The radius R is

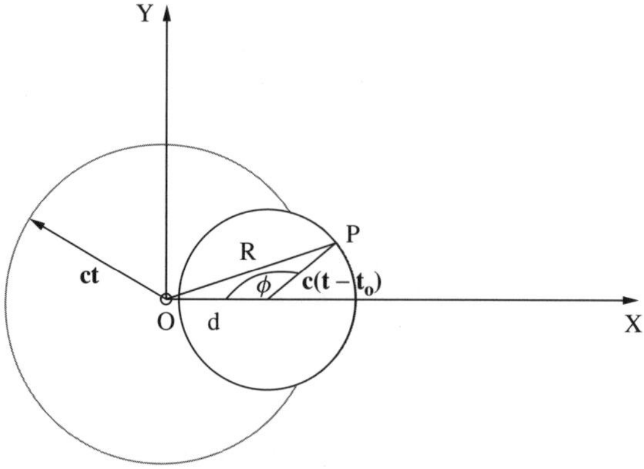

Figure 2.2 *The sketch shows two spherical wave fronts combining to produce a complicated surface in space. The source at O launched its wave at* t = 0 *while that at a distance* d *along the x axis was launched a time* t_o *afterwards. One can think of the* X − Y *plane shown being a cut at constant* t *so that the circles are actually spheres in three-dimensional space. In the text we only discuss the geometry of the two-dimensional section*

continuous at the points of intersection. Hence we can choose to write the equation of the complete front at fixed $t = R(x, y)/c$ as

$$H(x,y) \equiv \frac{R(x,y)}{\sqrt{1+\frac{d^2}{c^2(t-t_o)^2}-\frac{2d}{c(t-t_o)}\cos\phi}} + ct_o = ct. \tag{2.3}$$

In three dimensions, and considering a general source, this argument implies that we may choose the equation of an arbitrary composite electromagnetic wave front to be

$$H(x,y,z) - ct = 0 \equiv S(x,y,z;t). \tag{2.4}$$

The value of the wave field *on the front* is $\Phi_f(x,y,z) = \Phi(x,y,z,H(x,y,z)/c)$, where $H(x,y,z)/c$ is the current value of t. We can moreover assume that the first spatial derivatives of Φ together with the first time derivative are known on the front since they combine to give the electromagnetic field there. The second spatial and temporal derivatives should not be defined on the wave front, however, since otherwise they would allow fields ahead of the front to be calculated.

It is useful to follow Fock's argument [3] to determine in general the local propagation of a wave front. We may calculate

$$\nabla\Phi_f = \nabla\Phi + \partial_t\Phi\frac{\nabla H}{c}, \tag{2.5}$$

since the operator ∇ on the left operates on the entire spatial dependence of Φ_f as constrained to its value on the wave front. We may also take the divergence and the partial time derivative of Equation (2.5) to obtain respectively

$$\nabla^2\Phi_f = \frac{1}{c}\partial_t\Phi\nabla^2 H + \nabla^2\Phi + \frac{\nabla\partial_t\Phi}{c}\cdot\nabla H, \tag{2.6}$$

and

$$\nabla \partial_t \Phi + \frac{\partial_t^2 \Phi}{c} \nabla H = 0. \tag{2.7}$$

Eliminating $\nabla \partial_t \Phi$ between these last two equations and using the wave Equation (2.1) yields

$$\nabla^2 \Phi (1 - (\nabla H)^2) = \nabla^2 \Phi_f - \frac{1}{c} \partial_t \Phi \nabla^2 H. \tag{2.8}$$

We may find the same equation for $\partial_t^2 \Phi / c^2$ on replacing $\nabla^2 \Phi$ on the left by this quantity, according to Equation (2.1).

Consequently all terms on the right of Equation (2.8) can be regarded as known on a given wave front. Therefore to maintain the discontinuity in the second derivatives of the wave function so that we cannot calculate electromagnetic fields (i.e. the first derivatives; recall that Φ may stand also for any Cartesian component of the vector potential) ahead of the front, we must have from the left-hand side of Equation (2.8)

$$(\nabla H)^2 = 1. \tag{2.9}$$

This last result also allows us to calculate that the local wave front propagates according to (recall that $S/c \equiv t - H(x, y, z)/c$ so each of the following terms is equal to one)

$$\left(\frac{1}{c} \partial_t S \right)^2 - (\nabla S)^2 = 0. \tag{2.10}$$

The sign ambiguity implicit in this equation is due to the two possible directions of propagation, according to whether the wavefront is moving towards or away from the origin. Thus, following the wave front so that $dS = 0$, we find (e.g. 2.4)

$$c dt - \nabla H \cdot d\mathbf{R} = 0, \tag{2.11}$$

and hence using Equation (2.9) we obtain for the change in local position of the front normal to itself (that is $dR_n = \nabla H \cdot d\mathbf{R}$),

$$\frac{dR_n}{dt} = \pm c. \tag{2.12}$$

This is the physical meaning behind restricting the 'signal' or 'front' velocity to be c. Locally the 'news', that is the front, cannot advance from a localized source faster than the phase velocity. This is true whatever the spatial shape of the front may be. We note that the argument depends solely on the fact that the field satisfies the wave equation, so that c is simply the phase velocity of whatever wave is carrying the signal. However, we know of no signal propagating faster than the phase velocity of electromagnetic waves in a vacuum. Moreover, should such a field ever be discovered, the problem of the invariance of Maxwell's equations would return. The Lorentz transformations would then necessarily incorporate the new maximal velocity in order to retain the operational (that is measurable) universality discovered by Einstein. Hence they would lose the property of rendering Maxwell's equations invariant.

It is readily seen that the condition (2.10) is not invariant under a Galilean transformation $\mathbf{r}' = \mathbf{r} - \mathbf{u}t$. Although the spatial gradients are invariant under this transformation, the time derivatives are related through $(\partial_t S)_r = (\partial_t S')_{r'} - \mathbf{u} \cdot \nabla' S'$. Here $S' \equiv S(t, \mathbf{r}')$ and this is numerically equal to S. Thus the form fails to be generally invariant, but it is essential to note for what follows that it is nevertheless approximately invariant under the Galilean transformation if $u/c \ll 1$. We might thus expect the Galilean transformation to be a low velocity limiting form.

In fact, since $d\mathbf{R} = d\mathbf{R}' + \mathbf{u}dt$, Equation (2.11) becomes $(c - \nabla H \cdot \mathbf{u})dt = \nabla H \cdot d\mathbf{R}'$ and the velocity or propagation changes, contrary to the extended (Poincaré) principle of relativity. Explicitly, remembering Equation (2.9), $dR'_n/dt = c - u_n$ where n refers to the local normal to the front.

More directly, by using $\nabla S = (\partial S/\partial R_n)\hat{\mathbf{e}}_n \equiv \partial_n S\,\hat{\mathbf{e}}_n$, we obtain from Equation (2.10)

$$\left(\frac{1}{c}\partial_t S - \partial_n S\right)\left(\frac{1}{c}\partial_t S + \partial_n S\right) = 0. \tag{2.13}$$

Recalling that $\partial_n S = \partial_{n'} S'$ and $\partial_t S = \partial_t S' - u_n \partial_{n'} S'$ we find from this last expression the explicit transformation of (2.10) under a Galilean transformation to be

$$\frac{1}{c^2}(\partial_t S)^2 - (\partial_n S)^2 = \frac{1}{c^2}(\partial_t S')^2 - \left(1 - \frac{u_n^2}{c^2}\right)(\partial_{n'} S')^2 - \frac{2u_n}{c^2}\partial_t S' \partial_{n'} S'. \tag{2.14}$$

If we assume that in the inertial frame O where the front velocity is c that the local plane wave patch moves according to $\exp i(\omega t - kR_n)$, then $\omega/k = c$. However for an inertial observer O' moving with the speed u_n relative to the first observer along the normal to the local patch, we must assume a wave according to $\exp i(\omega' t - kR'_n)$ since the spatial derivatives (i.e. k) are invariant. But since $dR'_n = (c - u_n)dt$ we obtain for O' the phase velocity $\omega'/k = c - u_n$ by holding the primed exponential constant, whence eliminating k

$$\omega' = \omega(1 - u_n/c). \tag{2.15}$$

This is the classical or Galilean Döppler shift, and Problem 2.1 shows that this follows also from Equation (2.14).

Problem

2.1 Use Equation (2.14) together with $S \propto \exp i(\omega t - kR_n)$ and $S' \propto \exp i(\omega' t - kR'_n)$ to show directly that $\omega' = -ku_n \pm kc$. By letting $u_n \to 0$ argue that the positive sign may be chosen so that finally $\omega' = \omega(1 - u_n/c)$ as in the text.

It is not in fact difficult to find a transformation between inertial observers that maintains Equation (2.10) invariant. However, if it is to be linear, so that events are not multiply mapped between inertial observers, then it requires transforming the time coordinate as well as the spatial coordinate. This would appear to be mainly a curiosity to someone accustomed to Newtonian absolute time.

Nevertheless Lorentz [4] carried out such an analysis that we can duplicate simply here. Let us refer to Equation (2.13) and give the normal to the wave front the label z. It is quite clear that the coordinates x, y in the tangent plane to the wave front should be unchanged betwen inertial observers. This is because a body extended in this plane for O could be made to move together with O' along parallel rails that are a fixed distance apart for O. A more modern arrangement would have the body extremities following parallel laser beams. We assume of course that, at least locally, space is Euclidean.

However, in order to maintain a historical context we assume for the moment that

$$x' = \kappa x, \quad y' = \kappa y, \tag{2.16}$$

where κ is a scalar. We also assume that an observer O', moving along the z direction with speed u relative to the original inertial observer O, finds the coordinates of the 'point' (really an 'event') (t, z) to be linearly transformed according to

$$\begin{aligned} z' &= bz + at, \\ t' &= fz + et. \end{aligned} \tag{2.17}$$

In these relations as well as for those in (2.16), the coefficients can only be functions of the relative velocity u since this is the only parameter that distinguishes the inertial observers. Moreover, they must be such that as $u \to 0$ we will have

$$a/b = -u, \tag{2.18}$$

plus both $b, e \to 1$, $\kappa \to 1$ and $f \to 0$. Otherwise the transformations would not have the Galilean limiting form. In order to give the coefficients a coherent system of units, we choose units such that $c = 1$. Thus for example 1 second of time equals c seconds of space. Consequently u is measured in units of c, and the coefficients are all dimensionless.

We proceed by calculating the derivatives in Equation (2.13) in terms of the primed coordinates. We recall that S is a scalar and so the new functional form is $S(t', x', y', z')$ where $t' = t'(t, z)$ and $z' = z'(t, z)$. Subsequently we insist that the primed form should be proportional to the unprimed Equation (2.13) with factor κ^2. Because the law of the wave front propagation in each frame follows by setting the differential form for each observer to zero, we can still allow this arbitrary (non-zero) factor κ. This procedure yields three relations for the five coefficients as

$$\begin{aligned} a^2 - b^2 &= -\kappa^2, \\ e^2 - f^2 &= \kappa^2, \\ ea - fb &= 0. \end{aligned} \tag{2.19}$$

To obtain a solution, two additional assumptions are necessary. If for example we assume that Equation (2.18) holds for all u (if a/b is expandable in a Taylor series about $u = 0$, then more general dependencies would be non-linear), then we obtain the solution

$$\begin{aligned} e^2 &= b^2 = \kappa^2 \frac{1}{1-u^2}, \\ \frac{a}{b} &= \frac{f}{e} = -u. \end{aligned} \tag{2.20}$$

The positive root must be taken for b because of the Galilean limit wherein $z' \to z$ as $u \to 0$. The same positive root must be taken for e to avoid reversing time sequences between observers, which would create problems with causality.

Problem

2.2 Derive the results given in Equations (2.19) and (2.20).

Substituting this solution into Equation (2.17) and restoring normal units by dimensional analysis yields the transformations

$$\begin{aligned} z' &= \gamma\kappa\,(z - ut)\,, \\ t' &= \gamma\kappa\left(t - \frac{uz}{c^2}\right). \\ x' &= \kappa x, \\ y' &= \kappa y. \end{aligned} \tag{2.21}$$

Here the 'Lorentz' factor is $\gamma = 1/\sqrt{(1 - u^2/c^2)}$. Evidently by changing the sign of u and interchanging the primes we obtain the inverse transformations from O' to O. This can also be seen by a direct algebraic inversion if $\kappa = 1$. In fact the famous Lorentz transformations are obtained by setting $\kappa = 1$ in Equations (2.21). This is as we have expected on operational (experimental) grounds, since transverse rods may move on rails in the $X - Y$ plane.

This value for κ may also be established by applying these transformations to obtain the coordinates of an observer O'' moving with the velocity $-u$ along the z axis relative to O'. This observer then coincides with the observer O and so the double primed coordinates must be identical to the unprimed coordinates [5,6]. We are effectively applying the group inverse to the group operation and insisting that the result be the identity.

The latter operation requires $\kappa(u)\kappa(-u) = 1$ (most simply seen by considering the transverse transformations). Subsequently we invoke the symmetry of empty space to require that the effect on transverse lengths should be the same whether we move along the positive or negative z direction. Consequently $\kappa(u) = \kappa(-u)$ and each must therefore be equal to one, as was concluded operationally above. We have not escaped implicit assumptions by this argument, however, namely that space is isotropic and homogeneous. This is equivalent to the implications of our operational argument above.

The isotropy of space has recently been tested to very high precision [7] by fitting the Robertson, Mansouri and Sexl (RMS) parameters to the beat frequency of two orthogonal resonant cavities stimulated by a split laser beam of tunable frequency. The optical system is mounted on a stabilized rotating table and is maintained in a high vacuum. Modulation of the beat frequency is expected if there is anisotropy in the speed of light just as in the Michelson-Morley experiment. In fact the orthogonal cavities are in the form of a cross at the centre of the system, so that we do not encounter a Sagnac effect (see the next chapter). The RMS parameters are shown to be very small and $\Delta c/(2c)$ is less than 0.6×10^{-17} at the 1σ level.

The Lorentz transformations succeed in making the propagation of a light wave invariant between inertial frames, but at what price! The speed of light is now completely independent of the relative speed between source and observer. Thus Equation (2.12) must look the same in all inertial frames. Hence, considering a spherical wave front around a point source

$$c^2t^2 - R^2 = c^2t'^2 - R'^2, \tag{2.22}$$

or in a Cartesian reference frame ($\kappa = 1$)

$$c^2t^2 - (x^2 + y^2 + z^2) = c^2t'^2 - (x'^2 + y'^2 + z'^2). \tag{2.23}$$

And this only at the cost of abandoning a universal Newtonian time. It is not surprising that Lorentz and Poincaré hesitated over the physical meaning of t'.

Problem

2.3 Show by a direct application of the transformations (2.21) that Equation (2.23) holds ($\kappa = 1$).

Moreover the solution is not unique. If following Voigt [8] (already in 1887!) we maintain an absolute Newtonian clock rate by setting $e = 1$ in Equations (2.17), but still allow for a velocity dependent change in the zero point ($f \neq 0$). Moreover, if we still maintain the Galilean limit (2.18), then the solution for the transformation that leaves the wave equation invariant becomes

$$b = e = 1, \tag{2.24}$$
$$a = f = -u, \quad \kappa^2 = 1 - u^2, \tag{2.25}$$

and hence we obtain the Voigt transformations

$$x' = x - ut \qquad t' = t - ux/c^2. \tag{2.26}$$

Although these transformations do maintain the invariance of the speed of light and Newtonian time (and one at least seems to have been used by Poincaré, see discussion below), they are clearly not physically acceptable since $\kappa^2 \neq 1$, let alone the disagreement with experiment. This exercise merely indicates that other physical considerations must be applied in order to obtain a unique solution from the principle of relativity, that is the invariance condition.

Before leaving this section we should generalize the Lorentz transformations to arbitrary orientations of the axes of two different inertial observers. In order to do this easily, we introduce a matrix calculus just as we did in Chapter 1. We adopt the notation for a column vector of Equation (1.1), where now we wish the generalized coordinates to be simply the Cartesian quantities so that $q^i = x^i$. Recall that Latin indices $a \dots h$ may each take any values among $0, 1, 2, 3$, while those in the group $i \dots z$ may each take only values among $1, 2, 3$.

The number zero designates the time coordinate. We require homogeneity of units so that t will actually stand for ct, that is, we take units such that the velocity of light is one. If we measure time in seconds, then space must also be measured in 'seconds', which we call 'light-seconds'. One light-second of space is therefore c metres, roughly 80% of the distance to the Moon. This is convenient for astronomical discussions. In the laboratory one might prefer to choose a metre of space as a common unit more conveniently. These metres of time we might call 'light-metres' where now we divide by c, but there is no ubiquitous term. One light-metre is roughly 3.3 nanoseconds.

The form of the Lorentz transformation that we have given above is peculiar to the arrangement of axes relative to the relative velocity of the two observers. Both z axes are parallel to this relative velocity $\mathbf{u}$. This is called 'standard configuration' and the corresponding form of the transformations is called a Lorentz 'boost'. We can write this in matrix form as

$$\begin{pmatrix} t' \\ x' \\ y' \\ z' \end{pmatrix} = \begin{pmatrix} \gamma & 0 & 0 & -\gamma u \\ 0 & 1 & 0 & 0 \\ 0 & 0 & 1 & 0 \\ -\gamma u & 0 & 0 & \gamma \end{pmatrix} \begin{pmatrix} t \\ x \\ y \\ z \end{pmatrix}. \tag{2.27}$$

More concisely this becomes

$$x'^a = L^a{}_b x^b, \tag{2.28}$$

where as usual the upper index labels the row while the lower index labels the column in the 'Lorentz boost' matrix $\underline{\underline{\mathbf{L}}}$.

Suppose that we wish to transform to an arbitrarily oriented set of axes that are used by O' who is moving with velocity $\mathbf{u}$ relative to O as sketched in Figure 2.3.

It is evident that we can accomplish this transformation in two stages. We transform from O to O' using the Lorentz boost matrix, and then to the double primed axes by a spatial rotation. However, the Lorentz boost acts on both space and time, which causes us to use four-element column vectors[2] (and row vectors where necessary). Our familiar[3] spatial rotation $\underline{\underline{\mathbf{S}}}$ acts only on three-element vectors. But this is easily cured by creating

[2] One should emphasize here, however, that, unlike three-vectors, we have as yet assigned neither modulus nor dot product to four-vectors.

[3] See Chapter 1.

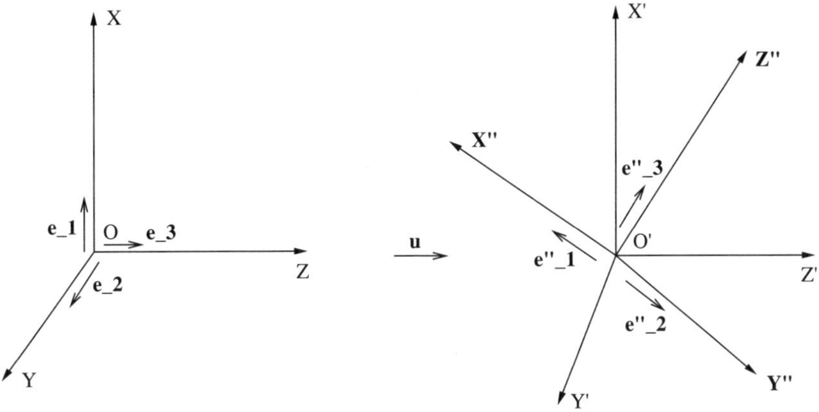

Figure 2.3 *The sketch shows the Cartesian axes of two inertial observers O and O′. O′ can choose to use either the primed axes that are in standard configuration relative to the unprimed axes, or O′ can choose arbitrarily rotated axes labelled″. The primed base vectors are the same as the unprimed base vectors (shown), but the double primed base vectors (shown) are not*

an extra dimension for $\underline{\underline{\mathbf{S}}}$ (i.e. extra row and column) that allows the time coordinate to be unchanged under the spatial rotation, even when the expanded $\underline{\underline{\mathbf{S}}}$ acts on four-element vectors. This is readily achieved by writing the expanded rotation matrix as

$$\begin{pmatrix} 1 & 0 & 0 & 0 \\ 0 & S^1_{\ 1} & S^1_{\ 2} & S^1_{\ 3} \\ 0 & S^2_{\ 1} & S^2_{\ 2} & S^2_{\ 3} \\ 0 & S^3_{\ 1} & S^3_{\ 2} & S^3_{\ 3} \end{pmatrix}. \tag{2.29}$$

We will also refer to this matrix as $\underline{\underline{\mathbf{S}}}$ or $S^a_{\ b}$ and let the indices tell us whether it acts on three or four vectors (after the letter h they imply three, before this letter there are four).

We know that it is sufficient to represent the elements of $\underline{\underline{\mathbf{S}}}$ in terms of the Euler angles. Thus a rotation through the angle ϕ about the z axis is easily realized by a four-matrix $\underline{\underline{\mathbf{S}}}_{\phi}$, modified as above. This angle has the merit of being defined by the ratio of lengths in the plane perpendicular to the relative velocity. Although rotations $\underline{\underline{\mathbf{S}}}_{\theta}$ and $\underline{\underline{\mathbf{S}}}_{\psi}$ may be similarly effected separately or in combination, it is important to remember that these angles are relative to the axes of observer O'. These angles necessarily transform between observers, while ϕ does not.

For our present purpose we need not simplify the rotation matrix. Thus we write the boost followed by a rotation in the index form

$$x''^a = S^a_{\ b} L^b_{\ c} x^c, \tag{2.30}$$

which is explicitly

$$
\begin{aligned}
t'' &= \gamma(t - uz),\\
x'' &= S^1_{\ 1}x + S^1_{\ 2}y + \gamma S^1_{\ 3}(z - ut),\\
y'' &= S^2_{\ 1}x + S^2_{\ 2}y + \gamma S^2_{\ 3}(z - ut),\\
z'' &= S^3_{\ 1}x + S^3_{\ 2}y + \gamma S^3_{\ 3}(z - ut).
\end{aligned}
\tag{2.31}
$$

We must now remember the representation of a rotation in terms of the direction cosines. In terms of the base vectors (recall that the unprimed and singly primed base vectors are parallel) this is

$$S^i_{\ j} \equiv \mathbf{e}^i \cdot \mathbf{e}''_j. \tag{2.32}$$

The index on the base vector gives its direction. An upper index indicates a column vector while a lower index indicates a row vector. This is all as in Chapter 1.

We may now write this result in a much more conventional and intuitive form. If $\mathbf{r}$ is the spatial position of an event for O, and $\mathbf{r}''$ the same for O' *but resolved along the double primed axes*, the spatial parts of Equation (2.31) become

$$\mathbf{r}'' = (\mathbf{r}_\perp)'' + \gamma\left((\mathbf{r}_\parallel)'' - (\mathbf{u})''t\right). \tag{2.33}$$

Here we have used parentheses with a double prime on the right to indicate 'resolved along the double primed axes', used by the inertial observer O'. This resolution is just what the various products with the components of $S^1_{\ j}$ do. The quantity $\mathbf{r}''$ on the left is the position vector defined by O' and resolved along the double primed axes. Moreover we have put

$$\mathbf{r}_\perp = \begin{pmatrix} x \\ y \end{pmatrix}, \tag{2.34}$$

and $\mathbf{r}_\parallel$ equal to z. That is, the perpendicular and parallel designation is relative to $\mathbf{u}$. Consequently Equation (2.33) becomes, writing the parallel and perpendicular components more generally

$$\mathbf{r}'' = \left(\mathbf{r} - \frac{\mathbf{u}\,(\mathbf{u}\cdot\mathbf{r})}{u^2}\right)'' + \gamma\left(\left(\frac{\mathbf{u}\,(\mathbf{u}\cdot\mathbf{r})}{u^2}\right)'' - (\mathbf{u})''\,t\right), \tag{2.35}$$

or rearranging

$$\mathbf{r}'' = (\mathbf{r})'' - \gamma(\mathbf{u})''t + (\gamma - 1)\left(\frac{\mathbf{u}(\mathbf{u}\cdot\mathbf{r})}{u^2}\right)''. \tag{2.36}$$

This is correct as it stands but we can write it more concisely by recalling that a vector equation can be resolved along an arbitrary set of axes. Hence writing the new three vector $\mathbf{r}''$ as $\mathbf{s}$ we have[4]

$$\mathbf{s} = \mathbf{r} - \gamma\mathbf{u}t + (\gamma - 1)\left(\frac{\mathbf{u}(\mathbf{u}\cdot\mathbf{r})}{u^2}\right). \tag{2.37}$$

[4] Note the similarity in this procedure with the construction of a new velocity vector in rotating coordinate axes as discussed in Chapter 1.

This expression gives the three-coordinate vector in arbitrarily oriented inertial axes moving with velocity $\mathbf{u}$ relative to the given inertial axes of O.

The general time component of Equations (2.31) can be written as a trivial generalization of the boost relation in the form

$$t'' = \gamma(t - \mathbf{r} \cdot \mathbf{u}) \equiv t'. \tag{2.38}$$

This combination of spatial rotation and Lorentz boost is the general operation in the 'homogeneous Poincaré group'. Evidently there are six group parameters, three from the boost operation and the three angles of spatial rotation. The operation always returns a member of the group, and there are obvious inverse and identity elements.

It is convenient to write this group operation in operational form by expressing it as a matrix operator $\mathcal{L}^a{}_b$, where as usual the upper index refers to the rows and the lower index refers to columns. It is left as a Problem to show that this matrix may be written using Cartesian coordinates as

$$\begin{aligned} \mathcal{L}^0{}_0 &= \gamma(u), \\ \mathcal{L}^0{}_k &= -\gamma(u)u_k = \mathcal{L}^k{}_0, \\ \mathcal{L}^k{}_\ell &= \delta^k_\ell + \frac{(\gamma(u) - 1)}{u^2} u_k u_\ell. \end{aligned} \tag{2.39}$$

We have used our normal index convention whereby letters after h in the alphabet refer only to spatial values $1, 2, 3$ or in this case x, y, z. This allows Equations (2.37) and (2.38) to be written in the compact form

$$x''^a = \mathcal{L}^a{}_b x^b. \tag{2.40}$$

Problem

2.4 Using Equations (2.36) and (2.38), show that the components of $\underline{\underline{\mathcal{L}}}$ given in component form in the text are equivalent to the matrix

$$\underline{\underline{\mathcal{L}}} = \begin{pmatrix} \gamma & -\gamma u_x & -\gamma u_y & -\gamma u_z \\ -\gamma u_x & 1 + \frac{(\gamma - 1)}{u^2} u_x^2 & \frac{(\gamma - 1)}{u^2} u_x u_y & \frac{(\gamma - 1)}{u^2} u_x u_z \\ -\gamma u_y & \frac{(\gamma - 1)}{u^2} u_y u_x & 1 + \frac{(\gamma - 1)}{u^2} u_y^2 & \frac{(\gamma - 1)}{u^2} u_y u_z \\ -\gamma u_z & \frac{(\gamma - 1)}{u^2} u_z u_x & \frac{(\gamma - 1)}{u^2} u_z u_y & 1 + \frac{(\gamma - 1)}{u^2} u_z^2 \end{pmatrix} \tag{2.41}$$

In the above expressions we have also followed the usual practice of writing the Lorentz boost between observers who coincide at $t = 0$. This is known as the homogeneous Lorentz boost. There is always some O observer who coincides with O' at some

t, so that it is no real restriction to write this standard simple form. Essentially we have made use of an extra four parameters of the Poincaré group.[5]

We shall discuss this again below, but should the coincident event between O and one of the friends of O' be at $\mathbf{R}_{\mathbf{o}}' = \{0, 0, Z_o'\}$ at time t_o' for our particular O', then the time transformation of Equations (2.31) must become (between O and the rotated axes of O')

$$t'' = \gamma(t - uz) + t_o', \tag{2.42}$$

and the spatial transformations must become

$$x'' = S^1_{\ 1}x + S^1_{\ 2}y + S^1_{\ 3}Z_o' + \gamma S^1_{\ 3}(z - ut), \tag{2.43}$$

$$y'' = S^2_{\ 1}x + S^2_{\ 2}y + S^2_{\ 3}Z_o' + \gamma S^2_{\ 3}(z - ut), \tag{2.44}$$

$$z'' = S^3_{\ 1}x + S^3_{\ 2}y + S^3_{\ 3}Z_o' + \gamma S^3_{\ 3}(z - ut). \tag{2.45}$$

But if the coincident event happens at t_o for O and for the coincident friend of O', then it can only happen at $t_o' = |Z_o'| + t_o$ for O' due to the light travel time. We shall see below that $Z_o = Z_o'/\gamma$ when Z_o is measured at a fixed time so that the above relations may be written finally in terms of Z_o. This means that Equation (2.37) will have a term $\gamma(\mathbf{u}(\mathbf{u} \cdot \mathbf{R}_o))/u^2$ added, while Equation (2.38) will add t_o plus the modulus of this last expression.

We have been seduced into treating the Lorentz transformation as applying to every event, not just those that are connected by light waves. For the moment we have not justified this, but have only derived the transformation for light waves. The justification awaits the next section.

This section has presented an analysis of the propagation of light in electromagnetic theory, and of the Lorentz transformations of coordinates between inertial observers. These are required so that electromagnetic theory conforms to the principle of relativity. Taken literally, this invariance implies that the speed of light is invariant under any change of velocity between source and observer. It is not yet clear why a universal relation between the coordinates of inertial observers should follow from the invariance of the wave equation for light *in vacuo*. For example, what does the invariance between inertial observers of other 'fundamental' principles such as Newton's second law require? Are all such requirements mutually compatible? This is certainly not the case for Schrödinger's wave equation in quantum mechanics, as an application of the arguments of this section will show. This last fact foreshadows a continuing interaction between the quantum world and that of relativity, which leads ultimately to the Dirac theory of the electron and to quantum field theory.

For the moment these bizarre transformations between the coordinates of inertial observers are a curiosity of James Clerk-Maxwell's electromagnetic theory, together with the Galilean principle of relativity as extended by Poincaré [6].

Poincaré discovered the group nature (the inverse found as usual by interchanging the primes and writing $-\mathbf{u} \leftarrow \mathbf{u}$ and the identity is the 4×4 unit matrix) of the Lorentz transformations, and enunciated a general principle of relativity.

[5] In the complete Poincaré group of inertial observers, there are three parameters of relative velocity, three of spatial rotation, and four of translation in time and space.

The complete group operator has been discussed above, but as one example we can use this structure to deduce the addition formulae for parallel velocities. Consider three inertial observers O, O_1 and O_2. They all use axes in standard configuration. The velocity of O_1 relative to O is u_1, and the velocity of O_2 relative to O_1 is u_2, always directed along the common z axis. The group operation is simplified to multiplication by the boost matrix. To demonstrate closure under this operation we have to show that the operation of the matrix in Equation (2.27) in sequence to go first from O to O_1 and then from O_1 to O_2 is equivalent to one boost (2.27) from O to O_2 in terms of some u and $\gamma(u)$. The velocity u must then be interpreted as the velocity of O_2 relative to O. It is left as Problem 2.6 to show that this requires $\gamma^2 = 1/(1-u^2)$ where u is given by

$$u = \frac{u_1 + u_2}{1 + u_1 u_2}. \tag{2.46}$$

This says that O must not use the Galilean $u_1 + u_2$ for u, but rather requires the strange velocity transformation as above. Strange because we observe that if $u_1 = 1$, then $u = 1$ no matter what u_2 is. That is, u_2 is never added to (or subtracted from) the velocity of light. Clearly this is an explicit corollary of the invariance of Maxwell's theory as alluded to above.

The general homogeneous group behaviour combines spatial rotations and Lorentz boosts. Remarkably the group operation couples boosts and spatial rotations in unexpected ways. For example, two successive boosts to inertial observers O' and O'' respectively relative to an inertial observer O, but with non-collinear boost velocities (hence some acceleration), result in a spatial rotation of the axes of O'' relative to the axes of $\hat{O}$. The observer $\hat{O}$ is boosted directly to the velocity of O'' from O in order to compare directly the O'' axial directions to those of O.[6]

This rotation occurs even though no spatial rotation has been imposed and there is zero torque. If we imagine a series of such rotations through a sequence of inertial frames, each tangential to the current $\mathbf{u}(t)$, then a direction fixed in the O world will be seen to 'precess' continually relative to the axes of O. The fixed direction might be that of the spin of a sub-atomic particle or of a small gyroscope. This phenomenon is known as "Thomas precession" [9] and we shall pursue it further in the next chapter. It is strictly a property of the Poincaré group and ultimately the postulates of special relativity.

Poincaré also understood how to transform forces and electromagnetic fields under these transformations (an excellent discussion of these matters is to be found in [10]). However, the true meaning of the time transformation appears to have eluded both Poincaré and Lorentz. In reference [10], Chapter 8, Poincaré is quoted as calculating a 'local time' by the clock synchronization method that is often attributed to Einstein. He finds $t' = t - uz$ in our notation, which is in fact the Voigt transformation discussed above (2.26) and not the Lorentz expression. The latter requires that Newtonian time be abandoned in favour of proper time, which introduces the Lorentz gamma factor.

[6] Observer O' could also be boosted directly from O' to O'' since O' differs from O only by a pure boost, for which there is no rotation of the spatial axes.

It was left to Einstein to resolve this question fully through the introduction of the theory of measurement, that is by an operational approach. Minkowski [11] created the full geometrical elegance by inferring a space-time manifold[7] from the Lorentz transformations, although this is not strictly necessary.

Problems

2.5 **(a)** Refer to Figure 2.3 and suppose that the double primed axes are the result of a single rotation θ' about the x' axis. First define the necessary extended $\underline{\underline{\mathbf{S}}}_{\theta'}$. Combine this with the standard Lorentz boost to show that

$$\begin{pmatrix} t'' \\ x'' \\ y'' \\ z'' \end{pmatrix} = \begin{pmatrix} \gamma(t - uz) \\ x \\ \gamma \sin\theta'(z - ut) + y\cos\theta' \\ \gamma\cos\theta'(z - ut) - y\sin\theta' \end{pmatrix}. \tag{2.47}$$

(b) Show that the previous result agrees with Equation (2.37). Note that $y'' \equiv \widehat{\mathbf{e}}_{y''} \cdot \mathbf{s} \equiv y\cos\theta' + z\sin\theta' - \gamma u \sin\theta' t + z\sin\theta'(\gamma - 1)$ and a similar argument yields the other spatial components.

2.6 Show that the requirement that the group be closed under the boost operation does lead to Equation (2.46) for the addition of parallel velocities.

2.3 Measurement Theory and the Lorentz Transformations

In this section we will make much use of 'space-time diagrams'. These are not physical representations of space-time, but are simply a diagrammatic means of keeping track of the spatial location at each time of a particle, or of a point 'event' (such as the encounter of a photon with a particle). This is rather like following a train or a bus as each encounters various destinations along a super highway. When we pretend that the 'space' is one-dimensional, these time lines are a complete measure of a particle's experience and may be called 'world lines'. Proper time is measured along the world line by a co-moving clock. An example is shown in Figure 2.4.

Einstein, inspired by Mach [12], introduced positivism and the theory of measurement to the discussion of the principle of relativity by postulating that *no 'signal' that has a localized source can propagate faster than the speed of light in a vacuum*. In this context 'signal' means either a causal influence or the transmission of information. Both aspects require the transmission of energy, and this is forbidden to be faster than that achieved by locally generated electromagnetic waves.

Positivist philosophy is attributed to Auguste Compte (a statue of whom is to be found in the Place du Sorbonne in Paris) and implies for us that only measurable quantities should have a place in physical theory. One must be careful of ignoring the time dependence of 'measurability', however. Mach refused to believe in atoms on the basis

[7] A mathematical manifold is a set of elements (here events) that is locally identical to Euclidean space but may not be globally.

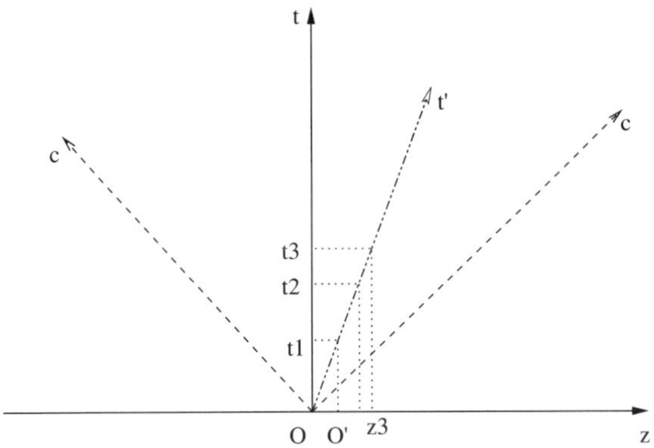

Figure 2.4 *The diagram shows time and one dimension of space for an inertial observer O. The line marked t′ is the track of an observer O′ who is moving along the positive z axis, that is, the 'world line' of O′ relative to O. Thus t′ is the proper time of O′. The world line of O is the t axis. The two observers are together at the origin. Two lines at 45° marked c are light rays moving left and right in our units. Several points along the world line of O′ are marked by coordinates in O*

of their 'undetectability'. Intriguingly, the positivist philosophy flowers most vigorously in quantum mechanics, a theory that Einstein could not accept. Admittedly he came to this view only after the triumph of his classical theory of gravity, in which an invisible space-time manifold (without matter) plays a major rôle.

We shall see shortly that this new postulate limits and transforms what one inertial observer and his 'world' can measure in the 'world' of another inertial observer in relative motion. The 'world' of an inertial observer O is a brief means of referring to the world line of O together with those of O's 'friends'. The friends are by definition rest observers at arbitrary positions in O's reference system, so that their world lines are all parallel to that of O. All 'events' measured at a time and place (including empty events free of particles or energy) by O and friends, form the world continuum of that inertial frame.

However, it is worth pausing to ask why electromagnetic waves should have this limiting property. Certainly the invariance of Maxwell's equations under the Lorentz transformations suggests such a thing. This invariance leads to Equation (2.46), which shows in turn that a source of light waves moving uniformly relative to an inertial observer does not propagate light faster or slower then when it is at rest. But why should this apply to all signals?

Consider that there does exist a physical field that propagates a wave described by the linear wave equation at a speed $v_x > c$ in a vacuum. Then the principle of relativity applied to that wave equation would yield, as in the preceding section, the Lorentz equations with c replaced by v_x. However, electromagnetic waves would now fail the principle of relativity *unless they were to propagate with the same speed*. One might attempt to change Maxwell's theory to accommodate the new speed, but experimental test argues against this. The simplest solution is to maintain c as the maximum for all physical fields, as is effectively done for gravitational waves.

But we seem obsessed with linear wave equations. Perhaps there is some propagating influence that is so rapid that we would consider it instantaneous. A simple way to model this is to assume an elliptic type diferential equation in space-time. We shall see that this amounts to assuming that the space-time diagram of Figure 2.4 is Euclidean, just as is the piece of paper on which it is drawn.

We might look for the transformations that leave invariant the equation

$$(\partial_t \phi)^2 + (\partial_z \phi)^2 = m^2, \tag{2.48}$$

where m is a numerical invariant (scalar), possibly zero. Using the linear transformations (2.17), we can again follow the procedure that renders the equation invariant. This yields (requiring once more $a/b = -u$, that time does not reverse direction and still measuring speeds in units of c)

$$\begin{aligned} t' &= \frac{1}{\sqrt{1+u^2}}(t + uz), \\ z' &= \frac{1}{\sqrt{1+u^2}}(z - ut), \\ y' &= y, \\ x' &= x. \end{aligned} \tag{2.49}$$

This transformation also forms a group (so that any two inertial observers are treated the same) but now the equivalent of (2.46) is

$$u = \frac{u_1 + u_2}{1 - u_1 u_2}, \tag{2.50}$$

and $\gamma^2 = 1/(1+u^2)$.

The change in sign in γ is highly significant. Light speed is no longer invariant. Thus if in Equation (2.50) we set $u_2 = 1$ for observer O_2 then $u = (1+u_1)/(1-u_1)$ for observer O, which may be arbitrarily large as $u_1 \to 1$ for observer O_1. Moreover, as $u \to \infty$, time intervals and space intervals are reduced to zero for the primed observer under the transformation! All this would be agreeable for O' off on an interstellar voyage, but unfortunately the transformations do not agree with experiment.

The unphysical behaviour is due essentially to the fact that the transformation (2.49) maintains $t^2 + x^2 + y^2 z^2 = t'^2 + (x'^2 + y'^2 + z'^2)$. This is in stark contrast to the Lorentz transformations of the linear wave equation that maintain

$$|t'^2 - (x'^2 + y'^2 + z'^2)| = |t^2 - (x^2 + y^2 + z^2)|. \tag{2.51}$$

The sign of the quantity under the modulus in this last equation differs in different parts of an observer's world.

Problem

2.7 Derive the transformations (2.49) and the group velocity addition formulae (2.50).

We proceed to discuss the momentous consequences that follow from assuming two things about electromagnetic waves (that we often call 'light' for brevity).

First:

the speed of light is independent of the relative velocity of source and observer,

and second:

it is the fastest signal by which events may make their consequences known.

The first assumption is explicit in Equation (2.46) (setting u_1 or u_2 equal to 1). The second assumption really requires that any other fundamental physical field that propagates a linear wave *in vacuo* from a local source does so *at the speed* c. Otherwise one such field or the other would not satisfy the principle of relativity, given the role of c in the Lorentz transformations. The speed of light *in vacuo* thus appears as the maximum speed at which the physical world may be 'connected'. It is satisfied by the modern theory of gravity due to Einstein and by quantum gauge fields. Empirically we have found no exceptions in our current science.

These assumptions are not quite identical to the argument from the invariance of the linear wave equation, however. There are non-linear equations that propagate signals (e.g. soliton solutions), and light might be a result of such an equation that is yet unknown. Thus in the original sense [12], these assumptions can be regarded as empirical and independent of the actual equation of light propagation. The remarkable result of this approach, however, is ultimately to confirm the importance of the invariance of the linear wave equation, for we find that the Lorentz transformations follow from these assumptions, and we know they are the invariance group of the linear wave equation.

In addition to the above, we take the speed of light to be isotropic in an inertial frame. If one accepts Clerk-Maxwell's theory, then it is rather space itself that we assume to be isotropic, since that theory has no intrinsic anisotropy. This is usually accomplished by assuming Euclidian space, at least locally. A background assumption that is always present in our considerations is the Galileo-Poincaré principle of relativity according to which:

All inertial frames are equally good for the description of physical phenomena.

The necessary abandonment of Newtonian absolute time is a truly radical result of these assumptions plus what might be called 'event positivism'. The latter is the positivist view that reality is what can be measured in principle.

In order for time to be absolute or universal, it is necessary that all observers can agree that two events are simultaneous, whatever their respective relative velocities. Measured time or 'chronometry' after all requires that an event in a system called a 'clock' or 'chronometer' (e.g. a specific phase of a periodic system such as the hands on an analogue watch) be simultaneous with an event whose 'time' (the clock phase or, in the case of a secular clock such as human civilization or the Julian Day count, the clock state) is being measured. But this is now generally impossible for separated events, as we proceed to show below by the application of our two principles.

Consider two events that are simultaneous for an observer O. This means in practice, since such events are in general not located spatially at O, that O must have a means of establishing simultaneity of events with all of the co-moving 'friends'. This may be achieved by having 'ideal' clocks that have been 'synchronized' to record the same measured time for instantaneous events.

'Ideal clocks' are those that will always run at the same 'rate'. That is we require ideally measured time to be 'homogeneous'. We are in 'Ouroboros mode' (one definition always implies another) here since a constant 'rate' must be measured by another ideal clock and so on to indefinite recursion. An amusing but evocative statement of this problem is described as 'Segal's Law' on the US naval observatory web site. This maintains that *A man with one watch knows what time it is. A man with two watches is never sure*. 'Watch' is a way of describing a precise clock that is nevertheless only approximately ideal.

In practice we use isolated (that is, undisturbed), normally periodic, systems in many realizations, and compare each time with the average time recorded for a given event. The periodic orbits of the Moon and planets together with the rotation of the Earth comprise the traditional set of ideal clocks, but now atomic vibrations and the regular pulses of the astronomical 'pulsars' are superior. These have permitted the measurement of the slowing of the Earth's rotation, which is henceforth no longer an ideal clock. All ideal clocks fail at some level of precision, but hydrogen maser clocks are at present accurate to at least one part in 10^{15} over macroscopic time intervals (up to 1000 seconds), which is ideal for most purposes. Atomic clocks also have the property of being rather insensitive to acceleration produced by external forces because of the enormous internal accelerations of the orbiting electrons. That is, they are relatively easy to isolate.

'Synchronization' of ideal clocks can be achieved in various ways between friends. One method is by arbitrarily 'slow' (depending on the accuracy required) clock transport. We use this method whenever we compare clocks in proximity in everyday life, or when someone asks us for the 'time'. However, the practical method between friends separated by distance is by the reflection of electromagnetic waves. This is often referred to as 'radar ranging' since distance is also obtained during the operation, but higher frequencies than radar frequencies are often used, such as by the optical lasers used by surveyors.

In Figure 2.5 a light ray is sent by O, at $t = 0$ according to a local clock, towards a friend O_1 (with parallel world line) at some distance d. It is received and immediately reflected at time t_1 according to a local clock of O_1, and received again on O's world line at t_2. Assuming Euclidean space, which is homogeneous and isotropic, it is clear that synchronization is achieved if O_1 adjusts the local clock so that $t_1 - t_2/2 = 0$, once t_2 is communicated at leisure by O to O_1. Since we are using electromagnetic waves, we also deduce that $d = ct_2/2$. In this way a $t = constant$ spatial slice (the z axis in one dimension) is constructed for O and friends on parallel world lines that may be referred to as a 'simultaneity'. It is the 'world space' for all O observers.

Problem

2.8 By considering the light travel times, show directly that the radar ranging of O' by O_2 leads to the results quoted in the caption of Figure 2.5; that is, $d = (c - u)T_1$, $d = (c - u)T_2/2$, and $d_1 = cT_2/2$.

We are now in a position to consider the lack of invariance of a simultaneity or world space under the change of inertial observers. Figure 2.6 shows events that are

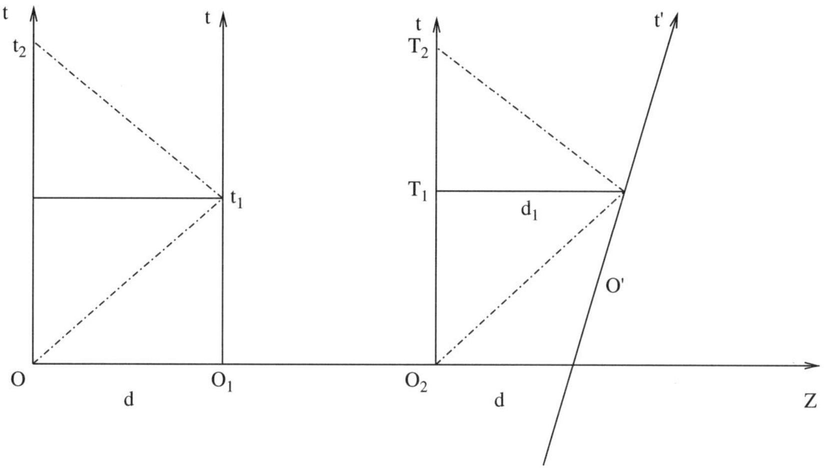

Figure 2.5 *This shows the world lines of* O*, two of* O*'s friends* O_1 *and* O_2*, plus the world line of another inertial observer* O′ *moving with the relative speed* u*. The* 45° *light paths of the radar ranging between* O *and* O_1 *show how synchronization is achieved when* $t_1 = t_2/2$*, and how a spatial slice may be mapped out for all* O *observers according to* $d = ct_2/2$*. The friend* O_2 *interacts with* O′ *also using radar ranging. It can be seen directly from this part of the figure that* $T_1 = T_2/2$*,* $d = (c - u)T_1$ *and* $d_1 = cT_2/2$

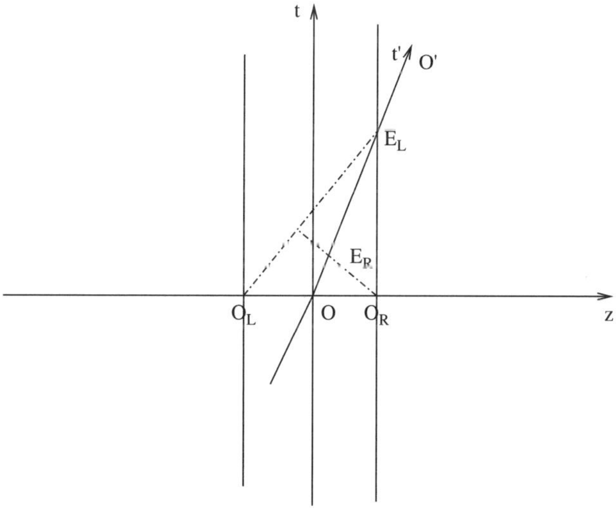

Figure 2.6 *The world line of* O′ *is shown passing through the world space of* O *observers at* O *at a speed* $u < 1$*. Two friends of* O*,* O_L *and* O_R*, emit photons (dashed-dot lines) towards* O′ *simultaneously (i.e. in the same world space or simultaneity). These are received by* O′ *at the distinct events* E_R *and* E_L*. There is no other way that these emission events can exist for* O′ *according to our basic principles*

simultaneous for two O observers O_L and O_R. These events comprise the emission of photons each directed towards the non-O observer O'. This observer is moving to the right with velocity u and is shown as passing through the arbitrary origin of the O observers. The figure is generic when we remember that in our units, electromagnetic waves propagate always at 45° to the axes and all relative velocities must be less than 1. This latter aspect requires that the line representing the world line of O' for O observers must be drawn closer to the t axis than 45°. The origin O can be any point in the world space of the O observers.

The figure shows without further effort that the photons do not arrive at O' simultaneously for $u \neq 0$, but rather at the events marked E_L and E_R. Recall that O' has no faster means of experiencing the original events in the O world space than through the reception of these photons. The positivist view is that these reception events have 'really' occurred for O' at the reception events E_L and E_R, with E_R before E_L. Should O' be moving to the left then the reverse time sequence, E_L before E_R, would hold. Thus for spatially separated events such as those at O_L and O_R, *neither simultaneity nor time ordering is invariant*.

Fortunately for our notions of causality, such events cannot be connected by an electromagnetic signal. They are absolutely separated in a manner we shall often call 'space-like'. Their simultaneous occurrence is not made possible by a triggering signal. It is due to the synchronization of O clocks in the world space, and by some previously agreed emission time as read on the ideal clocks of O_L and O_R. These two observers, or any number of O's friends, could agree at leisure on delay times between the emission of photons. If these were directed towards an unknowing O observer displaced, say, along the y axis, this observer might conclude that an arbitrarily fast signal was triggering the photon emission. But this arrangement is no different in principle from closing a pair of previously constructed very long scissors, in order to obtain an arbitrarily fast-moving intersection point. The source of the signal is not local, and the signal is not propagating as a causal wave.

It is interesting to remark that variable astronomical sources often show this acausal kind of 'superluminal' motion. In an important case it is due to electromagnetic radiation emanating from a distant stellar explosion. When this radiation strikes previously widely separated gas clouds, it excites the emission of photons towards us nearly simultaneously, creating an apparently superluminal causal connection. This is called a 'light echo' and will be discussed at length later. The delays programmed into Christmas tree lights from a central source produce a similar effect, although the delays creating the superluminal light echo are due to very different distances from the source.

We could readily calculate from the circumstance of Figure 2.6 the time difference between E_L and E_R *as measured by O observers at those events*. However, this time difference is not the proper time difference as measured on an ideal clock carried by O'. If different inertial observers cannot agree on the simultaneity of events, there is no reason to assume that they can agree on the clock phase to assign to an event. This clock phase is the ordinal time, proper time for O' and coordinate time for O inertial observers.

We prefer to deal with this question in the context of 'light clocks'. A light clock is an ideal clock comprising of a box extended along a fixed direction in space. Along this axis a single photon is directed. The box has perfectly reflecting mirrors exactly perpendicular to the axis of photon motion. Consequently if the length of the box at rest relative to the world space of an inertial observer is ℓ_o, then we have a periodic clock with time $2\ell_o$ between 'ticks'. The box may without loss of generality for these purposes be reduced to one dimension along the axis of the photon motion.

A slightly more practical means of constructing such a clock would be to use a phase coherent laser pulse in place of the photon. If the axis is the z axis, then a superposition of left- and right-going waves (say of the transverse electric field) with perfect reflection at each end would give a standing wave $Ae^{i\omega t} \sin kz$ for some constant A. To imitate our single photon, the frequency would be $1/2\ell_o$ in our units. The equivalent of measuring the bounce time is to measure a phase cycle at some z between the ends.

This is essentially the principle used in hydrogen maser clocks that are kept by national and international time services (e.g. National Research Council of Canada, US Naval Observatory, and for worldwide synchronization, Le Bureau International des Poids et Mesures in Paris). These clocks operate as suggested in the previous paragraph except that they count at the frequency of the 21 cm spin flip line of hydrogen, $\nu \approx 1420.40575_2$ MHz). The laser pulse is then actually a maser,[8] and it is produced by the coherent emission of spin-excited hydrogen atoms in a resonant evacuated cavity. The excitation is produced by microwaves applied at the correct frequency to low temperature hydrogen atoms. Cesium fountain atomic clocks are even more precise, although their stability is somewhat less.

Regardless of the technology, the universality of such a precise ideal clock should be manifest. The ordinal time given by any other clock in the same world space could be synchronized with the ticks of such a clock. By adjusting ℓ_o the ticks could be synchronized with those of *any* local clock. Then following the radar ranging procedure it could be synchronized with any clock over the entire world space. Of course, since we are measuring length in light seconds, an ideal clock that 'ticked' once a second would be very extended, nearly to the moon! It is greatly preferable to use the hydrogen maser frequency and to count precisely the ticks (phase cycles) in one second.

Let us consider such an ideal clock that is not at rest with respect to a set of inertial observers, but rather in rectilinear motion parallel to its length. Figure 2.7 shows the situation. At $t = 0$, O and O_1 are each coincident with an end of the light clock that is moving with speed u along the positive z axis. Since O and O_1 are in the same world space (have synchronized clocks), their separation L at $t = 0$ is known by radar ranging. Hence L is what O observers would declare to be the measure of the length of the moving light clock.

The clock phase is so arranged that a photon leaves the left end coincident with O at $t = 0$. It travels to the right end of the clock where it is reflected at time t_r and returns to the left end at time t_l, which is therefore the 'tick time' for O observers. Also for O observers, the reflection happens at coordinate z_r and the photon returns at z_l. The magnitude of the displacement along the z axis of the clock in time t_r (from O_1 to z_r) is marked ut_r, while in time t_l (from O to z_l) it is marked ut_l.

In the proper frame of the clock, which comprises O' observers for whom it is at rest, the total elapsed tick time is just $t'_l = 2L'$. This would be measured along the world lines

---

[8] A maser resonates at microwave frequencies rather than at visible light frequencies, as does a laser.

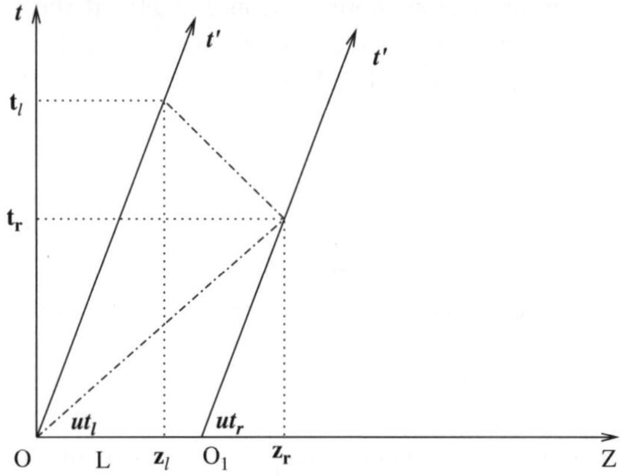

Figure 2.7 *The sketch shows a light clock moving to the right with speed* u *relative to* O *observers. It is of length* L *as measured by observers* O *and* O_1 *at* $t = 0$*. A photon emission event occurs at* O *and the photon is reflected at the event* $\{t_r, z_r\}$ *and received again at the event* $\{t_l, z_l\}$*. The coordinates are those of* O *observers. The tick time* t_l *and the length of the ideal clock for* O *observers* L *must be such that* $2L/t_l = 2L'/t_l'$*, where the primes indicate proper quantities. This is to maintain light speed invariant. In the text* t_l *is found as a function of* L *and Equation (2.56) is deduced*

of clock observers between the origin event and the photon return event. The length of the clock L' is measured at leisure by O' observers (the proper time of any O' observer is the coordinate time in the primed frame) since the clock is at rest relative to them. However, this proper tick time between emission and reception events is *not* the tick time measured by O observers along the t axis. This is equal to t_l measured along the t axis. We find this value as a function of u and L by noting that the distance travelled by the outgoing photon ($c = 1$ and distance may be measured in light-seconds) is t_r. Hence from Figure 2.7

$$t_r = L + ut_r. \tag{2.52}$$

On the return trip the photon travels the distance $t_l - t_r$ and, once again from the figure,

$$t_l - t_r = ut_r + L - ut_l. \tag{2.53}$$

Eliminating t_r from these last two equations yields the tick time for O observers as

$$t_l = \frac{2L}{1 - u^2} \equiv 2L\gamma^2, \tag{2.54}$$

which is also the distance travelled for O observers. However, by the principle of the invariance of the light speed we must have

$$\frac{2L'}{t_l'} = 1, \tag{2.55}$$

so that finally on employing Equation (2.54) for γ^2

$$\frac{L'}{t_l'} = \frac{1}{2} \equiv \gamma^2 \frac{L}{t_l}. \tag{2.56}$$

It is unlikely that $L' = L$ since the simultaneous events for O are not simultaneous for O'. Moreover, there is no reason that $t_\ell = 2\gamma^2 L$ should equal t_ℓ' for all γ, since the rest frame remains the rest frame. Consequently this last relation suggests that neither L nor t_ℓ are equal to their proper frame values.

This is a very general conclusion. The proper tick time of the ideal clock can be made to equal any time interval of the O' observers, and the proper length of the clock could be matched to any spatial interval measured by O' observers. We therefore suspect that any time interval, and any spatial interval parallel to the relative motion, must for O observers be different from their proper values for O' observers. Since the 'units' by which we measure time and space are simply standard intervals, we may refer to this measureable difference as a 'units transformation'.

We turn once again to a light clock in order to separate definitively the transformation of time and length intervals between inertial observers. For the moment we have only the ratio of these quantities in Equation (2.56). Consider a light clock that is moving perpendicular to its length for O observers as sketched in Figure 2.8.

In the figure, A and B are the world lines of the two ends of an ideal clock. At $t = 0$, O observers measure its length as L. This will also be the proper length L' since A and B could be moving on rails in the Y–Z plane. At the event where A crosses the $t = 0$ plane, a photon is emitted and directed towards the right end B of the clock. After a time T the photon bounces back to A where it is received after a time $2T$. Assuming

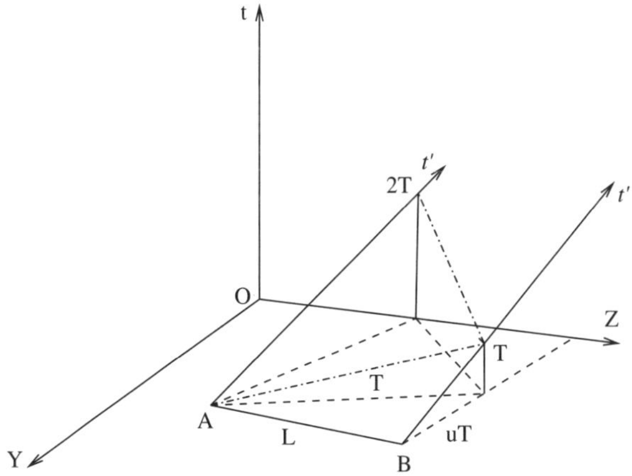

Figure 2.8 *This shows a light clock moving perpendicular to its extent along the Y axis. The size of the clock is exaggerated and ideally it is point-like. The mechanism of the clock is described in the text. It may be synchronized to any proper periodic process*

Euclidean geometry in the $Y-Z$ plane, we see from the figure that by Pythagoras

$$T^2 = L^2 + u^2T^2. \tag{2.57}$$

However, due to the perpendicular arrangement, $L = L' = T'$, where T' is the time that O' observers assign to the reflection event. We use once again that the speed of light is invariant and equal to 1 in our units. We have then by substituting $L = T'$ into the Pythagoras statement that

$$T' = \frac{T}{\gamma}. \tag{2.58}$$

We have extracted the positive square root to avoid inverting the time direction between inertial observers.

This time part of what we refer to as the units transformation is generally referred to as 'time dilation'. This is because, by Equation (2.58) and since $\gamma \geq 1$, the proper time interval T' of the moving observer O' is perceived by O observers to be 'dilated' by the factor γ. The measurement we have seen employs invariant 'radar ranging' in a light clock. Evidently O' observers will similarly perceive the proper time of an O observer to be dilated by the same factor, since no inertial frame is to be preferred. There is no logical difficulty in this unavoidable symmetry unless some observer travels between inertial frames by virtue of a controlled acceleration. In that case we shall have to study the world line or 'history' of this observer in detail in order to avoid what is known as the 'twin paradox' (see next chapter).

The transformation of parallel lengths that we anticipated as the spatial part of the units transformation is now found from Equation (2.56) if we set $t_l = T$ and $t_l' = T'$. This yields

$$L' = \gamma L. \tag{2.59}$$

Thus any 'proper length' (i.e. one measured at rest with respect to the observer) of a moving observer parallel to the apparent motion is 'contracted' by the factor γ as measured simultaneously by O observers. This is true for any proper spatial interval including a spatial unit, and is generally referred to as the 'Lorentz-Fitzgerald' contraction. Once again there is perfect symmetry between inertial observers so that O' observers also detect the Lorentz-Fitzgerald contraction of O observer proper lengths when they measure them simultaneously. However, in this instance, because of the non-invariance of simultaneity, it is clear that the events that define L' for O are not the same as the events that define L for O', so that the risk of paradox is removed.

The means that we have used to derive time dilation and the Lorentz-Fitzgerald contraction might be used to argue that they are only 'apparent' rather than 'real'. However, we must remember that if the speed of light is really maximal, there is no other way to perceive moving systems. According to the positivist philosophy then, the dilation and contraction are 'real' since they are true of any measurement of which we can conceive. Moreover, experiment (to be discussed below) detects them as part of our reality.

We turn finally in this section to show that the units transformation, plus invariant light travel time arguments and linearity, are together equivalent to the Lorentz transformations. The argument will derive them in terms of this theory of measurement. Henceforward these transformations will give the different coordinates of the

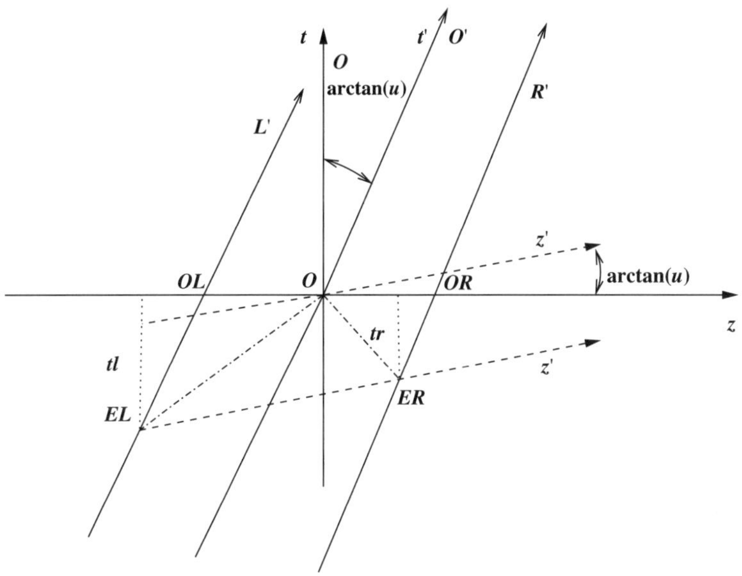

Figure 2.9 *The diagram shows a* t − z *slice of the world of* O *inertial observers. The world lines of* O′ *moving with the relative velocity* $u\hat{e}_z$, *together with two other* O′ *observers* L′ *and* R′ *who are equally spaced left and right of* O′ *by the proper length* $\ell'/2$, *are also shown on this diagram. These lines are lines of constant* z′. *The dashed oblique lines are lines of constant* t′. *The dashed-dot lines are light rays that arrive at the event where* O *and* O′ *coincide. Their backward extrapolations intersect the world lines of* L′ *and* R′ *at the events* EL *and* ER. *These emission events are simultaneous for* O′ *observers since the light rays arrive at* O′ *at the same origin event after travelling the same distance* $\ell'/2$

same event for different inertial observers, whether or not electromagnetic events are involved. Because of this universality, we will replace the preceding cumbersome type of argument concerning the different perceptions of events by algebraic manipulation of the Lorentz transformations. However, the physical justification is always as above.

We refer to Figure 2.9 as a graphical summary of the following thought experiment. A rod of proper length ℓ' for O' observers is moving relative to the world of O observers along the z axis with the speed u. The O' observers L' and R' are at rest on the ends of the rod. These observers emit photons in the direction of the central observer O' which arrive simultaneously at O' at the event where this observer coincides with O (this implies a choice of $t = 0$ for O observers). The photons also arrive at O simultaneously since O and O' are at the same event, but O must infer, by retracing the light rays, that they were emitted non-simultaneously at events E_L and E_R. Thus the dotted line through these two events and any line parallel to it are lines of constant t' represented in O's world. They are thus the space along which z' is measured. Similarly the lines marked O', L' and R' are lines of constant z' represented on O's world, and hence are the axes parallel to which t' is measured. An important feature of the experiment is that the two O observers OL and OR make a simultaneous measurement of the length of the rod. We know this to be $\ell = \ell'/\gamma(u)$by our previous arguments.

The representations of the axes of t' and z' in O's world should correspond to the linear Lorentz transformations when the correct units are used. This is easy to see for the lines of constant z'. We observe that $z' = 0, -\ell'/2, +\ell'/2$ on the world lines of O', L' and R' respectively. For O observers these world lines are respectively $z - ut = 0$, $z - ut = \ell/2$, $z - ut = -\ell/2$. Consequently we must have $z' = constant \times (z - ut)$ where the constant allows for the units change. At the instant $t = 0$ we know that $\ell'/2 = \gamma\ell/2$ by the units change transformation. Since $z' = constant \times z$ at this instant, we identify the constant as equal to $\gamma(u)$ by setting $z' = \ell'/2$ and $z = \ell/2$. Hence the equation for a line of constant z' in O's world becomes

$$z' = \gamma(u)(z - ut). \tag{2.60}$$

This is one of the Lorentz transformations that render Maxwell's equations invariant, but presently it depends only on the theory of measurement given a maximal, invariant speed of light.

The equation of a line of constant t' for O observers is slightly more involved but nevertheless instructive. We have indicated on the figure t_l and t_r as the times of the events E_L and E_R measured backwards from the present (so as to be positive) for O observers. These are readily calculated from the figure by setting the light travel time equal to the distance travelled along the z axis in each case. We find them to be $t_l = (\ell/2)/(1 - u)$ and $t_r = (\ell/2)/(1 + u)$ so that $t_l - t_r = \ell\gamma^2 u$. But this difference is just the rise in the line between E_L and E_R over the total distance along the z axis for O observers. From the figure this horizontal distance is equal to $(\ell/2 + ut_l) + (\ell/2 - ut_r)$, which is equal to $\ell(1 + u^2\gamma^2) = \gamma^2\ell$. Consequently the slope of the line E_L-E_R is simply $(\gamma^2\ell)u/(\gamma^2\ell) = u$. Given this slope, the equation for O observers of any line of constant t' (z' axis or simultaneity for O' observers) is $t' = constant(t - uz)$. This assumes the synchronization of coincident observers in order to set $t' = 0$ at the event $t = 0$, $z = 0$ where O and O' coincide. The constant allows for the units change in time.

On the world line of O', $z = ut$ and t' is the proper time. Hence our linear relation gives $t' = constant \times t(1 - u^2)$ for all slices of constant t' along the world line of O'. But we know that $t' = t/\gamma$ by our light clock argument that relates proper time t' to coordinate time t. We therefore require that the constant here be equal to γ. Consequently the equation of a line of constant t' on O's space-time diagram is

$$t' = \gamma(u)(t - uz). \tag{2.61}$$

This is the second Lorentz transformation.

We see by these arguments that the Lorentz transformations can be understood as requiring the grid of coordinate lines of an O' observer to be represented on the orthogonal grid of O observers as oblique lines, each of which make that angle $\arctan u$ with the corresponding O line. If an O observer is chosen to coincide with O' at $t = 0$ and taken to be the common orgin, then on the figure the dashed line marked z' and the solid line marked t' through the origin represent the axes of O'. The coordinate grid is formed by all lines parallel to these.

We emphasize that these figures are still diagrams and are not manifolds. This means that distances along the various axes have meaning either as temporal or spatial intervals, but distances in the diagrams that are oblique to the coordinate axes have no direct meaning. They would have such meaning only if space-time were regarded as a manifold with the equivalent of a Pythagorean metric. This is not the case as previously discussed. Even when we come to regard space-time as a manifold following Minkowski [7], the effective metric will be different from that of the plane of the space-time diagrams. Thus oblique distances in the diagrams continue to have no direct meaning.

We rewrite these linear equations in practical form by restoring units to find

$$z' = \gamma(u)(z - ut) \tag{2.62}$$

$$t' = \gamma(u)\left(t - \frac{uz}{c^2}\right), \tag{2.63}$$

but we must not forget that these are valid only in standard configuration. For arbitrary orientations we must use Equations (2.37) and (2.38).

Moreover, we have constructed these transformations so that the origin event $t = 0$, $z = 0$ coincides with the origin event $t' = 0$, $z' = 0$, such that one observer O and one observer O' coincide at the spatial origin. If one considered a friend of O' for whom the origin event was at $Z_{O'}(0)$ ($Z_{O'}$ positive or negative), then the new spatial coordinate of an O event t, z for this primed observer would be evidently

$$z' = \gamma(u)(z - ut) + \gamma(u)Z_O(0), \tag{2.64}$$

since the proper length $Z_{O'}$ is γZ_O where $Z_O(0)$ is the distance of the primed observer from O' for O observers at $t = 0$. At any subsequent t at $z = 0$, z' becomes correctly $\gamma(Z_O(0) - ut)$. If we are only interested in the difference in a coordinate, then this detail may be neglected.

The time transformation must also have a time $\gamma|Z_O|/c$ added to t' since the coincidence of O and O' cannot happen until then for the friend of O'. However, unlike the case for the spatial coordinate, we must take subsequently the current value of $Z_O(t)$ in this expression to allow correctly for the light travel time between O' observers. Hence the primed coordinates for this friend of O' become

$$z' = \gamma(u)(z + Z_O(0) - ut),$$
$$t' = \gamma(u)\left(t - \frac{uz}{c^2} + \frac{|Z_O(t)|}{c}\right). \tag{2.65}$$

We must emphasize that this latter discussion can be, and is usually, omitted; since one can always choose two world observers O and O' to coincide at a fiducial time taken to be zero. Thus if the coincident event for a train leaving the station is the front of the train exiting the station building, and if a passenger on the train is at a distance $Z_{O'}(0)$ from the head of the train, then for this passenger the coincident event has coordinates $Z_{O'}(0)$ and $Z_{O'}(0)/c$. But nothing prevents us from taking the coincident event to be at the passenger when the passenger crosses a fixed mark on the station. In that case the transformations in the standard form apply. Figure 2.10 reminds us of the structure of railway stations.

Figure 2.10 *This is a dramatic image of a railway station in Milan. The parallel tracks appear to converge as the distance between them subtends a smaller and smaller angle, but in fact in Euclidian space they never cross. This is not so on a curved surface. Much can be gained by loitering in railway stations. A train moving steadily and slowly along the track will give the illusion of a bystander moving in the opposite sense. The location of the spatial origins on the train and in the station are arbitrary.* Source: *Reproduced with permission from www.flickr.com/photos/paolomargari/2550814754. © 2008 Paolo Margari (See Plate 4.)*

We have in this section derived the transformations that leave Maxwell's equations invariant between inertial observers, by a theory of measurement. This approach is due to Einstein [5] and it has the merit of applying to all determinations of points in space and time (i.e. events) as they are compared between moving observers. The results would not change form if the role of c were usurped by some other speed, but we have seen that this would be incompatible with the principle of relativity as applied to Maxwell's equations.

The positivist philosophy that lies behind these transformations insists that we treat them as 'real' in every sense, since they are what we can measure. In this sense the philosophy is entirely compatible with quantum mechanics wherein nothing is known with certainty until it is measured.

Problems

2.9 Derive the expressions for t_l and t_r used in the text using Figure 2.9. Derive also the distance along the z axis (spatial separation for O observers) between EL and ER for O observers.

2.10 Draw the axes of O' on the space-time diagram of O if O' is moving in the negative z direction. One can take O' to coincide with O at $t = 0$, but consider also the representation if this is not the case.

References

1. Robertson, H.P. (1949) *Reviews of Modern Physics,* **21**, 378.
2. Mansouri, R. and Sexl, R.U. (1977) *General Relativity and Gravitation*, **8**, 809.
3. Fock, V. (1964) *The Theory of Space, Time and Gravitation* (2nd edn), MacMillan, New York, p. 12.
4. Lorentz, H.A. (1904) *Proc. Acad. Sci., Amst*., **6**, 809.
5. Einstein, A. (1905) *Annalen Phys*., **17**, 891.
6. Poincaré, H. (1905) *Comptes Rendus Acad. Sci. Paris*, **140**, 1504.
7. Eisele, Ch., Nevsky, A. Yu and Schiller, S. (2009) *Physical Review Letters*, **103**, 090401.
8. Voigt, W. (1887) *Nachricten Ges. Wiss. Gottingen*, 41.
9. Thomas, L.H. (1927) *Philosophical Magazine*, **3**, 1.
10. Samueli, J.J. and Boudenot, J.C. (2005) *H. Poincaré (1854–1912)*, Ellipses, Paris.
11. Minkowski, H. (1909) *Phys. Zeitshrift*, **10**, 104.
12. Mach, E. (1883) *Die Mechanik in ihrer Entwicklung historischkritisch dargestellt*. Leipzig.

3

Implications: Using and Understanding the Lorentz Transformations

Non nova, sed nove.
Nothing new except the style.

3.1 Prologue

We have argued in the last chapter that the Lorentz transformations between any two inertial reference frames must replace the familiar Galilean transformations. It is only true at high velocities since the Lorentz transformations (2.63) reduce to the Galilean ones to first order in u/c. This is evident for the spatial transformation, but we remark that in the time transformation we really require that $z \leq O(ut)$ ($O()$ means of the order of the argument) for this reduction to occur. For arbitrarily large distances z, the second term in the time transformation may not be neglected even at low velocity. It allows for significant light travel time between the appreciation of 'events' by widely separate observers, as we shall see.

The concept of 'event' is one that we have introduced casually in previous chapters. Now it will play an increasingly important rôle in our considerations. An 'event' is simply a set of four coordinates, one temporal and three spatial. This was discussed at length in the first chapter. On our space-time diagrams it is simply a point. Once an event is defined in one inertial frame, it can be known in any inertial frame by the Lorentz transformations. As discussed at the end of the last chapter, we normally choose temporal and spatial origins so that the standard forms apply for observers coincident at the origin, but this is not necessary. Most conceptual problems that arise in the application of these transformations are due to the absence of a consistent set of events for one inertial observer.

Practical Relativity: From First Principles to the Theory of Gravity Richard N. Henriksen

We can well imagine that this 'punctuality' (zero extent in time and space) associated with an event is not compatible with uncertainty in quantum mechanics. It would seem that more structure should be associated with an event. This will not concern us here, but efforts in this direction have been made under the heading of non-commutative algebra, or in terms of 'atoms' of space-time considered as an emergent manifold. A major proponent of non-commutative geometry is Alain Connes (Collège de France, Paris), while T. Padmanabhan (IUCAA, Pune, India) has considered irreducible units of space-time (see e.g. [1]).

3.2 Kinematic Applications

The complete Lorentz transformations must be used in general to map the coordinates of events between observers, and so to understand their different perceptions. However, there are cases when one coordinate is of more interest to us than the others, and we shall start with the important case of time.

3.2.1 Time

We have built into the Lorentz transformations the dilation of proper time, as perceived by another inertial observer, by the light clock argument. Now we can turn the tables and use the time transformation (2.38) to emphasize this point. The transformation gives the proper time of an event on the world line of O' ($\mathbf{r}' = \mathbf{0}$) that occurs at $(t, \mathbf{r})$ for O. Recall that these two observers coincided at the origin and agreed on the coordinates of this event.

Thus $\mathbf{r} = \mathbf{r}_\parallel = \mathbf{u}t$,[1] and so applying the transformation between two closely spaced events on the world line of O' we find $dt' = \gamma(dt - \mathbf{u} \cdot d\mathbf{r})$ whence, replacing $d\mathbf{r}$ by $\mathbf{u}dt$ and recalling the definition of γ,

$$dt' = \frac{dt}{\gamma}. \tag{3.1}$$

This is the relation between a proper time interval dt' and the coordinate time interval assigned to this proper interval by O. The coordinate time interval is always longer than the proper time interval, since the Lorentz factor is always greater than 1.

An even easier way to obtain the same result is to use the inverse transformation of (2.38) in the form $t = \gamma(t' + \mathbf{u} \cdot \mathbf{r}')$. This yields immediately the O coordinate time interval associated with the proper time interval at O' by setting $d\mathbf{r}' = 0$. Hence $dt = \gamma dt'$ as above.

This time dilation is real between relatively moving inertial observers according to our positivist philosophy. It implies that any unit of time on any proper clock (including a heartbeat or a lifetime) becomes an arbitrarily large time interval of O coordinate time as u approaches c arbitrarily closely, and so $\gamma \to \infty$. Can such a radically non-Newtonian result be tested? Frequently the arrival of excessive numbers of muons at the surface of the Earth, even though their 'half-life' is only about 2.2 microseconds (a light

[1] This may also be inferred from Equation (2.36) when $\mathbf{r} \parallel \mathbf{u}$ and $\mathbf{r}' = \mathbf{0}$.

microsecond is ≈ 0.3 km), is cited as a test. Since they are created by cosmic rays near the top of our atmosphere, this observation is indeed an indication of a 'dilated' lifetime. However, the measured half-lives of fast-moving muons in accelerator experiments is now even more convincing, as we discuss below.

First we emphasize that any pair of inertial observers together with their respective worlds are completely interchangeable, that is 'symmetric'. There is nothing that O can perceive about the world of O' that will not be also perceived by O' about the world of O. This includes time dilation, the Lorentz-Fitzgerald contraction, and indeed the entire content of the Lorentz transformations. The apparent change in sign when taking the inverse is due purely to an observer-specific choice of the positive direction of z. It could quite well be chosen so that the relative motion is positive for O' rather than O.

There is no difficulty with this symmetrical situation so long as the two inertial frames remain forever distinct, that is, moving relatively and interacting electromagnetically. It is only when at least one of the respective inertial observers contrives to change state by means of local acceleration, that two observers who were moving relatively can be brought back together in the same reference frame. At this point the total number of clock ticks or heart beats for each observer can be compared directly, given that there was an inital coincident origin at which the clocks were synchronized.

If the previously affirmed symmetry holds despite the acceleration history, then we do indeed confront a paradox, since each proper observer claims that the other proper clock runs 'slow', that is, displays less time than the local proper clock. This is commonly referred to as the 'twin paradox', since there seems to be no way to decide which twin has aged more. We discuss one resolution of this question below, but which twin is older continues to be a question of debate if strong gravity is present [2]. Ultimately it requires that we extend our non-Newtonian considerations to accelerations and real forces, which we reserve for later.

Nevertheless, accelerator based tests of the dilation of the muon 'half-life' confirm our time dilation expectations. The half-life of an unstable sub-atomic particle is simply the time over which half of any initial number will have decayed on average. This is a statistical value and it is subject to $1/\sqrt{N}$ fractional fluctuation in a sample of N muons. Thus it is best to have a large number N. This is achieved, for example, as a byproduct of the measurement of the anomalous magnetic moment of the muon [3,4].

One should recall that the half-life of low-velocity (relative to c) muons has been measured to be ≈ 2.2 microseconds. The mass of the muon also measured at low velocity is, in energy units, 105.7 MeV.[2]

In this illustration we make use of the famous equivalence $E = mc^2$ between 'rest mass' m (low-velocity inertia) and 'rest mass energy' E, which we will have to infer from dynamics later. The total energy of a rapidly moving body of rest mass m is actually $E = \gamma mc^2$, rather than the more famous preceding expression. However, if we expand this expression to first order in $(u/c)^2$ we obtain $E = mc^2 + mu^2/2$, and so the rest mass energy does appear as the major qualitative change to the classical energy. Accepting the general expression, we see that the ratio of the total energy to the rest mass energy gives directly the Lorentz γ.

[2] An MeV is one million electron volts, and an electron volt is the energy acquired by the electronic charge falling through a potential difference of one volt $\approx 1.6 \times 10^{-19}$ joule. The mass of the hydrogen atom is roughly one GeV.

In the course of the anomalous muon experiments, a beam of many thousands of muons was established at an energy of 3.09 GeV. This implies, according to our comments above, that $\gamma = 3090/105.7 = 29.2$. Consequently, according to time dilation each muon lifetime, and therefore the half-life of the ensemble of muons, is dilated to $2.2 \times 29.2 = 64.2$ microseconds. This was incidental to these experiments since they really wanted sufficient time to measure the precession rate of the muons in a magnetic field before they decayed. This 'proper time engineering' is how it was achieved. Suffice to say that the number of muons N was observed [4] to decay with exactly this half-life! Moreover the statistical errors increase as one would expect as the number of muons declines.

Example 3.1

The Lorentz factor for almost all macroscopic velocities that we encounter practically is only very slightly different from unity, since these velocities are very sub-light speed. For such examples we have by expansion in a Maclaurin series to first order that $\gamma \approx 1 + u^2/(2c^2)$. Then neglecting possible gravitational effects on clocks, we see that jet plane travel at 1000 km/h or roughly $u = 0.28$ km/s makes each of the traveller's seconds equal to only $1 + 4.4 \times 10^{-13}$ times the second of a terrestrial observer. After a return voyage of a million years the traveller would have gained only $\approx$ 14.2s over the fixed observer. This is not the 'fountain of youth', but atomic clocks do have the accuracy to detect this divergence [5].

Photons travel at the speed of light in a vacuum. Formally they require zero proper time even to traverse the Universe. They are in this sense 'delocalized' (existing only as infinite wave trains) and we shall encounter this peculiarity again. Fortunately they have no rest mass since then their energy would be infinite.

We are all astronomical voyagers. Our 'star ship' is quite literally the Sun and its planetary system. The equatorial speed of the Earth is about 0.5 km/s so that clocks and residents on the equator will run 'slow' relative to polar clocks and residents by a factor only slightly larger than that above. The average orbital speed of the Earth is about 30 km/s, so that relative to the Sun each second is slow by the factor $1 + 5 \times 10^{-9}$ and is easily detectable. But none of these frames is truly inertial. Relative to the cosmic microwave background radiation that may reflect the mean Universe, the sun is known to be moving at about 370 km/s. Hence we are ageing more slowly than the mean Universe by the factor $1 + 7.6 \times 10^{-7}$. This is still small, but over the lifetime of the 'Big-Bang' Universe ($\approx 4.1 \times 10^{17} s$) we would be younger by about $10,000$ years! More realistically, the Earth is perhaps only one third of the age of the Universe, and humans less that 10^{-4} this age, so that 'racial time' is within one year of mean Universe time.

Such verifications occur frequently in high-energy accelerator experiments. They confirm the positivist view that the dilation must apply to any 'clock', since the internal mechanism of muon decay is not invoked here. However, there is another accessible means of confirming time dilation.

The Döppler shift is familiar to everyone as it occurs in sound waves. A source of sonic frequency that is moving (subsonically) relative to an observer is shifted to lower frequency when moving away and to higher frequency when approaching. The amount of this 'Döppler shift' depends only on the frequency and the relative velocity. Train whistles and ambulance sirens behave in this familiar fashion.

Sound waves are very different from electromagnetic waves in that they require an 'aether', that is, a medium in which to propagate (the fact that they are longitudinal waves rather than transverse waves is of no essential importance here). Moreover, we do not normally encounter sonic sources moving at high enough velocities (approaching c, the sound speed in a medium c_s is much less than this) to justify taking into account time dilation. That is, we treat the 'ordinal' (i.e. the usual cycle count from some reference phase) proper time of the source to be the same as the ordinal coordinate time. This coordinate time would be measured by a instantaneously coincident local friend of a distant observer O, for whom the source is moving. The distant observer does not see the same ordinal time for the emission of a wave front by the source, however, due to finite sound travel time. It is useful to consider why this is so in this familiar example, since the electromagnetic Döppler shift depends on finite light speed in a similar fashion.

Figure 3.1 shows a space-time diagram where the speed of sound in a still medium (for some inertial observer O whose world is filled by the medium) is set equal to unity by our choice of units. Hence sound waves move at 45° to the axes in the diagram. The source is moving towards the observer with the speed u *relative to the medium*, which is necessarily subsonic if the sound is to arrive at O before the source. A supersonically moving source only emits sound waves inside a trailing cone that has

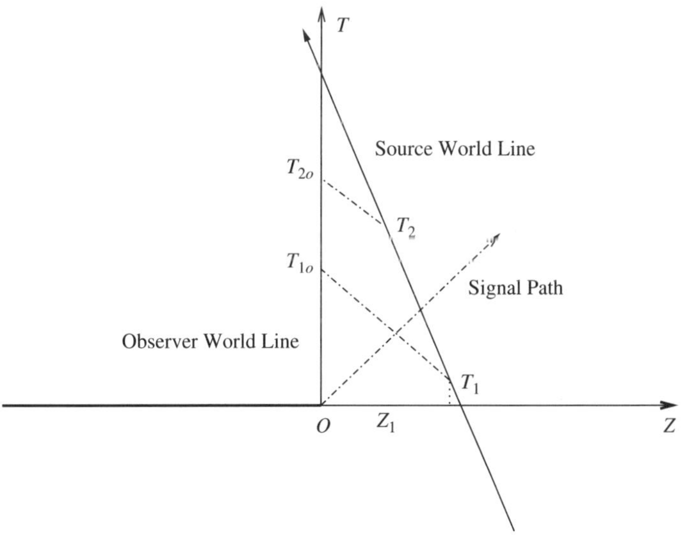

Figure 3.1 *The figure shows the world of O observers with a source of sound waves moving towards the origin observer with speed u < 1 in the medium. We have taken units such that the signal speed (sound or electromagnetic) is unity. The O observers are at rest in the medium for sound waves (general inertial observers for light). Wave fronts are emitted at* T_1 *and subsequently at* T_2 *in coordinate time and received at O at* T_{1O} *and* T_{2O} *respectively. All times are ordinal coordinate times for O observers*

a vertex angle $\arcsin(c_{s/u})$ at the source, which is called the Mach angle (our familiar positivist). The cone surface is a sonic wave front or pressure discontinuity. Once past an observer the source eventually produces a 'sonic boom' as the front passes over the observer. The same phenomenon occurs for a source of electromagnetic waves, but only in a medium with an index of refraction greater than unity. In that case a source may be faster than the light propagation speed in the medium, and the consequent 'light boom' is called Çerenkov radiation.

Returning to Figure 3.1, the source is shown to emit a sound wave front at proper time T_1 that is received at O at time T_{1O}. One period later it emits a second front at proper time T_2 that arrives at O at time T_{2O}. The times T_1 and T_2 may be taken to be the same as those times read on O coordinate clocks held by O observers that coincide with the source at those times. That is, all times cited are O coordinate times. This is the neglect of time dilation. In any case the interval $T_2 - T_1$ between successive wave fronts is the reciprocal of the local source coordinate frequency $1/\nu_c$. Similarly the interval $T_{2O} - T_{1O}$ is the reciprocal of the sound frequency observed by O, namely $1/\nu_O$. By our neglect of time dilation for sound sources, the local coordinate frequency ν_c is identified with the proper source frequency ν_s.

Consulting the Figure once more, we see that since the distance of the source from O at emission of the first front is Z_1, we have $T_{1O} = T_1 + Z_1$. Remember that the signal speed is unity in our units. Moreover it is clear that $T_{2O} = T_2 + (Z_1 - u(T_2 - T_1))$. Hence by subtraction $T_{2O} - T_{1O} = (T_2 - T_1)(1 - u)$. Consequently, remembering the definitions of our various frequencies,

$$\nu_O = \frac{\nu_s}{1 - u}. \tag{3.2}$$

This argument displays the signal travel time dependence very clearly and is exactly the same if our diagram is taken to be for light signals, except that the speed is then relative to the observer not the medium. Effectively the sound speed is invariant for a (inertial) world in which the medium is at rest, since the sound propagates always in the medium regardless of the source speed. Should the medium itself be moving along the line of sight relative to the observer; then this speed must be added to or subtracted from u as the case may be. The signal speed is relative to the medium and so the medium speed is added to u when they are parallel and subtracted when they are anti-parallel. Note that if the source is moving away subsonically with speed u, then we change the sign of u in the denominator above.

Another way of understanding the sonic Döppler shift (from the spatial part of the space-time diagram) is to realize that because of the source motion, the observed spatial interval between wave fronts that we call λ becomes $(1 \pm u)/\nu_s$. Hence the observed frequency $1/\lambda$ is indeed $\nu_s/(1 \pm u)$. It should be clear from these derivations that only the source velocity relative to the observer acts (the medium may be moving relative to the inertial frame of the observer). Moreover, it is only the component of this relative velocity along the line of sight of the observer that affects the observed frequency shift, at least when time dilation is ignored.

We can summarize the situation for a generally moving source by introducing the radial unit vector $\widehat{\mathbf{e}}_r$ drawn as usual *from* the observer along the line of sight to the source. Then the sonic Döppler shift is for general relative velocity **u** (restoring the conventional units)

$$\nu_O = \frac{\nu_s}{1 + \frac{\widehat{\mathbf{e}}_r \cdot \mathbf{u}}{c_s}}, \tag{3.3}$$

and u may include a medium velocity relative to O, added vectorially.

We have thus understood a principal part of the Döppler shift for light. Our singular omission was to ignore the difference between local coordinate time T and proper time, say τ. This difference is given by our familiar time dilation formula $T = \gamma\tau$ for coincident observers at $T = 0$. Consequently we now obtain from the analysis of Figure 3.1, $T_{2O} - T_{1O} = (T_2 - T_1)(1 - u) = \gamma(\tau_2 - \tau_1)(1 - u)$.

The transformation to vector velocity is done just as for sound waves. However, we must remember that for light sources the relevant velocity **u** is always relative to the observer, since there is no aether medium for light. Thus the Döppler frequency shift for light is

$$\nu_O = \frac{\nu_s}{\gamma(1 + \frac{\widehat{\mathbf{e}}_r \cdot \mathbf{u}}{c})}, \tag{3.4}$$

where ν_O is observed in an inertial frame for which the source velocity is **u**, and ν_s is the proper frequency measured by inertial observers for whom the source is at rest. Since the speed of light is invariant, a similar equation follows for the observed wavelength in terms of the proper wavelength from the definition $\lambda \equiv c/\nu$.

We remark in passing that should there ever be a medium in which c_s approached c, then a moving source of sound waves could be both subsonic and yet have a speed comparable to that of light. In that case the time dilation gamma factor would need to appear in Equation (3.3).

One of the few cases when our previous discussion of the coordinates of a remote observer (see Equation (2.65)) applies is to the Döppler shift. Let us take the origin observer O in that discussion (i.e. the one with world line coordinates $\{t, 0\}$) to be the source. Observer O transmits one wave front at t and another, one period later, at $t + \Delta t$. Since $\Delta z = 0$, the second of Equations (2.65) gives $\Delta t' = \gamma\Delta t + \gamma u \Delta t/c$, since $\Delta Z_O = u\Delta t$. Explicitly then

$$\Delta t' = \gamma(1 + u)\Delta t. \tag{3.5}$$

Hence, now with $\nu_s = 1/\Delta t$ and $\nu_O = 1/\Delta t'$, one obtains again Equation (3.4) by using the same directional convention. This argument allows us to emphasize the relative nature of the Döppler shift. The first of Equations (2.65) gives correctly $\Delta z' = -\gamma u \Delta t$ for the difference in location of the emission events as *seen* by this remote observer O'.

The Lorentz factor $\gamma(u)$ that appears in Equation (3.4) implies a frequency shift even when the source velocity is wholly transverse to the line of sight. It is known as the 'transverse Döppler shift' and is characteristic of sources moving close to the speed of light. This frequency shift is always a 'red shift' to lower frequency since $\gamma \geq 1$, but it is of 'second order' as it depends on the value of $(u/c)^2$. It offers a means of testing the prediction of time dilation, but it is much smaller in practical cases than the first-order effect already recognized in sound waves. It is necessary for the test to eliminate the first-order frequency shift to the desired accuracy.

Example 3.2

The transverse Döppler shift is as difficult as time dilation to detect, but the first-order effect due to the line of sight motion is readily apparent. The speed of sound at a temperature of 20°C in dry air is about 343 m/s. So according to Equation (3.3) a sound source approaching the observer at 30 m/s changes from high to low by a total fraction ≈ 0.175 of the rest frequency. Since the notes in an octave are separated by the fraction $2^{1/12} - 1 \approx 0.059$, this change is easily detected even by those of us without perfect pitch! In fact, even at much lower relative velocity the effect is noticeable.

For electromagnetic waves we must replace c_s by c and use Equation (3.4). At 30 m/s an approaching source has a higher frequency by the fraction $u/c \approx 10^{-7}$. This is readily detectable by modern frequency standards that equate to time-keeping. In fact, even angles of approach (or recession) that approach 90° to the line of sight are readily detectable. At 80° the effect is reduced by only 0.174. The magnitude of the effect is doubled in radars. All this permits traffic monitoring at multiple angles. The true second-order transverse electromagnetic Döppler effect is minute at these speeds.

Problem

3.1 The planet Mercury has no atmosphere and its surface reflects radar signals quite efficiently. The rotational period of the planet is approximately 58.65 days. Note that there are no second-order relativistic relative velocities among the planets, and that the reflection of a radar signal from Mars happens so rapidly that the relative configuration of the Earth and Mercury is fixed before and after the reflection. Hence show that the maximum frequency spread in a radar signal reflected from Mercury back to the Earth, after having been transmitted to Mercury in a narrow band at 10^{10} Hz, is ≈ 403 Hz. The radius of mercury may be taken as 2439 km and the radar signal should be regarded as absorbed by the planet and subsequently re-emitted.

Experimental tests were first performed on spectral lines from thermal ions where the width of the line was produced by the first-order Döppler effect, but the net red shift of the line centre was due to the second-order time dilation factor γ. The accuracy was improved with the advent of laser resonance and particle storage rings. A remarkable test was performed [6] in which a beam of lithium ions with speed $u = 0.064c$ in the laboratory frame was circulated in a storage ring. Parallel and anti-parallel collinear laser beams were tuned (frequencies ν_p and ν_a) to resonance with ionic transitions, which were known to produce optical lines of frequency ν_1 and ν_2 in the laboratory frame. It then follows from Equation (3.4) that if the second-order factor is indeed correct, then (see Problem 3.2)

$$\nu_a \nu_p = \nu_1 \nu_2. \tag{3.6}$$

Problem

3.2 Recall that the lithium ions are circulating in a storage ring at speed u. Show that Equation (3.6) is true. The parallel laser resonant frequency ν_p is tuned to an emission line of rest frequency ν_1 and the anti-parallel resonant laser frequency ν_a is tuned to an emission line of rest frequency ν_2. Note the crucial dependence on the transverse factor.

Corrections were made for the rotation and translation of the Earth to the cosmic inertial frame. By parameterizing possible deviations from the relativistic Döppler law, they were able to exclude deviations from the predicted second order at about the one part in 10^6 level.

A more direct measure of the transverse Döppler shift would be the frequency shift between a central source of radiation and an identical source on the edge of a rotating object. Such experiments with various physical techniques continue to look for undetected discrepancies [7].

When the motion is along the line of sight of the observer, Equation (3.4) takes a simpler form, namely

$$\frac{\nu_O}{\nu_s} = \sqrt{\frac{1 \mp u}{1 \pm u}}, \tag{3.7}$$

where the upper sign in each of the numerator and denominator on the right is for a receding source and the lower sign is for an approaching source. The factor on the right is exact and is often referred to as the 'K factor' between source and observer in standard configuration. We use K_+ for an approaching source and K_- for a receding source.

There is a type of elementary but practical (eventually one hopes!) problem involving interstellar travel that the reader may now consider. We present this in the following problem set. Few calculations are involved and space-time diagrams help enormously.

Problems

3.3 A starship travels to α Centauri at a speed of $u = 0.8c$. The star is at a distance of four light-years from the Earth, which may be considered to be an inertial frame.

- **(a)** How long does the trip take as measured by a clock on the Earth?
- **(b)** How long does the trip take as measured by a clock on the starship?
- **(c)** What is the distance between Earth and α Centauri according to starship observers? How can this be measured?
- **(d)** A radio signal is sent from the Earth to the starship every six months as measured by an Earth clock. What is the time interval between reception of the signals by the starship?
- **(e)** A radio signal is sent from the starship to Earth every six months as measured by a starship clock. What is the time interval between reception of the signals at the Earth?
- **(f)** If the wavelength of the radio signal sent from the Earth is 15 cm, to what wavelength must the starship receiver be tuned in order to receive it?

3.4 Two spaceships A and B leave Earth (taken to be an inertial origin) at $t = 0$ travelling in opposite directions along the z axis (say) with velocities $-u\widehat{\mathbf{z}}$ and $u'\widehat{\mathbf{z}}$ respectively. When the clock on ship A reads T the crew sends a light signal to ship B. Show that the time read by a clock on ship B when the signal is received is $T\sqrt{(1+u)(1+u')/(1-u)(1-u')}$ (if $c = 1$). Hint: The coordinates of the reception event should be found for A and hence B if the relative velocity of B relative to A u_{BA} is found.

We turn now to the famous twin paradox. Figure 3.2 illustrates the thought experiment. The world of O is shown with an observer O' who travels away from the origin (where O and O' coincide at $t = 0$) along the positive z axis with relative speed u. This continues until the event E_1 which has the coordinates $(T/2, Z)$ in the world of O. The dashed line marked Z_o' and the parallel line labelled t' through E_1 are lines of constant proper time for the outward bound observer. At E_1, O' reverses direction and returns to intersect the world line of O at event E_2, which has coordinates $(T, 0)$ for O. Neither at the origin nor at event E_2 does O' change inertial frames, but this does occur at E_1.

The paradox is supposed to reside in the symmetric diagram that we could draw that shows the perspective of O'. In that diagram E_1 would lie symmetrically on the negative z' axis and would mark the turning point of O. The symmetry is illusory, however, since

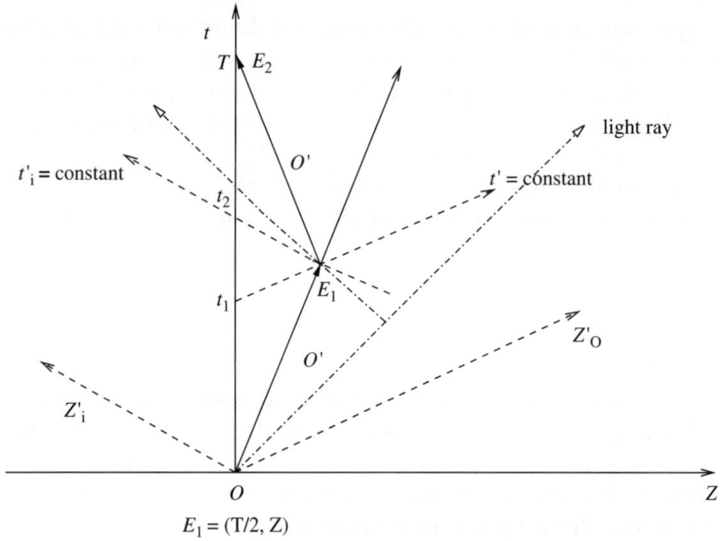

Figure 3.2 The figure shows the world line of an inertial observer O′ who coincides with the inertial observer O at the spatial origin at $t = 0 = t'$. This observer changes inertial frame at the event E_1 due to an acceleration that may be considered punctual on the scale of the figure. This causes O′ to intersect the world line of O once again at the event E_2. Thus E_2 has coordinates $(T, 0)$ for O and $(T/\gamma(u), 0)$ for O′. Each ordinal time is in the units of the respective observer. Light rays and dashed lines representing the lines of constant t' on both the outward and inward journey are also shown. There are two such lines through the event E_1 that reduce the interval $t_2 - t_1$ to zero there

O does not actually change inertial frames. Only O' suffers a measurable acceleration. It must be the acceleration required to change the inertial frame of O' that is responsible for the physical asymmetry. This must be so even though the acceleration can be made to occupy a negligible fraction of the journey as indicated on the figure and used below. Let us try to understand this.

We do not yet know how to describe the period of acceleration at E_1. But we can arrange that this takes place in a time much shorter than the time $T/2$, so that the process can appear to take place at the punctual event E_1. In this way we make the time in which the O' journey takes place in either one of two inertial frames, much longer than the time to reverse direction. The second inertial frame appears during the return, and the dashed line marked Z_i' and the parallel dashed line through event E_1 are lines of constant proper time for the inward bound observer. Thus the inward and outward dashed lines through the event E_1 correspond to the same proper time of O'. They do not, however, correspond to the same proper time for O. In fact the extrapolated lines of constant t' meet the world line of O at the separate times t_1 and t_2. The interval of proper time for *O has no corresponding interval of O' proper time*. This is therefore the time that O says is 'missing' from the world line of O'.

There are thus two ways in which this missing time may be calculated. The time dilation formula (e.g. 3.1) does not depend on the sign of the relative velocity, so that we infer that $T = \gamma(u)\tau$, where τ is the elapsed time along the world line of O'. The missing time should be calculated consistently in units of O proper time. The event E_2 has coordinates $(\tau, 0)$ for O' and $(\gamma\tau, 0)$ for O, but the ordinal time τ is only τ/γ in units of O. This is because each unit of O is larger than each unit of O' by the factor γ. Consequently the missing time is $\Delta t = \gamma\tau - \tau/\gamma \equiv T - T/\gamma^2 = u^2 T$.

The second method requires recalling the equation of a line of constant t' for O as summarized in Equation (2.61). We drop the factor γ there in order to obtain units of O proper time. Then the line of constant t' through t_1 has the equation $t - uz = T/2 - uZ = t_1$ and that through t_2 has the equation $t + uz = T/2 + uZ = t_2$. Hence $\Delta t = t_2 - t_1 = 2uZ$. But Z, the location of the turn event E_1, is $uT/2$ for O observers, so that once again $\Delta t = u^2 T$.

A similar calculation could be made for the gap in O' time caused by the apparent turn of O. However, this turn has not really happened for O. As is frequently the case, not all apparent motion is physical. Thus it is acceleration that is required to effect any change of inertial frames by an observer, which produces the asymmetric time dilation. It is important to note that the physical effect is really the difference in elapsed time in the proper units of each observer. *Thus T ticks or heart beats of O correspond to only $T/\gamma(u)$ ticks or heart beats of O'*. We have converted the missing time to units of O simply to compare with the 'gap' on the world line of O, which gap is necessarily in units of the proper time of O. The physical nature of inertial frames and the measurable effects of changing them is the real conclusion of this discussion.

Example 3.3

An interesting astronomical application of the electromagnetic Döppler shift occurs in the analysis of apparent superluminal motion in active galaxies [8] and in the so-called

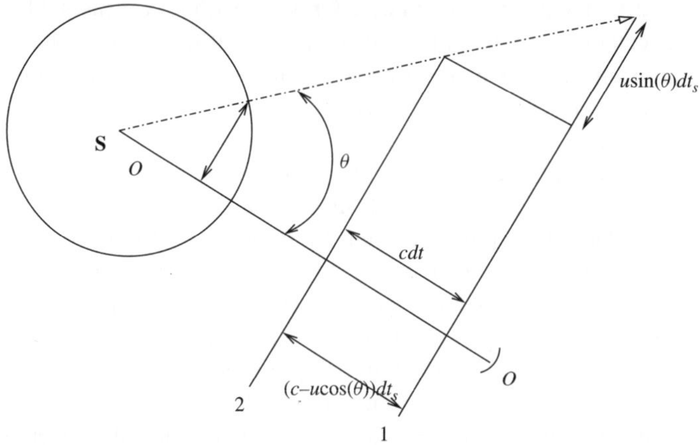

Figure 3.3 *The figure shows the observer O on the Earth and a friend near the source. The friend measures the coordinate time at the source* t_s. *Observer O also uses coordinate time to measure the distance between two received wave fronts* cdt. *This is equal to* $(c - u\cos\theta)dt_s$ *as shown in terms of the emission coordinate interval. The source expands transversely on the 'sky' in this time by the distance* $u\sin\theta dt_s$

microquasars in our own galaxy [9]. Consider a jet of material that is ejected at relativistic speeds at an angle θ to the line of sight of an Earth observer (Figure 3.3). We may take the Earth observer to be inertial since our speed with respect to the mean Universe is small relative to c. We also suppose that an observer at rest relative to the source of the jet is also approximately a 'friend' of the Earth observer. We shall refer to the friend at rest near the source as the 'remote' Earth observer. This observer keeps coordinate time, assigning the ordinal value t_s to a local event. This supposes that any relative velocity between the Earth observers is very small compared with c or the jet velocity.

The remote Earth observer measures a velocity transverse to the line of sight to the Earth as $d\ell/dt_s \equiv u\sin\theta$, where ℓ is the perpendicular displacement from the line of sight and u is the jet velocity. The velocity measurement would have to be made by observing the displacement of a luminous spot on the jet using friends of the remote Earth observer. The Earth observer, however, can only observe wave fronts directed along the line of sight. Just as in our discussion of the Döppler shift, the time interval dt for the Earth observer between line of sight wave fronts that are emitted in interval dt_s for the remote Earth observer is (reciprocal of Equation (3.4) without the time dilation) $dt = (1 - \beta\cos\theta)dt_s$. Here $\beta \equiv u/c$. Consequently the transverse motion for the Earth observer is measured to be $d\ell/dt = d\ell/dt_s(dt_s/dt) \equiv u\sin\theta(dt_s/dt)$. Finally, then, the apparent motion on the plane of the sky for the Earth observer is

$$u_{app} = \frac{u\sin\theta}{(1 - \beta\cos\theta)}. \tag{3.8}$$

The angular motion on the sky in radians per second is just this last result divided by D, the distance to the source. We explore the very important practical consequences of this result in the Problems.

Problems

3.5 Show that the maximum apparent velocity on the sky due to relativistic jet motion occurs when $\cos\theta = u/c$ and that the maximal speed is $\gamma(u)u$. Thus for motion close to the Earth observer's line of sight, the apparent motion may be very superluminal.

3.6 In 1994 the first microquasar in the galaxy was discovered [10] using a radio interferometer. It was interpreted in terms of the apparent motion generated by an approaching jet and a receding jet, both assumed similarly close to the line of sight to the Earth. Using the theory summarized in Equation (3.8), derive the formula for $\beta\cos\theta$ in terms of the apparent angular velocities of the two jets. The result is

$$\beta\cos\theta = \frac{\mu_a - \mu_r}{\mu_a + \mu_r}. \tag{3.9}$$

We have used μ_a for the approaching jet (normally brighter because of relativistic aberration) and μ_r for the receding jet, measured normally in arcsec per unit time.
Show also that the distance to the source D is given by the same theory, as

$$D = \frac{\mu_a - \mu_r}{\mu_a\mu_r}\frac{c\tan\theta}{2}. \tag{3.10}$$

3.7 One measures $\mu_a = 17.6 \pm 0.4$ mas/day, and $\mu_r = 9.0 \pm 0.1$ mas/day, and $D = 12.5 \pm 1.5$ kpc. Calculate β and θ. The units are 'mas' which is milliarcsec and 'kpc' which is kiloparsec or 3.086×10^{21} cm ($\theta \approx 70°$, $\beta = 0.92$).

3.2.2 Time and Rotation

In 1915 Sagnac [11] published an experimental result that tested directly the idea that the speed of light was independent of the relative velocity of source and observer. Using the device of a half-silvered mirror, a beam of monochromatic light mounted on a rotating disc (angular velocity Ω about its axis) was split into two beams while maintaining their coherence. One beam was directed with the sense of rotation of the rotating disc and one in the contrary sense. Reflecting mirrors mounted on the disc caused the two beams to be recombined again at the half-silvered mirror, and subsequently their interference could be measured. We shall deal with the theory again in a more sophisticated way, but because of its relevance to clock synchronization on a rotating Earth and to the construction of laser gyroscopes, we introduce it here in an idealized version [12].

We suppose that there are an infinite number of reflecting mirrors encompassing the circumference of the disc, so that each of the contra-directed light beams travels exactly tangentially to the disc.[3] Then after a complete circuit ($2\pi r$ for background inertial O observers if r is their measured radius) the beam moving in the sense of the rotation will have taken the O world time t_+ where $ct_+ = 2\pi r + \Omega r t_+$. Similarly the beam

[3] Such an arrangement can actually be realized using fibre optics or by using ring laser cavities. However, the index of refraction may not be unity as is assumed here.

moving against the rotation of the disc will have taken the world time t_- for one circuit, where $ct_- = 2\pi r - \Omega r t_-$. This is the usual synchronization argument between relatively moving observers when the speed of light is independent of this relative motion.

We conclude then that the two beams have experienced a path difference $c(t_+ - t_-) \equiv \Delta s$ which may be written as

$$\Delta s = \gamma^2 \left(\frac{4A\Omega}{c} \right), \tag{3.11}$$

where $A = \pi r^2$ is the area interior to the light path, equal to the area of the disc in this idealized calculation. The Lorentz factor $\gamma = 1/\sqrt{(1 - (\Omega r/c)^2)}$ may be set equal to unity here, since $(\Omega r/c)^2$ is of second order and very small in practice. The resulting fringe shift ($\Delta N \equiv \Delta\Phi/\Phi$ for phase Φ) in the recombined beams as a function of disc angular velocity is thus predicted by our hypothesis to be

$$\Delta N = \frac{4A\Omega}{c\lambda}, \tag{3.12}$$

where λ is the O world rest wavelength of the beam. We have derived this result in an idealized experiment, but the result is in fact general, provided that A is interpreted as the area enclosed by the light path and Ω is perpendicular to that area. The predictions have been verified in repeated experiments, and an excellent review is to be found in reference [13] as of 1997.

An intriguing practical application is to ring laser gyroscopes that are commonly used in aircraft, although now they are secondary to the Global Positioning System (GPS, which can itself be used to measure the Sagnac effect, e.g. [12]). In the gyroscope application, an optical system detects the fringe shift of an interference pattern due to the rotation of the aircraft away from a predetermined direction. Lasers on closed paths provide the necessary phase coherence, and when mounted on three independent axes they provide all components of Ω.

The fringe shift in the Sagnac effect is closely related to the problem of synchronization of clocks on a closed loop in a rotating frame [12]. Consider an observer O_1' riding with the edge of the disc. We suppose that two events E_1 and E_2 in the O background world are separated by the arc $rd\phi$, where r and the central angle ϕ are determined by O observers. The events are punctual flashes of light and at a given O coordinate time, O_1' coincides exactly with the mid-point of the arc $rd\phi$.

By our usual arguments based on the light travel time (see e.g. the Sagnac argument above for the whole disc), O_1' sees E_2 at the O coordinate time $t_- = rd\phi/(2c(1 + \Omega r/c))$ and E_1 at the O coordinate time $t_+ = rd\phi/(2c(1 - \Omega r/c))$. That is, for O_1', E_2 occurs before E_1 by the O coordinate time interval $t_- - t_+ = -rd\phi\gamma^2\Omega r/c^2$. Moreover we assume that time dilation holds for O_1' relative to O coordinate time, despite the acceleration of this observer. This is because we can imagine a set of inertial observers who move linearly with disc edge speed, each one of which coincides with O_1' for an instant. Then in the units of these inertial clocks the ordinal interval between E_2 and E_1 is $-\gamma\Omega r^2 d\phi/c^2$. Because at every instant the ordinal time on coincident clocks can be made the same, we take this also to be the interval on the clock of O_1'.

But this argument can be extended to n disc observers $\{O_j'\}_{j=1\ldots n}$ who are spaced equally around the circumference of the disc such that O_n' is again adjacent to O_1', but behind it in the sense of the rotation. At a given coordinate time instant, n flashes of light $\{E_j\}_{j=1...n}$ are arranged similarly around the circumference of the disc, each pair separated by the arc $rd\phi$. At the instant of these simultaneous events the disc observers are all opposite the mid-points of the corresponding arcs E_j to E_{j+1}. Then each O'_j will determine that E_{j+1} happens before E_j by the interval calculated above.

The fact that the O' observers do not agree with the simultaneity of the events $\{E_j\}_{j=1...n}$ does not surprise us, for we have seen that this is an inevitable consequence of the limiting speed of light. However, normally we compare two inertial 'worlds', each of which have an agreed coordinate time between friends. Here we see that this cannot be so for the disc observers, since O'_1 says that E_2 is before E_1 and O'_2 says that E_3 is before E_2 and so on, each step in time being the interval calculated above. Finally O'_n is forced to say that E_1 is before E_n by the cumulative amount $2\gamma\Omega\pi r^2/c^2$, since now the total change in angle is 2π. So according to disc clocks, E_1 is both the last event, being after E_2 which is after all of the other events, and the first event, being before E_n which is before all the other events!

There is thus no unique ordinal disc time that can order the simultaneous events in the O world. Rather a cut (i.e. a discontinuity) in disc time of order $2\gamma\Omega A/c^2$ is inevitable. This cut is just one half the Sagnac effect, but counter-rotating beams would each encounter this step in the opposite sense giving the usual difference. Hence not only do the disc observers not see the n flashes as simultaneous, but they cannot agree on the ordinal time of the series of events among themselves. This is characteristic of rotating observers, and we conclude that it implies that no radar synchronization is possible around a closed loop of such observers.

A circle of constant latitude on the Earth is a practical example of this thought experiment, as is indeed any small circle on the rotating Earth. The Sagnac phase difference has been measured in such cases using satellite links.

Nevertheless terrestrial clocks are synchronized and the speed of light is still held to be isotropic in inertial space. This is possible because terrestrial clocks may be radar synchronized relative to a clock on the axis of the Earth's rotation, or by slow clock transport over the surface of the Earth. To avoid gravitational corrections, all clocks would best be located on the Earth's 'geoid' (equipotential) at all times. So the solution is not to regard the circle of latitude as a closed space, but to use the properties of the space in which it is embedded. We shall encounter this point again in the context of more general relativity.

3.2.3 Time and the Lorentz Transformation

The Lorentz transformations transform the space and time coordinates from one inertial observer to another that is in relative motion with respect to the first, by assuming that the speed of light is invariant and maximal. As such they allow correctly for the light travel time, and this can lead to vast differences in coordinates even for slow relative motion if the events are distant.

An amusing example is to consider two pedestrians, O at rest and O' moving with speed u relative to the Earth (taken inertial). Let us suppose O' passes O while moving directly away from the red star Betelgeuse in Orion that is just rising. For O the distance to Betelgeuse at that moment is $z_B \approx 600$ light years. If an explosion happens at Betelgeuse for O at time t, then by the Lorentz time transformation $t' = \gamma(t + uz_B) \approx t + uz_B$. A comfortable walking pace is one metre per second or $u = 4/3 \times 10^{-9}$ so that $t' \approx t + 24s$, a detectable deviation from absolute time. However, suppose that it was a distant galaxy rising in the same configuration with $z_B \approx 3 \times 10^6$ light years. Then $t' = t + 1.2 \times 10^5 s$, or just under a day and a half.

This example emphasizes that the time transformation does not imply Newtonian absolute time in the limit $u \to 0$, if distance is large enough. One should note, however, that the answer only holds between these two inertial frames that are forever identified. Observer O' has to keep walking until the signal arrives so that the extra distance explains the 24 s. That is for 24 light-seconds, more than ten times the distance to the moon! Otherwise, of course, O' reverts to being an O observer.

Problem

3.8 In the text example showing non-absolute time even at walking speed for distant objects, we used the Lorentz transformation to compare coordinates. Given the finite and maximal light speed, what is the positivist explanation of the time difference?

A deeper question is to ask under what conditions the order of occurrence of two events may be reversed under a change of inertial observer. By the Lorentz time transformation if the interval $\Delta t' < 0$ even if $\Delta t > 0$, then we require $\Delta t < u\Delta z/c^2$ or equivalently $\Delta z > c\Delta t/(u/c)$. Since $u/c < 1$, this requires that the separation between the events Δz be greater than the distance that light can travel during the interval Δt between the events. Consequently causality (the cause must precede the effect) is safe, since by the hypothesis that c is the maximum signal speed, no causal link can exist between such events.

Graphically, the line through two such events on a space-time diagram has a slope $c\Delta t/\Delta z$ less than unity so that it has a larger projection on the z axis than has the light line. The separation between a pair of such events is said to be 'space-like' since one can always find an inertial observer (u) such that $\Delta t' = 0$ if $u/c = c\Delta t/\Delta z < 1$. By contrast, setting $\Delta z' = 0$ would require (from the spatial Lorentz transformation) $u/c = \Delta z/c\Delta t > 1$, which is not possible.

Conversely, it is easy to show from the Lorentz transformations (when applied to the intervals between events as above) that with $\Delta z < c\Delta t/(u/c)$ such intervals are 'time-like'. That is, one can find the inertial frame in which $\Delta z' = 0$ namely $u/c = \Delta z/c\Delta t < 1$, but the time interval in this case cannot be set equal to zero by any physical choice of u/c (it would have to be > 1). The slope of the line between such events $c\Delta t/\Delta z > 1$, and so it has a larger projection on the t axis than has the light line. Evidently all causally linked events, including those on the world line of a particle, are time-like.

3.2.4 Space

Recall that the Lorentz-Fitzgerald contraction refers to the extent of an object in the direction parallel to its velocity $\mathbf{u}$ relative to an inertial observer O. It is rigorously prescribed as the length measured simultaneously by two O observers who coincide with the parallel extremes of the object at that time. If the parallel extent is $\Delta z' = \ell'$ for O' observers for whom the object is at rest, then this prescription plus the Lorentz transformation for spatial interval gives $\ell' = \gamma(u)\Delta z = \gamma(u)\ell$. So the predicted contraction factor is indeed $\gamma(u)$, which is once again a second-order effect.

It is very difficult to carry out this defining experiment in any fashion except possibly optically, and then the light travel time serves to obscure the issue as we shall see under the heading of optics below. It is more difficult than the verification of time dilation since relativistic objects that we know are sub-atomic particles. For these we can measure lifetimes at least statistically but, for quantum mechanical reasons, we cannot measure particle sizes even in principle. We can measure lengths statistically as is done for example to small fractions of a nucleus in experiments designed to detect gravitational waves, but objects in these experiments are not moving relativistically.

Mostly we content ourselves with verifying the absence of contradiction by pondering a multitude of thought experiments. Consider for example the traditional 'barn and pole' experiment. A more modern context would be in terms of a space capsule and an airlock, but in any case it involves a pole and a box with a door. The pole is considered to be moving parallel to its length at relativistic speed relative to the box, while the box is aligned with this motion so as to capture the pole. We do not need to consider either the commotion that will ensue when the pole strikes the end of the box or the strength of the respective materials, since nothing will occur globally more rapidly than can occur at the speed of light.

The problem consists of resolving the following apparent paradox. The box observer can see a pole, whose proper length is longer than that of the box, sufficiently contracted that the whole length of the pole may be in the box simultaneously. This could be signalled by shutting the door at this instant. The pole observer however sees an even smaller contracted box such that the box may be much smaller than the pole. How can this observer explain the door closing on the complete pole? Moreover, at a given relative speed, is there a minimum proper length of the box relative to the proper length of the pole for which this trick holds? Such a limit may be referred to as the 'just possible' condition.

Figure 3.4 shows the space time diagrams for the box observer O and the pole observer O' in a 'fits easily' condition, rather than the limiting case. Everything is implicit in these diagrams, but it is instructive to discuss them at various levels of analysis. The figure reveals qualitatively by inspection the resolution of any possible paradox in the 'fits easily' case. Neither the pole nor the box can be considered rigid over times comparable with their lengths in (light) seconds. Hence the pole may accordian into the box until E_2, and it is only the distance between E_1 and E_2 that must be less than the proper size of the box. Similarly on the right of the figure we see that the box may stretch over the pole until the event E_2 and it is only the distance between E_1 and E_2 that must be larger than the proper size of the pole. In neither reference frame are events E_1, E_2 simultaneous. We turn then to the question of the limiting condition.

To find the minimum ratio ℓ_b/ℓ'_p for which this trick is possible, we must allow a maximum lack of rigidity for both the pole and the box. The general argument in one reference frame is sufficient, but it is instructive to consider both perspectives. Thus starting with the pole reference frame, the light distance ct'_2 must equal the stretched

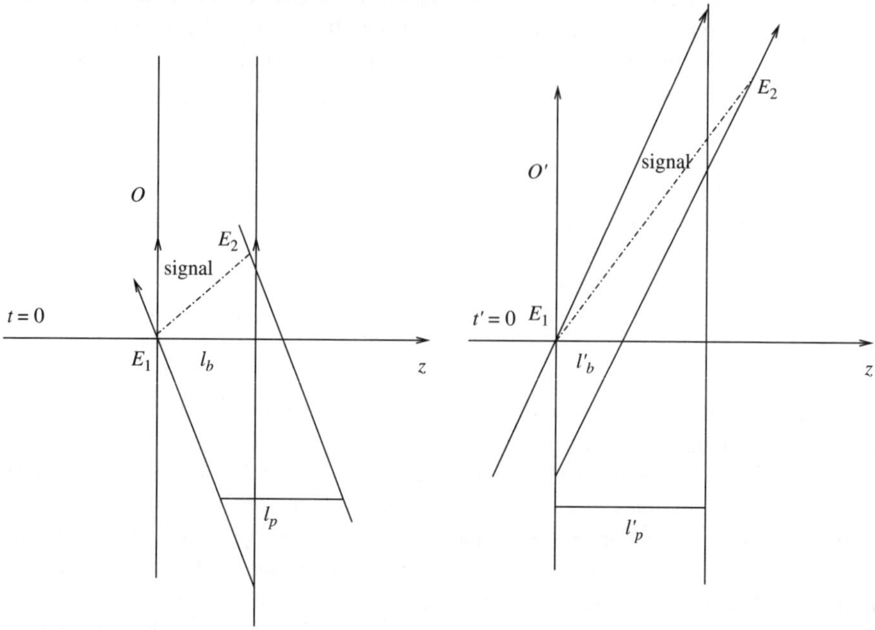

Figure 3.4 *On the left side the two vertical lines are the world lines of the ends of the box in the world of O. The oblique lines are the world lines of the ends of the pole that is moving at speed u along the negative z axis. At t = 0 the length of the pole measured by two O observers may be greater than ℓ_b, but it is the distance between E_1 (the collision with the back of the box) and E_2 that counts. The pole accordians into the box until it is stopped at E_2 by the arrival of a light signal from E_1. The right side of the figure shows the sequence from the perspective of the pole observer O'. The vertical lines are the world lines of the ends of the pole. The box is moving to the right and event E_1 is again where the back of the box and the left end of the pole collide. The contracted box length ℓ'_b stretches to contain the pole, until it is stopped at E_2 by the arrival of a light signal announcing the collision*

length of the box $\ell_b/\gamma + ut'_2$. This gives $t'_2 = (\ell_b/c)/(\gamma(1 - u/c))$. Therefore the maximum stretched length of the box is $\ell_b/\gamma + ut'_2$, which is $\ell_b\sqrt{(1 + u/c)/(1 - u/c)}$. We must have this $\geq \ell'_p$ for the box to stretch over the pole, which condition requires

$$\ell_b \geq \sqrt{\frac{1 - u/c}{1 + u/c}}\,\ell'_p. \tag{3.13}$$

Taking the equality gives the minimum box length possible as desired. For $\gamma = 2$ as an example, we have $u/c = \sqrt{3}/2$ and so we find $\ell_b(min) = \ell'_p/(2 + \sqrt{3})$, substantially less than $\ell'_p/2$ which would be the naive conclusion from the Lorentz-Fitzgerald contraction alone.

But what of the Lorentz transformation to events in the frame of O? We have that at the minimum condition $t'_2 = \ell'_p/c$ since the stretched length of the box is then just equal to the proper length of the pole. Moreover (O')'s coordinate for event 2 is

$z_2' = \ell_p'$. Hence by the inverse transformations $(u \leftarrow -u)$, $t_2 = (\gamma \ell_p'/c)(1 - u/c)$ and $z_2 = \gamma(\ell_p' - \ell_p' u/c) = \ell_p' \sqrt{(1 - u/c)/(1 + u/c)}$. But Equation (3.13) shows that this is $z_2 = \ell_b$ in the minimum condition. In addition $t_2 = z_2/c = \ell_b/c$, which is just the time for light to travel the proper length of the box. So the Lorentz transformations show that the events separated by a light travel time in one frame of reference are separated by a light travel time in the other frame of reference, and a mutual lack of rigidity is thus permitted.

However, how is is this perceived for the box observer? The left side of Figure 3.4 indicates correctly that the pole will accordian into the box until the receipt of the light signal at E_2. This signal will arrive at $t_2 = (\ell_p'/c)(\gamma(1 + u/c))$ (the familiar light travel time argument). Consequently the pole will be compressed to the length $\ell_p'/\gamma - ut_2$, which is equal to $\ell_p'/(\gamma(1 + u/c))$. This must be less than ℓ_b so that once again

$$\ell_b \geq \sqrt{\frac{(1 - u/c)}{(1 + u/c)}}\ell_p'. \tag{3.14}$$

But now we may use the Lorentz transformations to find the events in the frame of reference of the pole. Applying the minimum condition we have that $t_2 = \ell_b/c$ also, and $z_2 = \ell_b$. Hence the forward Lorentz transformations give $z_2' = \gamma(z_2 + ut_2) = \ell_b \gamma(1 + u/c)$, which Equation (3.14) shows to be ℓ_p'. Moreover from the time transformation, $t_2' = \gamma(t_2 + uz_2/c^2) = \gamma(\ell_b/c)(1 + u/c)$, which is just ℓ_p'/c by Equation (3.14). So the light travel time between events is again preserved between observers. This is the maximum time over which a maximally rigid object (sound speed equal to the light speed) can be treated as a fluid.

Thus we have a reassuring consistency between the light travel time arguments and the Lorentz transformations, as one should expect. We have only had to sacrifice the classical concept of rigidity. The resultant fluidity is completely independent of the composition of the object, which is a general property of relativistic effects. Time dilation does not require knowledge of the construction of clocks and the Lorentz-Fitzgerald contraction does not require a study of the structure of matter.

It is worth remarking that this absence of a mechanistic explanation for time dilation and length contraction has disturbed some physicists [14], just as quantum 'entanglement' has done. Lorentz, and others subsequently [14], attempted to give a physical explanation based on a model of structure of rods and clocks. This required a preferred inertial reference frame (that of the 'aether') so that velocity relative to it was a physical effect on absolute time and space. However these arguments are all contingent on structure, and once the speed of light is recognized as invariant and maximal, there is no escaping the universality of the preceding arguments.

An old and famous problem concerning space is that of the nature of a relativistically rotating disc (see [15] for a historical discussion of the early protagonists). Einstein remarked several times [16] that the rotating disc demonstrated the physical interdependence of an 'effective' gravitational field[4] and spatial geometry.

[4] Here the centrifugal field; Einstein invoked the 'principle of equivalence' by which locally an acceleration and a gravitational field are indistinguishable.

The simplest form of the argument assumes that a circular disc (which may be thought of as a sequence of rings of different radii; e.g.[12]) may be placed into relativistic rotation about its axis of symmetry with angular speed Ω, without being destroyed. This means that it remains a circular disc in rotation in which every particle follows a helical world line. Therefore in the inertial Euclidian space, the elementary arcs ($rd\phi$) of every ring add to give a circumference of $2\pi r$.

Considering once again that the inertial observer who is instantaneously tangent to each arc of the ring is equivalent to the proper disc observer at that instant, we infer that the proper arc length is $\gamma(\Omega)rd\phi$. Here $\gamma \equiv 1/\sqrt{1-(\Omega r/c)^2}$. Consequently these appear to add to $2\pi\gamma r$ for all disc observers, and we would have to conclude with Einstein that the geometry on the disc is hyperbolic (negative curvature: i.e. the circumference of a circle greater than $2\pi r$) rather than Euclidian (zero curvature). This dramatic transformation would presumably be due to the effective gravitational field.

However, we have seen in our discussion of the Sagnac effect that rotating observers cannot agree on a single time. There must be a cut in the amount of $2\pi\gamma(\Omega)r^2/c^2$ around every ring. Actually any closed path on the disc will display such a temporal cut of order $(\Omega/c^2)\oint\gamma(r)r^2(\phi)d\phi \approx 2A\Omega/c^2$. This implies that the notion of a geometry for disc observers is suspect on any closed path, since the path does not exist at any one agreed disc time.

One could synchronize disc clocks on a ring with inertial clocks (e.g. by emitting and reflecting a cylindrical wavefront from the axis of the disc and correcting for time dilation), but then we would no longer have a self-contained reference frame. It appears then that this particular argument for the dependence of geometry on effective gravitation is not warranted. Rather, restricting our considerations to open paths, we conclude that particles in the disc have separated (by the factor γ on the arc of a ring). Thus the disc cannot be a rigid body just as for the pole and the box in our previous example.

This cannot be the whole story, however, since we have taken the disc to remain rigid in the inertial frame. We might have assumed arbitrarily that it remained rigid and Euclidian in the rotating frame (e.g. [15], pp. 130–132). Then we would infer that the circumference of a ring was $2\pi r/\gamma$ in the inertial frame. Implicitly we would be insisting on an agreed ring instant for which the ring geometry existed. This leads to the bizarre conclusion [12] that there would be a cut in the inertial time on a closed path by the same amount $2\gamma\Omega A/c^2$ (when measured in disc time units). This allows the resolution of the inertial geometric question just as for the disc observers, since there would be no one inertial time for the disc in the inertial frame. Particles would be compressed by the factor γ on an open arc of each ring, rather than being separated by the same factor.

Neither of these assumptions is liable to be strictly true, although the assumption of rigidity in the inertial frame is more consistent with special relativity since we know how to establish an inertial 'world time'. The asymmetry, much as in the twin problem, is due to the absolute acceleration of one set of observers. It is interesting that, just as the gap in time in the twin problem is due to the jump in proper time of the travelling observer at the turn, we might make a similar argument here about a gap in space. For the 'spatial gap' on a ring *between the two points of view* is $2\pi\gamma r - 2\pi r/\gamma = (\Omega r)2\gamma\Omega A/c^2$. This is, however, just the ring rotational velocity multiplied by the cut in time, which is required in either point of view. It is likely that a ring circumference is not defined for either set of observers to within this spatial gap. There is a kind of 'uncertainty' due to our measurement limitations and the actual structure of matter.

Problem

3.9 Derive the expression for the time discontinuity around any closed loop on a disc rotating with the angular speed Ω about its axis of symmetry as quoted in the text. Note that the integration in angle is carried out at a fixed inertial time. The approximation omits second-order terms.

3.2.5 Space and Time

In this section we make full use of the Lorentz transformations to discuss the space-times (that is, the 'worlds') of different inertial observers. After the measurements of space and time, a natural feature of any world is the velocity of an object. By this we mean the velocity of a point object, or of an object in which every part moves with the same velocity. The discussion of rotation is postponed until later.

Let us begin by recalling the general space and time transformations in the forms (2.37) and (2.38). Then the velocity for the O' observers (we need not distinguish time between the axes in standard configuration and the co-moving rotated axes) is $\mathbf{v}' = d\mathbf{s}/dt'$. A straightforward differentiation gives ($c = 1$)

$$\frac{d\mathbf{s}}{dt'} = \frac{dt}{dt'}\left(\frac{d\mathbf{r}}{dt} - \gamma\mathbf{u} + (\gamma - 1)\widehat{\mathbf{e}}_u\left(\widehat{\mathbf{e}}_u \cdot \frac{d\mathbf{r}}{dt}\right)\right), \tag{3.15}$$

where $\widehat{\mathbf{e}}_u$ is the unit vector in the direction of the relative velocity.

But for the O observers $\mathbf{v} = d\mathbf{r}/dt$, and Equation (2.38) in differential form gives dt/dt' so that after rearrangement we obtain

$$\mathbf{v}' = \frac{\mathbf{v}_\perp}{\gamma(1 - \mathbf{u}\cdot\mathbf{v})} + \frac{\mathbf{v}_\parallel - \mathbf{u}}{1 - \mathbf{u}\cdot\mathbf{v}}. \tag{3.16}$$

Explicitly we have set the components of $\mathbf{v}$ perpendicular and parallel to $\mathbf{u}$ as

$$\mathbf{v}_\perp \equiv \mathbf{v} - \mathbf{v}_\parallel, \tag{3.17}$$

and

$$\mathbf{v}_\parallel \equiv \widehat{\mathbf{e}}_u(\widehat{\mathbf{e}}_u \cdot \mathbf{v}). \tag{3.18}$$

To restore conventional units we need only multiply $\mathbf{u}\cdot\mathbf{v}$ by $1/c^2$.

We observe first that the inverse of the parallel part of the transformation ($-u \leftarrow u$ and exchange the primes on $\mathbf{v}$) rederives the Lorentz/Poincaré group condition (2.46) when $\mathbf{v}_\parallel \equiv u_2$ and $u \equiv u_1$. It is worth remarking that the Lorentz factor γ does not appear in this parallel transformation of velocity. This is due to the mutual cancellation of the time dilation and the Lorentz-Fitzgerald contraction. The only remaining non-classical effect is the light travel-time factor. In contrast, the non-classical part of the transverse or perpendicular velocity transformation is due to time dilation plus the light travel-time factor, since the Lorentz-Fitzgerald contraction does not apply transversely.

If we introduce a spherical polar coordinate system by taking the common axis of the relative velocity $z - z'$ as polar axis, then we may write explicitly in each inertial frame $\mathbf{v}_\perp = \mathbf{v}\sin\theta$ and $\mathbf{v}_\parallel = \mathbf{v}\cos\theta$. We mean that θ' replaces θ in the frame of O'. Hence the perpendicular and parallel velocity transformations become respectively

$$\mathbf{v}'\sin\theta' = \frac{\mathbf{v}\sin\theta}{\gamma(u)(1 - u\mathbf{v}\cos\theta)},$$
$$\mathbf{v}'\cos\theta' = \frac{\mathbf{v}\cos\theta - u}{1 - u\mathbf{v}\cos\theta}, \tag{3.19}$$

with the inverse found by changing the sign of u and interchanging the primes.

An interesting effect of these transformations is seen by writing them in inverse form and taking their ratio to obtain

$$\tan\theta = \frac{\tan\theta'}{\gamma(u)(1 + u/\mathbf{v}'_\parallel)}. \tag{3.20}$$

Thus suppose that u is the speed of a relativistic beam of particles along the z axis and θ' is the angle that a particle makes with this axis in the primed frame moving with the beam. Then we see from this last expression that the beam is better 'collimated' (the angle with the axis is smaller) for the O observers to the extent that the beam is relativistic. This is true for all particles except possibly those very close to satisfying $1 + u/\mathbf{v}'_\parallel = 0$. These latter particles are directed against the beam in such a way as to effectively cancel their longitudinal motion since $\theta \approx \pi/2$ (see the inverse of the second of Equations (3.19)).

This phenomenon is commonly observed in the relativistic beams of particle physics and of astrophysics. It is a kind of 'velocity aberration' between relatively moving observers. Even in the limiting case of the backward moving particles, when $\theta = \pi/2$ one sees that $\sin\theta' = \mathbf{v}/\mathbf{v}'$. This is only possible for $\mathbf{v}' > \mathbf{v}$.

Such a phenomenon is contingent on the particular particle velocities. However, suppose that the particle is a photon moving in a vacuum at speed c. That implies that $\mathbf{v} = \mathbf{v}' = 1$ and thus Equations (3.19) become

$$\sin\theta' = \frac{\sin\theta}{\gamma(u)(1 - u\cos\theta)}$$
$$\cos\theta' = \frac{\cos\theta - u}{1 - u\cos\theta}. \tag{3.21}$$

Thus there is no longer any contingency beyond the usual relative motion of the inertial observers. Consequently these formulae give the change in direction of a light ray or photon between inertial observers. This is known as the 'aberration of light'. The two formulae are not of course independent, as they must satisfy $\sin^2\theta' + \cos^2\theta' = 1$. Their ratio suffices in many circumstances in the form (we use the inverse transformations now, since we normally identify with O observers experimentally)

$$\tan\theta = \frac{\sin\theta'}{\gamma(u)(\cos\theta' + u)}. \tag{3.22}$$

A succinct and conclusive expression of light aberration (e.g. [15,17]) is found easily by using the trigonometric identity $\tan\theta/2 \equiv (1-\cos\theta)/\sin\theta$. One substitutes for $\cos\theta$ and $\sin\theta$ from the inverse of Equations (3.21) to obtain

$$\tan\left(\frac{\theta}{2}\right) = \sqrt{\frac{1-u}{1+u}}\tan\left(\frac{\theta'}{2}\right). \tag{3.23}$$

The inverse may be found as usual, but even more directly by rearrangement.

This delightful expression will assist us greatly in the optics section below. However, it is amusing to note before leaving it, that starships as imagined in various TV dramas must obey this equation on approaching the speed of light.[5] Note that for O observers all electromagnetic radiation emitted by the starship is focused into $\theta \approx 0$ as $u \to 1$. The ship thus vanishes as seen from any angle except directly ahead of its motion, where it is essentially seen as a point source. Such observers are in trouble at impact for more than one reason, however, since Equation (3.7) for the Döppler factor K_+ shows that the radiation emitted by the ship (including any exhaust) is boosted in energy by an arbitrarily large factor! Fortunately this is also the factor by which the solid angle of the radiation is reduced, so that total energy may be conserved.

The case for O' (ship) observers is also of interest. Any emitting source in the O world will, by Equation (3.23), be seen by these observers as approaching with the angle π to the direction of motion. That is, the whole luminous Universe is focused into a small angle directly ahead of the starship. The rest of the sky is dark.

3.3 Kinematic Acceleration

Although we cannot yet consider dynamically accelerated observers relativistically, we can consider how two different inertial observers describe the same accelerating particle kinematically. This follows from a simple differentiation of the components of Equation (3.16) together with Equation (2.38) written for dt/dt'. One finds (see Problems) for the parallel acceleration ($\mathbf{a}_\parallel$ in either frame of reference) that

$$\mathbf{a}'_\parallel = \frac{\mathbf{a}_\parallel}{\gamma(u)^3(1-\mathbf{u}\cdot\mathbf{v})^3}. \tag{3.24}$$

The perpendicular acceleration is slightly more complicated, but may be written ultimately as

$$\mathbf{a}'_\perp = \frac{\mathbf{a}_\perp(1-\mathbf{u}\cdot\mathbf{v}) + (\mathbf{u}\cdot\mathbf{a})\mathbf{v}_\perp}{\gamma(u)^2(1-\mathbf{u}\cdot\mathbf{v})^3}. \tag{3.25}$$

Once again $\mathbf{a} = \mathbf{a}_\perp + \mathbf{a}_\parallel$, where the reference direction is $\mathbf{u}$. The inverse is found in the usual way.

Care must be taken in using these equations not to assume in general that $\mathbf{a} \parallel \mathbf{v}$ in both frames. It is in fact possible to show that assuming this is true in both frames is generally incompatible with the angle transformations (3.20) (see Problems). This

[5] The technology of the 'warp' drive eludes us at present, so we do not discuss the trans-light possibility.

is possible when **a** and **v** are both parallel to **u**, of course, but the magnitudes of the acceleration are not equal: see Equation (3.24).

Problems

3.10 Derive Equations (3.24) and (3.25) following the procedure outlined in the text.

3.11 Show that if one assumes the acceleration to be parallel to the velocity in both inertial frames, the resulting angle transformation is incompatible with Equation (3.20).

3.12 Using Equation (3.25), show that

$$\mathbf{a}_\perp = \frac{\mathbf{a}'_\perp}{\gamma(u)^2}, \tag{3.26}$$

if; (i) O' is co-moving with the particle; or if (ii) $\mathbf{a}'$ and $\mathbf{v}'$ are both perpendicular to $\mathbf{u}$.

We do not consider here the dynamic cause of such a particle acceleration. In the Newtonian view the particle must be subject to a force $\mathbf{F}$ according to $\mathbf{a} = \mathbf{F}/m$, but it is easy to see that this will not be a Lorentz invariant definition if it holds, say, for the O' observers. It is necessary to determine separately the transformation law for the inertial mass m and the force. This will be postponed until the dynamics discussion below.

However, there is a case that allows us to discuss the resulting accelerated motion rather precisely. We adopt the idea, used frequently above, that an accelerating observer may be followed through a sequence of instantaneously co-moving inertial frames. Moreover, we take the particle to be instantaneously at rest for an O' observer, who may very well be taken to coincide with the particle of interest. The acceleration of this particle-observer is assumed to be entirely parallel to its velocity at any stage, which is therefore the instantaneous velocity $\mathbf{u}$ relative to an O observer.

It follows by the inverse of Equation (3.24) with $\mathbf{v}' = 0$ that $a_\parallel \equiv du/dt = a'_\parallel/\gamma(u)^3$. Here we are explicitly applying the sequence of Lorentz transformations to the instantaneously co-moving observer. Subsequently we shall drop the parallel designation on $a'_\parallel$.

This quantity a' is the acceleration that the moving particle feels, and should be given according to Newton by a force divided by the inertial mass. It is the co-moving or 'proper' acceleration. In any case it can be prescribed to be constant. If it were to be produced by the thrusters of a starship it might be conveniently set at g, the mean gravitational acceleration at the Earth's surface.

By taking a' as constant we have a well-defined equation for the instantaneous velocity $u(t)$ as seen by an O observer. This equation takes the convenient form

$$\gamma(u)^3 \frac{du}{dt} = a'. \tag{3.27}$$

There is a useful algebraic identity that allows immediate integration of this equation, namely (see Problem 3.13)

$$\gamma(u)^3 \frac{du}{dt} \equiv \frac{d(\gamma u)}{dt}. \tag{3.28}$$

Consequently, assuming that the particle starts from rest in the O world, one finds

$$u(t) \equiv \frac{dz}{dt} = \frac{a't}{\sqrt{1+(a't/c)^2}}, \tag{3.29}$$

where we have retained c in conventional units. This last equation is readily integrated to give

$$z(t) = \frac{c^2}{a'}\left(\sqrt{1+(a't/c)^2} - 1\right) + z_o, \tag{3.30}$$

where z_o is the starting location in the O frame on the axis of the relative velocity. This is an arbitrary value because of the arbitrary location of the origin in the O world.

By taking $z_o = c^2/a'$, the solution can be written very simply as

$$z^2 - c^2t^2 = \frac{c^4}{a'^2}, \tag{3.31}$$

which shows why this motion is referred to as 'hyperbolic motion'. It is sketched in the O world in the space-time diagram in Figure 3.5.

Problem

3.13 Derive Equations (3.28), (3.29) and (3.30).

We see from Figure 3.5 that the motion is asymptotic to the line $z = ct$. We are mainly interested in the upper half of the figure which displays the constant proper acceleration from zero velocity. The lower half describes a particle decelerating with constant proper deceleration until it comes to rest at z_o. If we let $\sinh\tau = a't/c$ then the velocity is given by $\mathbf{u}/c = \tanh\tau$. Therefore from the form of γ and the time dilation formula $dt' = dt/\gamma$ we find $dt' \equiv dt/\cosh\tau$. But from the definition of the τ symbol, $d\tau/dt = a'/(c\cosh\tau)$, so that $dt'/d\tau = c/a'$ and hence $t' = (c/a')\tau$. Consequently, again from the τ definition, we have the coordinate O time expressed in terms of proper time as

$$t = \frac{c}{a'}\sinh(a't'/c). \tag{3.32}$$

Thus after $10c/a'$ for the traveller, some $11,000c/a'$ has passed for the O world, and the discrepancy continues to grow. Relative to the co-moving observer, an O observer appears to behave kinematically in the same fashion. We know from our discussion of the twin 'paradox', however, that the result of real acceleration will be discovered only

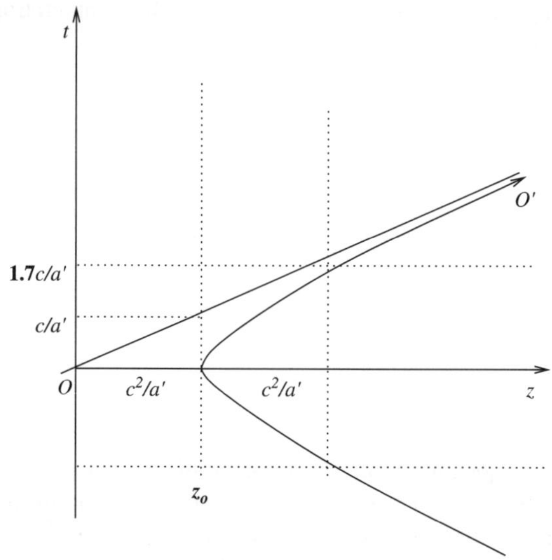

Figure 3.5 *This shows the hyperbolic world line of a particle* O′ *subject to constant proper acceleration in the world of* O *observers. The part below the* z *axis is shown merely to emphasize the hyperbolic character. The straight line is the asymptote* $z = ct$ *to the motion. The choice* $z_o = c^2/a'$ *is indicated as the vertex. Other points serve mainly to approximate the hyperbola, except for the time* c/a' *which is discussed in the text*

when the two observers coincide once again. In the meantime the traveller has time enough to explore vast tracts of our Universe, even if 'you can't go home again'.

Optically there is a certain symmetry, however. The last signal that the co-moving observer O' can receive from the origin O observer is the asymptote shown. No light signal launched subsequently can cross the asymptote. Hence no signal sent by O after the time c/a' can ever reach O'. Conversely, no signal sent from O' can reach O later than the time c/a'. This is the first time we meet the idea of a 'horizon', beyond which no contact is possible. It is an example of a 'Cauchy' horizon for the origin event and an 'event' horizon for the particle. These are more commonly associated with 'black holes' or gravitationally collapsed objects. These have the advantage of occurring 'naturally', but they are not necessary to horizons, as this example shows.

The other effect associated with a horizon is the Döppler shift. An object vanishes even as it approaches the horizon according to Equation (3.7), since every emitted frequency will be 'red-shifted' to zero. The same is true for any signal directed towards the accelerating particle.

3.3.1 Thomas Precession

The topic of this section is to compute the spatial rotation associated with the composition of two Lorentz boosts in non-parallel directions. We explore it in some detail, as this is seldom done in the literature (although see [18] for a similar discussion). Its successful application to hyperfine splitting of atomic spectra [5,19–21] represents, moreover, another test of our relativistic concepts.

Let us consider a boost from O to a frame O' in standard configuration with velocity $u\widehat{\mathbf{e}}_z$. Relative to the primed frame, a second boost is applied in an arbitrary direction relative to the common z axis of O and O'. We may always rotate the axes about the common z direction until the second boost velocity lies in the $x' - z'$ plane of O' (and the $x - z$ plane of O). It will thus have the form $\mathbf{u}' = (u'_x, u'_z)$. This second boost must therefore take the general form of Equation (2.40). The two boosts take us finally to an inertial observer O'', whom we may regard as instantaneously co-moving with an accelerating particle as we wish.

Symbolically the net result of these two boosts must be a coordinate transformation given by

$$\mathbf{x}'' = \underline{\underline{\mathcal{L}}}(\mathbf{u}')\underline{\underline{\mathbf{L}}}(u)\mathbf{x}. \tag{3.33}$$

A tedious but straightforward calculation yields the result for the matrix operator

$$\underline{\underline{\mathcal{L}}}\,\underline{\underline{\mathbf{L}}} = \begin{pmatrix} \gamma\gamma' + vv'_z, & -v'_x, & 0, & -(\gamma' v + \gamma v'_z) \\ -\gamma v'_x - \frac{vv'_x v'_z}{1+\gamma'}, & 1 + \frac{(v'_x)^2}{1+\gamma'}, & 0, & vv'_x + \frac{\gamma v'_x v'_z}{1+\gamma'} \\ 0, & 0, & 1, & 0 \\ -\gamma v'_z - v\left(1 + \frac{(v'_z)^2}{1+\gamma'}\right), & \frac{v'_z v'_x}{1+\gamma'}, & 0, & vv'_z + \gamma\left(1 + \frac{(v'_z)^2}{1+\gamma'}\right) \end{pmatrix} \tag{3.34}$$

The notation we use here for brevity is $v = \gamma(u)u$ and $\mathbf{v}' = \gamma'(u')\mathbf{u}'$.

One way to proceed is to compare this matrix with that of the direct Lorentz boost from O to O''. To achieve this we must know the velocity of O'' relative to O, say $u(nr)$ to indicate the non-rotating step from O to O''. The first row of the matrix operation (3.34) on x^a gives

$$t'' = (\gamma\gamma' + vv'_z)t - v'_x x - (\gamma' v + \gamma v'_z)_z.$$

But this must, by Equation (2.38), be equal to $\gamma(nr)(t - u(nr)_x x - u(nr)_z z)$. Hence

$$\begin{aligned} \gamma(nr) &= \gamma\gamma' + vv'_z, \\ \mathbf{u}(nr) &= (v'_x/\gamma(nr))\widehat{\mathbf{e}}_x + ((\gamma' v + \gamma v'_z)/\gamma(nr))\widehat{\mathbf{e}}_z, \end{aligned} \tag{3.35}$$

must give the desired direct boost. The two resulting sets of double primed coordinates may then be compared in order to detect the spatial rotation.

It is easier to follow a procedure similar to that used in [18]. We observe that our matrix operator (3.34) is symmetric neither in the purely spatial part nor between the row and column temporal vectors. This is not the case for a pure general homogeneous boost, which is always symmetric, as can be seen in the form (2.41). The temporal asymmetry does not concern us here, since after all we have mixed the times of two observers in the two-step process. However, the spatial asymmetry is a puzzle, since no explicit transformation has been applied to the spatial axes. Such an asymmetry can be expected if nevertheless there has been in some fashion a rotation of the axes (see e.g. the Euler angle rotation matrix or (1.7)). We must remember that spatial rotations and boosts together form the homogeneous Poincaré group. It appears that the Lorentz subgroup is not a closed subgroup with respect to the subgroup of rotations.

We therefore try to symmetrize the spatial part of the composite matrix by applying an inverse rotation that achieves symmetry among the spatial components (as would be the case for the direct boost $u(nr)$). The task is simplified by realizing that the required rotation must be about the y axis, since it has been left undisturbed in the two-step process. The necessary inverse rotation must therefore be of the form $\underline{\underline{\tilde{\mathbf{S}}}}_\theta$ (see Chapter 1; but the rotation is about the y axis in four-space), which becomes here

$$\underline{\underline{\tilde{\mathbf{S}}}}_\theta = \begin{pmatrix} 1, & 0, & 0, & 0 \\ 0, & \cos\theta, & 0, & \sin\theta \\ 0, & 0, & 1, & 0 \\ 0, & -\sin\theta, & 0, & \cos\theta \end{pmatrix}.$$

A direct calculation using this inverse rotation applied to $\underline{\underline{\mathcal{L}}}\,\underline{\underline{\mathbf{L}}}$ shows that the only off-diagonal spatial elements are the matrix elements (13) and (31). Setting these equal to restore the symmetry of the direct non-rotating boost $\mathbf{u}(nr)$ requires

$$\begin{aligned}\cos\theta\left(vv_x' + \frac{\gamma v_x' v_z'}{\gamma'+1}\right) &+ \sin\theta\left(vv_z' + \gamma\left(1 + \frac{(v_z')^2}{\gamma'+1}\right)\right)\\ &= -\sin\theta\left(1 + \frac{(v_x')^2}{\gamma'+1}\right) + \cos\theta\left(\frac{v_x' v_x'}{\gamma'+1}\right).\end{aligned} \tag{3.36}$$

After realizing that

$$1 + (v_x')^2/(\gamma'+1) \equiv \gamma' - (v_z')^2/(\gamma'+1), \tag{3.37}$$

there is no difficulty in showing that the required rotation angle is given by

$$\tan\theta = -\frac{vv_x' + \frac{\gamma-1}{\gamma'+1}v_x'v_z'}{\gamma + \gamma' + \frac{\gamma-1}{\gamma'+1}(v_z')^2 + vv_z'}. \tag{3.38}$$

This is the spatial rotation that is coupled in the Poincaré group to two successive non-collinear boosts. It is given here relative to the non-rotated O' axes since these are also parallel to the non-rotated direct boost axes.

This effect is most readily detected by following a single accelerating particle coincident with O''. We will seek the rate of change of this angle as the particle changes direction. To this end it suffices to treat the components of $\mathbf{v}'$ as infinitesimal $\delta\mathbf{u}'$ and so to retain them only in the first order. The angle θ then becomes the infinitesimal $\delta\theta$ and formula (3.38) gives (restoring γu and $\gamma' u'$ in place of v, v' and noting that $\gamma' = 1$)

$$\delta\theta = -\frac{\gamma u \delta u_x'}{\gamma + 1 + \gamma u \delta u_z'}. \tag{3.39}$$

We can express this result in terms of the velocity components for O, who observes the particle accelerate through the successive inertial frames O' and O'', by using the velocity transformation. From Equation (3.16) we find

$$\begin{aligned}\delta u_x' &= \frac{\delta u_x}{\gamma(1 - u\delta u_z)}\\ \delta u_z' &= \frac{\delta u_z - u}{1 - u\delta u_z},\end{aligned}$$

so that on substituting into Equation (3.39) we obtain

$$\delta\theta = -\frac{\gamma u \delta u_x}{\gamma(1 - u\delta u_z) + 1}. \tag{3.40}$$

Finally, on dividing by δt we obtain for the apparent angular velocity of the O'' axes relative to O axes (using vectors $\boldsymbol{\theta} \equiv \theta\widehat{\mathbf{e}}_y$, $\mathbf{a} \equiv (du_x/dt)\widehat{\mathbf{e}}_x$, $\mathbf{u} \equiv u\widehat{\mathbf{e}}_z$ to emphasize the generality, and restoring conventional units)

$$\boldsymbol{\omega} \equiv \frac{d\boldsymbol{\theta}}{dt} = \frac{\gamma \mathbf{a} \wedge \mathbf{u}}{c^2(\gamma(1 - \mathbf{u}\cdot\delta\mathbf{u}) + 1)}. \tag{3.41}$$

With our particular geometry this is explicitly

$$\omega = -\frac{\gamma a_x u}{c^2(\gamma(1 - u\delta u_z) + 1)}. \tag{3.42}$$

Here $a_x \equiv du_x/dt$, the transverse acceleration for O, and ω is in the negative y direction which is the direction of $\mathbf{a} \wedge \mathbf{u}$.

The same result can be obtained by expressing $\delta u_x' = a_x'\delta t'$ before the velocity transformation on $\delta u_x'$ and then using the acceleration transformation (3.25) together with the time transformation for two observers following a moving particle.

Equation (3.40) gives the general expression for the Thomas precession, which is an internal property of the Poincaré group. Formally, one says the Lorentz subgroup is not an invariant subgroup of the Poincaré group, unlike the group of spatial rotations and translations. This is a startling result, and is an early indication that space and time are bound together in more than just a diagrammatic sense. We shall see that in fact space-time can be regarded as a manifold, but of a rather different character relative to the curved manifolds that are familiar in Euclidian space.

The result is most frequently seen in the low velocity limit when u is also of first order. In general vector form this becomes

$$\boldsymbol{\omega} = \frac{\mathbf{a} \wedge \mathbf{u}}{2c^2}. \tag{3.43}$$

In this form it applies to the spin-orbit coupling or hyperfine structure, and the factor 2 was long sought.

An exact expression applies when the acceleration is always perpendicular to the velocity so that $\mathbf{u}\cdot\delta\mathbf{u} = 0$. Hence

$$\boldsymbol{\omega} = \frac{\gamma}{\gamma + 1}(\mathbf{a} \wedge \mathbf{u}), \tag{3.44}$$

and in the ultra-relativistic limit one loses the factor 2.

Consider now the International Space Station in orbit about the Earth. It is in an approximately circular orbit at an altitude of ≈ 347 km so that the acceleration is perpendicular to its orbital velocity. We assume for the moment that Newtonian gravity and dynamics apply, since the velocity is very non-relativistic. Then the Thomas angular change per orbit (see Problem) is

$$\Delta\boldsymbol{\theta} = \widehat{\mathbf{e}}_\phi \wedge \widehat{\mathbf{e}}_r \frac{2\pi GM}{c^2 R}, \tag{3.45}$$

where R is the radius of the orbit measured from the centre of the Earth and M is the Earth mass. The tangent vector $\widehat{\mathbf{e}}_\phi$ is in the direction of the orbital motion. Hence the precession angular velocity is about an axis perpendicular to the orbit, in the left hand sense (i.e. retrograde) for an O inertial observer. It is in the reverse sense relative to the space station since θ was defined as positive for O observers. This precession would be seen in a gyroscope whose spin lies in the plane of the orbit. Its direction would be different relative to the station after one complete orbit by the amount $\Delta\theta$, in the prograde sense of the orbit.

Numerically this becomes for the space station 8.5×10^{-4} arcsec per orbit or about 4.9 arcsec per year. This is about $2/3$ the value of the 'geodetic precession' due to Einsteinian gravity, which includes this effect plus the effect of 'curved space'. The geodetic quantity was the principal objective of the gravity probe B experiment, which is still in the analysis stage at the time of writing. Without a relativistic theory of gravity our calculation of this precession is heuristic. It holds only for the ensemble of Lorentz frames tangential to the orbit, since gravity does not act in the frame of the freely-falling orbit.

Problems

3.14 Show that the operation of $\underline{\underline{\widetilde{\mathbf{S}}}}_\theta$ on $\underline{\underline{\mathcal{L}}}\,\underline{\underline{\mathbf{L}}}$ yields the (13) and (31) components given in the text, as the only off-diagonal spatial components.

3.15 Starting with Equation (3.39) in the text, use the time transformation together with the transformation of perpendicular acceleration (3.25) to obtain Equation (3.41).

3.16 Derive Equation (3.45) assuming Newtonian gravity and dynamics. Verify the numerical estimates in the text.

3.4 Geometrical Optics

In special relativity, straight lines are defined by light rays, which have an absolute definition. Their directions, however, are not observer independent, as we have seen in Equations (3.21) and (3.23). This is the phenomenon of aberration that we wish to discuss in more detail in this section.

Our first topic is really about the appearance of a point source of 'light' as detected by a non-proper observer (i.e. an inertial observer for whom the source is moving). One way of summarizing the apparent 'focusing' of the light rays that we discussed previously is to consider the transformation of the solid angle between observers. This is defined as $d\Omega = \sin\theta d\theta d\phi$ for any inertial observer, and in standard configuration where the common z direction is the polar axis, we may treat $d\phi$ as invariant.

A direct calculation using the cosine version of Equation (3.21) yields

$$d\Omega' = \frac{d\Omega}{\gamma^2(1 - \widehat{\mathbf{k}} \cdot \mathbf{u})^2}, \tag{3.46}$$

where $\widehat{\mathbf{k}}$ is the unit vector in the direction of the light ray about which we compute the solid angle. Hence $\widehat{\mathbf{k}} \cdot \mathbf{u} = \cos(\theta)u$. The angle θ and the solid angle $d\Omega$ for the O observer correspond to the angle θ' and the solid angle $d\Omega'$ for the O' observer.

An important and useful application of this formula is to the radiation pattern of a moving point source, as 'seen' (the radiation would have to be captured by a sphere of O observers close to the source) in the world of an inertial observer O. In this case the O' observer is the co-moving inertial observer, which as usual is only instantaneous if the source is accelerated. Suppose that in the world of O' the point source emits an amount of energy per unit solid angle per unit time in the direction θ' according to

$$\frac{d^2E'}{d\Omega' dt'} = I'(\cos\theta', \phi'). \tag{3.47}$$

We may think of this ray as a stream of narrowly directed photons, each carrying an energy proportional to their frequency according to the Planck law $E = h\nu$. Consequently by the Döppler formula (3.4) we have $dE' = \gamma(1 - u\cos\theta)dE$. As always we have the time dilation formula $dt' = dt/\gamma$, and the transformation of solid angle is given above.

We may now calculate for the O world that ($c = 1$ until needed)

$$\frac{d^2E}{d\Omega dt} \equiv I(\cos\theta, \phi) = \frac{1}{\gamma^4(1 - u\cos\theta)^3} I'(\cos\theta', \phi), \tag{3.48}$$

where we have explicitly recognized that $\phi' = \phi$. This gives the 'appearance'[6] of the radiating source in the O world, if only we know what it is doing instantaneously in the co-moving frame.

In some cases we do know the co-moving radiation pattern. An isotropic emission from a black-body source would be focused dramatically into the forward direction for relativistic motion. In the limit it disappears for all O observers except those along the direction of motion.

Another example is that of a charge q undergoing acceleration parallel to its motion. This is dipole radiation, and we know from the Larmor formula that in the world of O' (we use Gaussian units here)

$$I' = \frac{q^2}{4\pi c^3}(a'_\parallel)^2 \sin^2\theta'. \tag{3.49}$$

Consequently by Equation (3.48) and the first of Equations (3.21)

$$I = \frac{q^2}{4\pi c^3} \frac{(a'_\parallel)^2 \sin^2\theta}{\gamma^6(1 - (u/c)\cos\theta)^5}. \tag{3.50}$$

[6] Strictly we should consider our quantity I per unit area normal to the direction and per unit time for a distant O observer, which would introduce additional transformation factors. Here we take the source point of view and calculate the energy loss rate.

However, just as in our discussion of hyperbolic motion, the parallel co-moving acceleration $\mathbf{a}'_{\parallel}$ is equal to $\gamma^3 \mathbf{a}_{\parallel}$ in terms of the acceleration in the O world. Therefore finally we have for any inertial O observer, that the emitted radiation pattern from a moving dipole is

$$I(\theta) = \frac{q^2}{4\pi c^3}(\mathbf{a}_{\parallel})^2 \frac{\sin^2\theta}{(1-(u/c)\cos\theta)^5}. \tag{3.51}$$

In the co-moving frame the dipole pattern is the familiar double sine lobe with peaks centred on $\theta = \pi/2$ and $\theta = -\pi/2$ respectively. However, in general the peak emission occurs at $\theta = \pm\theta_c$, where

$$\cos\theta_c = \frac{-1+\sqrt{1+15(u/c)^2}}{3(u/c)}. \tag{3.52}$$

At low velocities the cosine is zero as expected, but at relativistic speeds θ_c is small and becomes approximately equal to $1/(2\gamma)$. The peak angular factor in Equation (3.51) is then about $2.6\gamma^4$, but the beam is becoming narrower in proportion to θ_c, and the emission time is becoming longer.

One can apply this calculation to the case of hyperbolic motion calculated above. There is no difficulty if the constant acceleration continues for only a finite time [22], although there has been much confusion in the past based on the direct calculation of the electromagnetic fields [15].

An important historical discovery of aberration was by the English astronomer James Bradley in 1728 [23]. He observed that the apparent position of a star on the celestial sphere varied over a year. The path on the sky was elliptical in general, becoming circular for a star near the pole of the Earth's orbit.

This could be explained classically by assuming that the starlight travelled in the aether frame with the velocity $\mathbf{c}$ directed towards the Earth. An Earth observer moving with the relative velocity $\mathbf{u}$ transverse to the line of sight would, according to the aether theory, add $-\mathbf{u}$ to $\mathbf{c}$. From the resulting vector triangle, one infers that a telescope must be pointed at an angle $\tan\theta = u/c$ to the true line of sight in order to observe the star. As the Earth traverses its nearly circular orbit, this direction will trace out a circle of this angular radius for a pole star. A generally located star will trace out an ellipse of this angular semi-major axis. The ellipse collapses to a line of twice this angular length for a star in the orbital plane. Of course, a complete trajectory is only measurable for a given observer if the star in question remains visible over the year.

Historically this was taken to be a confirmation of Copernicus before stellar parallax was discovered, since the amplitude requires a relative speed of some 30 km/s for all stars. It was also an argument against an 'aether boundary layer' moving with the Earth that might have explained the lack of anisotropy in light propagation at the Earth [24]. For such a case there would have been no relative velocity between the Earth and the light ray.

According to special relativity the Equation (3.23) applies. Consider standard configuration between the distant star and the Earth observer with the z axis parallel to the orbital velocity of the Earth. Let the inertial frame of the star be the world of O and that of the Earth observer be that of O'. Then $\theta = \pi/2$ and Equation (3.23) gives $\tan(\theta'/2) = \sqrt{(1+u)/(1-u)}$ when we let u be the orbital speed of the Earth.

But trigonometrically $\tan\theta'/2 \equiv \sqrt{(1-\cos\theta')/(1+\cos\theta')}$ so that $\cos\theta' = -u$. Hence $\tan\theta' = -1/(\gamma u)$. But this angle is measured relative to the orbital direction of the Earth, and so it is equal to $\pi/2 - \theta_b$, where θ_b is measured from the line of sight. Consequently the Bradley result is now (restoring c)

$$\tan\theta_b = \frac{\gamma(u)u}{c}. \tag{3.53}$$

This is the same as the classical result to first-order. The first-order angle is about 20 arcsecs, and over six months would yield a position shift twice that value and nearly resolvable by the human eye. The next order is third order in u/c that requires a precision to be detectable of better than $\approx 2 \times 10^{-7}$ arcsec, which is not currently attainable.

3.4.1 Pictures of Moving Objects

The subject of this section is full of surprises, and an excellent although brief summary is provided in [17]. Moreover, one cannot rival in print the images found on web sites that offer animations of the visual appearance of the world in various states (e.g. [25]). However, we can hope to gain the insight required to believe these images.

In general a moving object is distorted and rotated (e.g. [26,27] shown explicitly in Figure 3.6) to the camera or the eye. This is because these optical devices operate on photons that arrive simultaneously and always at the invariant speed of light. Thus they cannot have left the object simultaneously if it has a significant extent in space. The set of earlier times at which photons left each point of the body so as to arrive together at the observer is the interval of 'retarded time' corresponding to the current observer instant. Hence a photograph or a 'sight' of an object is always viewed extended in retarded time. A moving object is seen stretched and distorted in space due to its motion during this difference in photon travel time. This effect is perfectly classical, given that the object is 'seen' through the simultaneous collection of photons. The non-classical effects are due to the Lorentz contraction and time dilation.

In addition to the distortion due to finite light travel-time, there is a Döppler shift that will ultimately determine whether the object is actually 'seen' or not (i.e. photons detected in the optical band). In general there will be a non-uniformity of a large object's perceived 'colour', due to the angular dependence in Equation (3.4).

We begin by considering limiting cases wherein nothing unusual is perceived. This is the case of the 'life-size' photograph. That is when the image is received using only parallel rays on a life-size detector. The key principle [26] is that parallel rays are parallel for all inertial observers. This is easy to prove by using our aberration formulae.

Thus consider two light rays 'moving' (it is a plane wave front that is moving, effectively a photon) along the z axis for O observers so that $\theta = 0$ for each ray. Then the aberration formula (3.23) shows that each ray will also move parallel to the z' axis for O' observers, since θ' will also be zero. The component of relative velocity parallel to the wave front can have no effect. This simple fact implies that life-size images made with parallel rays will appear the same in all inertial frames. The 'camera', however, must have an aperture as large as the actual size of the object. Hence this is not strictly a single observer point of view, except for objects smaller than the aperture of the observer.

Figure 3.6 A few cubes are set in a row (bottom). A second row of cubes on top moves along the first row from left to right at 90% of the speed of light. All cubes, whether moving or at rest, have the same orientation: the face with three dots is in front while that with four dots is on the trailing side. The fact that we can see the trailing sides of the moving cubes is a consequence of the finiteness of the speed of light. Source: *Reproduced with permission from www.spacetimetravel.org/galerie/galerie.html. Copyright 2002 Ute Kraus. Universität Hildesheim (See Plate 5.)*

The aberration formulae also show that paraxial rays (that is, those that differ only slightly from parallel rays) will deliver images which differ uniquely by a scale factor between two inertial observers (these rays can be received by one 'eye'). This will be the case for all rays coming from objects that subtend a small solid angle at the aperture of the optical detector, be it an eye or a camera. Sufficiently distant objects (depending on their size) will satisfy this criterion. Thus let the negative z axis along which an object is moving in the positive direction, make a small angle θ to a ray directed from the object to the observer. The received ray will make this same angle with the z direction. Assuming that the rays are also paraxial in the frame of the object (O'), the Equation (3.23) becomes

$$\theta = \sqrt{\frac{1-u}{1+u}}\theta', \tag{3.54}$$

so that an image formed from these rays for O will be a scaled version (smaller) of the image formed for O' from the same rays. This assumes the same image-forming device in each case. It also assumes that u is not very close to 1. In that case θ may be small for all θ' as discussed previously.

Should we regard the object from the side, however, the rays are all close to parallelism with the y axis as in Figure 3.7. Hence $\theta = \pi/2 - \alpha$, where α is the angle to the y axis.

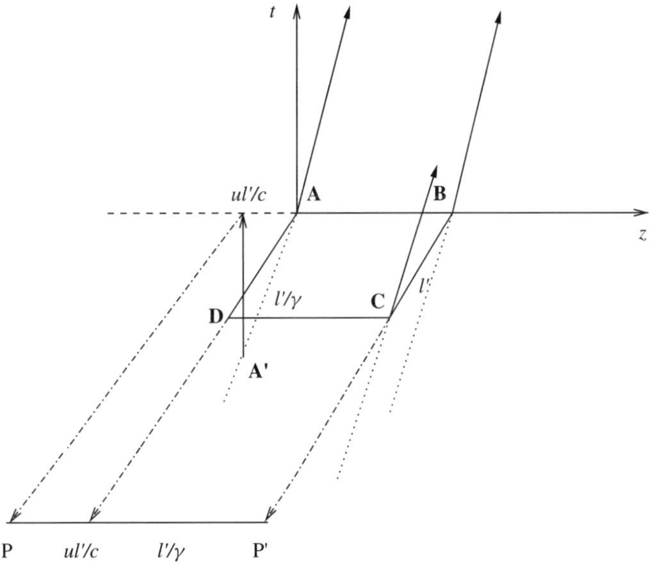

Figure 3.7 *The figure shows the four corners of a square section of a cube as the points A,B,C,D. The world lines for the corners A,B,C are shown for O observers, while that of D is omitted for clarity. A′ is the event where light is emitted from A that arrives at the aperture PP′ at the same time as does light from D and C. O observers perceive DC as the length ℓ'/γ and CB or DA as ℓ'. The length projected parallel to the t axis, $u\ell'/c$, is the extra transverse length of the side DC due to the time difference between A and A′. The image is perceived as life-size by an extended aperture or by a point observer using paraxial rays. The actual size of the image depends on the nature of the camera*

Now the formula (3.23) gives exactly

$$\sin\alpha = 1 - \frac{\cos\alpha}{\cos\alpha'}\sqrt{\frac{1-u}{1+u}}(1-\sin\alpha'). \tag{3.55}$$

Hence in this case when α and α' are small, we only have $\alpha \propto \alpha'$ when $u \ll 1$. In effect this is the condition for the object to remain 'small' under the distortion due to its motion in retarded time. We deal with such perspectives more explicitly below.

This discussion of nearly life-size images does not mean that every observer sees what is expected, merely that the image is similar under the right conditions. Consider for example Figure 3.7. This shows a section of a cube seen passing by a symmetrically located observer O at $t = 0$. The observed spatial location shown is not the current location of the section, since it will have moved on during the light travel-time to O. Nevertheless the image of interest is formed when the section is in the position shown.

Suppose as above that the image is constructed from purely parallel rays for O so that it is life-size and the same in all inertial frames. The rays from corners C and D will have left the object at the same instant to arrive at the aperture PP′ of O simultaneously, but the ray from corner A must have left earlier by the time $\ell/c = \ell'/c$ in order to arrive at O simultaneously with rays C and D. This is because the transverse length of the square is $\ell' = \ell$ for either observer, and the A ray has this extra distance to travel.

Each point on the side DA will be delayed proportionately to its distance from D. The side CB is projected to the corner C, and the side DC is Lorentz contracted to ℓ'/γ since it is seen simultaneously by the O observer. Consequently the plane wave front that arrives at the life-size aperture shows a total transverse extent of the square equal to $\ell'/\gamma + u\ell'/c$. That is, the side of the square is stretched by the first-order amount $u\ell'/c$ from the Lorentz-Fitzgerald contracted value. This first-order effect is due to the classical differential time travel and is *much more important than the second-order contraction*.

However, our brains will not construct the image this way, even if a camera incapable of constructing three-dimensional images would. In the first instance we notice that the parallel rays from DC for O were emitted for O' at an angle $\sin\theta' = 1/\gamma$ to the z axis according to the first of Equation (3.21). This is because the rays were emitted at the angle $\theta = \pi/2$ for O, at the instant shown in the figure. The angle θ' implies the angle $\alpha' \equiv \pi/2 - \theta'$ to the line of sight (y axis) according to O'. Hence in this frame $\cos\alpha' = 1/\gamma$. But these rays have left the edge DC parallel to all the other rays in the O world. Hence observer O, *able to see along the edge DC but not the edge AB*, must interpret the square section as turned away from the direction of motion (see Figure 3.8). It will be turned through the angle α' for O, since such an angle aberrates to $\pi/2$ in the world of O.

This construction of a rotated square fits all of the observed facts very nicely (again Figure 3.8 and sections of the cubes in Figure 3.6). The projection of the side DC (of length ℓ') on the z axis is then $\ell' \cos\alpha' = \ell'/\gamma$ as the Lorentz-Fitzgerald contraction requires. Moreover, the extra transverse length due to the projected side DA is $\ell' \cos(\pi/2 - \alpha') = \ell' \sin\alpha' \equiv \ell' u/c$, as was found above by the light travel-time argument.

This argument has used parallel rays in the O world to image the cube section from the side, but our argument regarding paraxial rays argues that they would form a similar (although scaled) image. This applies to a nearly head-on view of an object (also tail-on, but the image is bigger not smaller: see Problem) and to a general view of low-velocity objects. Objects obeying these criteria would also be seen as slightly rotated in the world

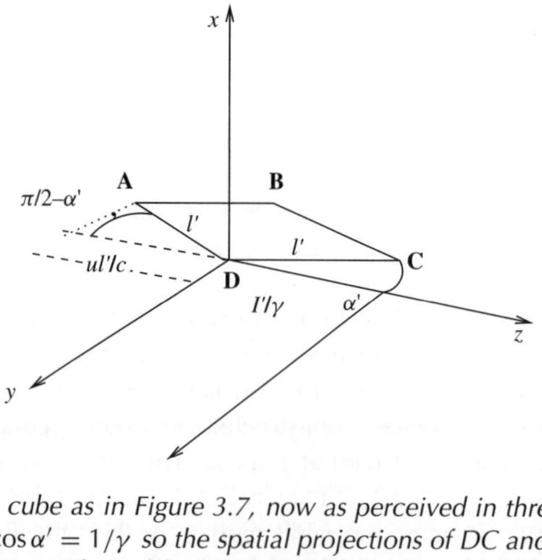

Figure 3.8 *The same cube as in Figure 3.7, now as perceived in three-dimensional space. The angle α' satisfies $\cos\alpha' = 1/\gamma$ so the spatial projections of DC and AD on the z axis are correctly indicated. This is either a life-size image or one taken by paraxial rays at a distance by a 'point' observer*

of O' (i.e. an O' observer instantaneously coincident with O), because of the similarity of paraxial images.

Problems

3.17 For a distant object receding along the positive z axis, show that the angle to the line of sight α of an emitted ray is $\alpha = \pi - \theta$. Using this relation in both the object world O' and the observer world O together with Equation (3.23), show that for paraxial rays $\alpha = \sqrt{\frac{1+u}{1-u}}\alpha'$.

3.18 From the diagram of Figure 3.9, show that ($c = 1$ and Euclidian geometry)

$$\left(\frac{\ell'}{2\gamma} + u\tau_A\right)^2 + d^2 = \tau_A^2$$
$$\left(\frac{\ell'}{2\gamma} - u\tau_B\right)^2 + d^2 = \tau_B^2,$$

and hence deduce the expressions for τ_A and τ_B given in the text. It is important to note that the lengths AD and DB are each $\ell'/(2\gamma)$ in the world of O by the usual length contraction.

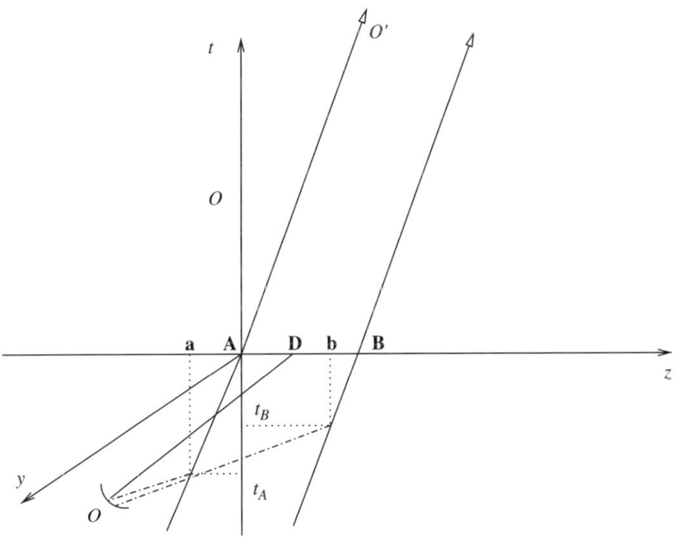

Figure 3.9 *The figure shows a rod AB currently crossing the surface t = 0, and the world lines of the ends of the rod indicate the motion along the z axis. The length AB is ℓ'/γ if ℓ' is the length for O′, and D marks the mid-point of the rod. The perpendicular distance OD is d. The retarded times t_A and t_B indicate the times of light emission from A and B such that the rays arrive at O as indicated (dash-dot) when the rod is currently in the position shown. The projected length aA is thus $u\tau_A$ while that of the length bB is $u\tau_B$. The intervals τ_A and τ_B are the magnitudes of the times t_A and t_B respectively*

So far we have implicitly assumed that the surface of our object emits in accordance with Lambert's law, that is, that no direction from the surface is favoured in emission. If the perpendicular direction were favoured in the co-moving frame, then this would assist us in identifying those perpendicular rays in the observer's frame. Hence the reconstruction of the rotated configuration from the optical data would be re-inforced. There should also be slight colour shifts across the image that would provide additional clues, assuming sufficient sensitivity.

We have in the above arguments considered only a plane section of a cube. Different sections would extend in the third spatial direction, and the light received from them in the configuration of the figure would start at different times. A cube would then appear differently rotated at different sections and thus distorted (Figure 3.6). Colour and brightness data would then be even more useful.

We have thus seen how an object can appear rotated. But what of Terrell 'rotation' for a specific observer? This requires us to follow the passing of the object across the line of sight, while allowing it to subtend large solid angles before and after the passage. In this case, each point on the surface of an object must be treated separately in order to establish its retarded time, so the complete image becomes a computational problem. We restrict ourselves to the simple case of a moving rod, as in Figure 3.9. By letting the observer be at an arbitrary perpendicular distance d from the rod along the y axis, we can think of the rod as an arbitrary line section of the cube section shown in Figure 3.7.

In order to study the evolution of the image as the rod passes the symmetrically placed observer, we construct the image *as seen* when the rod is currently in the configuration of Figure 3.9. In our previous discussion we constructed the image of the cube section using parallel rays *as it was formed during this passage*, rather than the retarded image. The retarded image of the rod is formed by the rays arriving at O from the ends A and B, which were emitted at the earlier (retarded) times t_A and t_B. If we use τ_A and τ_B as the magnitudes of the time intervals $[0, t_A]$ and $[0, t_B]$, then these must satisfy

$$\tau_A = \gamma\left(\frac{u\ell'}{2} + \sqrt{d^2 + \frac{\ell'^2}{4}}\right),$$
$$\tau_B = \gamma\left(-\frac{u\ell'}{2} + \sqrt{d^2 + \frac{\ell'^2}{4}}\right). \tag{3.56}$$

Consequently, the total length transverse to the line of sight is

$$u(\tau_A - \tau_B) + \frac{\ell'}{\gamma} = \gamma\ell'. \tag{3.57}$$

As in the discussion of the cube section, we will not interpret this as a linearly stretched rod. We observe from Equations (3.56) that a point at a fraction f of $\ell'/(2\gamma)$ (replace ℓ' by $f\ell'$ in the formulae) from D is not stretched linearly with f. This implies that any small section of the rod will deviate from a straight line between the end points, and together the sections will comprise an arc. Let us see how this works.

The rays from A and B make angles θ_A and θ_B with the z axis such that $\sin\theta_A = d/\tau_A$ and $\sin\theta_B = d/\tau_B$. According to the first of Equations (3.21), the angles that these rays make with the z axis in the frame of the rod are

$$\sin\theta'_{A,B} = \frac{1}{\gamma}\frac{d}{\tau_{A,B} - \sqrt{\tau_{A,B}^2 - d^2}}. \tag{3.58}$$

Once again we see that these angles vary non-linearly with f, just by replacing ℓ' by $f\ell'$ in the formulae (3.56) and letting A,B stand for L,R, the left and right ends of the small section. The angle in the frame of the rod is generally smaller than the angle in the O world according to Equation (3.58), so that each small section is turned away from z, but with its own particular angle. Hence the rod is really an arc of total length $\gamma\ell'$.

If we stand back and regard the rod from a distance large compared to its size, we can expect to see it as nearly straight. When d is much larger than ℓ' we are in the realm of paraxial rays. Moreover, $\sin\theta_{A,B} = 1/\gamma$, since $\tau_{A,B} \to \gamma d$ according to Equation (3.56). The rays are paraxial since we may not have γ too different from 1 (since otherwise all rays from the object are directed forward). This means by Equation (3.58) that in the world of O'

$$\sin\theta'_{A,B} = \frac{1}{\gamma^2(1-u)} = \bigcirc(\frac{1}{\gamma^2}). \tag{3.59}$$

These angles are smaller than the corresponding angles in the O world so that we regard the whole rod (nearly straight) as being turned away from the direction of the motion through the angle $\cos\alpha = 1/\gamma^2$. The projection of the 'arc' on the z axis is thus $\gamma\ell'/\gamma^2 = \ell'/\gamma$, consistent with the expected contraction.

To consider the change in the orientation of the rod in time, one must compare the retarded image of the configuration in Figure 3.9 to that of an earlier configuration. For simplicity we choose the earlier configuration of the rod when the right end B is at the origin, rather than the left end. In this earlier position the retarded times are

$$\begin{aligned}\tau_A &= \tfrac{3\ell'}{2}\gamma u + \gamma\sqrt{d^2 + \left(\tfrac{3\ell'}{2}\right)^2}\\ \tau_B &= -\tfrac{3\ell'}{2}\gamma u + \gamma\sqrt{d^2 + \left(\tfrac{3\ell'}{2}\right)^2}.\end{aligned} \tag{3.60}$$

The rod again appears curved by the arguments above. Moreover, taking the large d limit the object would appear the same as in the subsequent position shown in the figure. This is as expected in the paraxial limit. However, if we remain in the large solid angle domain, we find that the arc length (see Problem) is now $\gamma\ell'(1+2u^2)$, with no restriction on γ or u. Since this is larger than the arc $\gamma\ell'$ that appears in the subsequent position, we would interpret the arc as being at a larger average angle to the z axis. The average angle to the direction of motion therefore changes with the position of the object. It increases from a small value when the rod is seen at a distance to a maximum, and then returns to the same small value as it again passes into the distance.

Problems

3.19 Sketch a space-time diagram that generalizes Figure 3.9 to show three contiguous positions of the rod on the z axis at different times. The first has the end B at the origin and the end A ℓ'/γ to the left. The second position is that shown in the figure and may be taken to occur at $t = 0$. The third position has the end A concident with end B in the second position, with the end B ℓ'/γ to the right. Show the space-time position of the rod that corresponds to these three z axis positions. Show also the y axis and the world line of the observer D symmetric to position two at a distance d along the y axis.

3.20 Derive the retarded times for signals arriving at D at $t = 0$ from A and B when they are in the first position of the previous Problem. These are given in Equation (3.60) of the text.

In this section we have confined ourselves to discussing the effects on the image of a moving body in simple cases. However, most of the important issues have been discussed. One can combine the principles computationally to yield the image of a body in (moderately) relativistic motion. Ultrarelativistic motion leads to images confined to the forward and backward directions, as we have discussed in the section on aberration.

3.4.2 Light Echoes

In the previous section we learned that the appearance of a moving object has more to do with the finite and invariant speed of light than with second-order effects such as the Lorentz-Fitzgerald contraction. The consequent distortion is an important effect only for relativistically fast objects, and does not find application in the solar system. Even distant astronomical objects such as the discs of galaxies are not significantly distorted by this effect. Relativistically fast substructures in galactic jets, or fast substructures in gaseous discs orbiting black holes, may be our only practical examples.

In this section we prefer to consider the first-order light travel-time effect in a rather important astronomical context. This is the phenomenon of light echos from material ‘neighbouring’ a luminous outburst from a point source. These echoing rays have taken an indirect path to a terrestrial observer, and so will appear at a time delayed relative to the direct ‘discovery’ rays. Let us consider Figure 3.10.

For either indirect path shown in the figure, a time delay $t > 0$ relative to the direct path is given by ($c = 1$) $t = z + \sqrt{z^2 + r^2}$, where z is positive behind the source and negative in front. This yields by re-arrangement the equation of the curve C as $z = (t^2 - r^2)/(2t)$. Thus for a path leading directly away from O and reflected or re-radiated directly toward S and O, we find with $r = 0$ that $Z = t/2$. Along $z = 0$, we see that $r = t$.

The curve C is better described by shifting the origin from S to the apex at Z. Thus we set $\tilde{z} = z - Z = z - t/2 = -r^2/2t$. All points on C have negative $\tilde{z}$ from this origin, and we see in fact that the curve is a section of a parabolic surface with the source at the focus and the latus rectum equal to t. The parabola thus broadens with the delay time. Scattering material anywhere on this surface thus acts as part of a parabolic reflector at

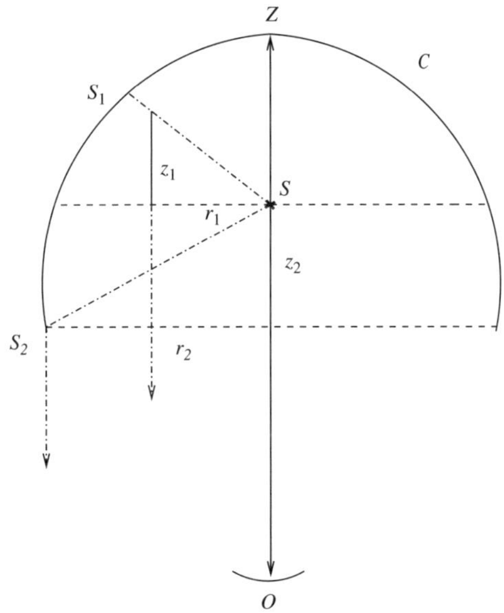

Figure 3.10 *A source S suddenly emits rays in all directions. The direct path is from S to the very distant observer O, while two possible indirect paths are labelled by their cylindrical coordinates r_1, z_1 and r_2, z_2 at the scattering points. All parallel rays are received by the distant observer. The parabolic curve C is the locus of all reflection (echo) points that are delayed by the time t for O, relative to the arrival of the direct ray. There may or may not be material present on this locus to reflect or re-radiate the incident rays. The Z axis is positively directed away from the observer*

time t. We may think of this as a parabolic 'dish', however incompletely 'covered' it may be.

A classic application of this idea is to supernova 1987A. We use this first as an illustration of 'reverberation mapping' by which at least relative distances internal to the neighbourhood of the source may be found. Two years after the initial discovery by a Canadian (Ian Shelton) at the Las Campanas observatory [28] on 23 February 1987, multiple scattered light echoes appeared (e.g. [29]).

The echoes were roughly circular about the line of sight. This might suggest the intersection of the 'two-year' parabola with a sphere of gas ejected from the star. However, in order to form multiple echoes the sphere would have to be of radius smaller that the distance to the vertex Z of the two-year parabola, which is only one light-year. Moreover, such a sphere centred on the star should be stellar gas ejected before the explosive event. Thus rather than simple scattering as observed, one would expect fluorescence in various emission lines.

A successful model [29] assumes clouds of dust distributed across the line of sight to the star at various distances. These must be broad enough to contain the relevant parabola at their distance from the explosion, in order to form a complete ring. The observations also show evidence for isolated clouds on the echo parabola, but we shall only consider the two-year ring.

If the distance to an echoing dust cloud is d_c from the observer, then $r = d_c\theta$, where θ is the angular radius of the ring echo. With d_c in light years, we have therefore from the equation of the echo parabola $\tilde{z} = -d_c^2\theta^2/(2t)$. Moreover, we may assume that $d_c \approx d$ for each cloud, where d is the distance from the observer to the star. Otherwise if $\tilde{z}$ were comparable to d, then by the preceding $d_c = \sqrt{4d}/\theta$ for the two-year echos. But the explosion was in the Large Magellanic cloud, which is a satellite galaxy of our own at a distance of about 166000 light-years. For the observed angles quoted below, this would put the clouds at the quite unreasonable distance of nearly one million light years (they should rather be between the observer and the source!).

On the two-year delay parabola there were two ring echos. The outer ring had a radius of about 4.4×10^{-4} radian (about 1.5 arcminute) and the inner ring had a radius about 2.3×10^{-4} radian (about 0.8 arcminute). With $d = d_c$ in each case we see that the ratio of the distances of the dust clouds from the vertex (a negligible one light-year behind the star) is the ratio of the squared angular diameters, namely 3.6. Thus without knowing the actual distance to the star-dust cloud system, we may find relative distances. We only have to find the absolute distances of one echo to find them all. These are 'reverberation distances' and are very important for the relative mapping of the surroundings of galactic nuclei, if there is a variable source of emission in the nucleus.

More recently, the colossal explosions that produce gamma ray bursts which can be seen across the Universe also produce echoes that give relative distance measures [30]. Their 'neighbourhood', however, can be far away from both the source and the observer. When their 'red-shifts' are measured, their cosmological distance is known. Measured delay times and angular sizes then provide both relative and absolute distances.

Another constraint that might be applied is the apparent rate of expansion of the light echo in delay time t. If we assume that the dust cloud is at a fixed $\tilde{z}$ then it is readily shown (see Problem) that

$$\begin{aligned}\left(\frac{dr}{dt}\right)_{\tilde{z}} &= \sqrt{\frac{-\tilde{z}}{2t}}\\ &= \frac{-\tilde{z}}{r}.\end{aligned}$$

Consequently a measurement of the expansion velocity and either t or r yields $\tilde{z}$. But the measurements of r and the expansion velocity require the distance d to the system to be known. Using the approximation $r = \theta d$, the two relations implied above become (see Problem)

$$\begin{aligned}\frac{d\theta}{dt} &= \frac{\theta}{2t}\\ \frac{-\tilde{z}}{d^2} &= \frac{\theta^2}{2t}.\end{aligned} \tag{3.61}$$

The rate of expansion is thus not independent in this approximation, and we obtain only a constraint on the combination $\tilde{z}/d^2$.

In the case of supernova 1987A, we know from other methods (including the progressive shocking of material ejected from the star by the explosion seen in fluorescence [31], that $d \approx 165{,}750$ light-years. This allows us to calculate $\tilde{z} = 1330$ light-years for the outer echo cloud and 370 light-years for the smaller echo cloud. These distances are still internal to the Large Magellanic cloud, which is consistent with our approximation that $d_c = d$. The constraint (3.61) is satisfied only to within a factor of 2 with these trial numbers.

Our calculations are only for illustration, and a more careful treatment of the data is required and has of course been carried out. Light echos are quite common and very relevant to modern astronomy. One remarkable consequence is that echos of supernovae in our galaxy that happened many centuries ago have been recently detected [32]. With such delay times we are literally 'present' at the explosion long ago and we may record and study its spectral signature.

The scattering of light by clouds occurs in our own atmosphere, especially at sunrise and sunset. There is usually no convenient symmetry, but occasionally rings are seen around the Moon and the Sun. This is the other extreme where $\tilde{z} \approx d$, and there is no perceptible delay time since it is due only to light travel-times on terrestrial scales of several hundred kilometres.

Problems

3.21 Show that the apparent expansion velocity in delay time is given by either of the two expressions given in the text.

3.22 Show that the expressions in the previous Problem reduce to the expressions (3.61) under the approximation $r = \theta d$, and state the resulting constraint explicitly.

References

1. Alimi, J.M. (2009) Proceedings of the Invisible Universe International Conference, Palais de l'UNESCO, June 29–July 3, Paris.
2. Abramowicz, A.A. and Bajtlik, S. (2009) Adding to the paradox: the accelerated twin is older. arXiv:0905.2428v1.
3. Brown, H.N., *et al.* (2001) *Physical Review Letters*, **86**, 2227.
4. Farley, F.J.M. (2001) *Europhysics News*, **32**, 5.
5. Hafele, J.C. and Keating, R.E. (1972) *Science*, **177**, 166, 168.
6. Greiser, R., *et al.* (1994) *Applied Physics B (Lasers and Optics)*, **59**, 127.
7. Stanwix, P.L., *et al.* (2006) *Physical Review D*, **74**, 081101.
8. Rees, M.J. (1966) *Nature*, **211**, 468.
9. Mirabel, I.F. and Rodrigues, L.F. (1999) *Annual Reviews of Astronomy and Astrophysics*, **37**, 409.
10. Mirabel, I.F. and Rodrigues, L.F. (1994) *Nature*, **371**, 46.
11. Sagnac, G. (1915) *Comptes Rendus de l'Academie des Sciences*, **157**, 708, 1410.
12. Henriksen, R.N. and Nelson, L.A. (1985) *Canadian Journal of Physics*, **63**, 1393.
13. Stedman, G.E. (1997) *Reports of Progress in Physics*, **60**, 615.
14. Ohanian, H.C. (2008) *Einstein's Mistakes*. W.W. Norton & Co., New York.

15. Pauli, W. (1981) *Relativitätstheorie*, English Translation, Pergamon Press (1958), Dover Books on Relativity.
16. Einstein, A. (1912) *Annalen der Physik*, **38**, 355.
17. Rindler, W. (2006) *Relativity*, Oxford University Press, New York.
18. Ferraro, R. (2007) *Einstein's SpaceTime*, Springer, New York.
19. Thomas, L.H. (1926) *Nature*, **117**, 514.
20. Frenkel, J. (1926) *Zeitshrift für Physik*, **37**, 243.
21. Pauli, W. (1927) *Zeitshrift für Physik*, **43**, 601.
22. Röhrlich, F. (1965) *Classical Charged Particles*, AddisonWesley, Reading, MA.
23. Bradley, J. (1728) *Phil. Trans. Royal Society London*, **35**, 637.
24. Michelson, A.A. and Morley, E.W. (1887) *Philosophical Magazine*, **24**, 449.
25. Kraus, U. Universität Hildesheim, http://www.spacetimetravel.org/ (accessed 27 April 2010).
26. Terrell, J. (1959) *Physical Review*, **116**, 1041.
27. Taylor, E.F. and Wheeler, J.A. (1963) *SpaceTime Physics*, W.H. Freeman and Company, San Francisco.
28. IAU Circular # 4316, 1987A.
29. Malin, D. and Allen, D. (1990) *Sky and Telescope*, **January**, 22.
30. Irwin, J.A. (2007) *Astrophysics: Decoding the Cosmos*, John Wiley & Sons Ltd., Chichester, p. 156.
31. Panagia, N., Gilmozzi, R., Macchetto, F., Adorf, H.M. and Kirshner, R. (1991) *Astrophysical Journal*, **380**, L23.
32. Rest, A. *et al.* (2008) *Astrophysical Journal*, **681**, L81.

4

The Measure of Space-Time

Man is the Measure of all things: of things which are, that they are and of things which are not, that they are not.

Protagoras, 485–421 BCE

4.1 Prologue

In this chapter we follow Minkowski and find that Lorentz invariance imposes a 'measure' on space-time. That is, henceforth we shall see that position four vectors no longer merely summarize the location of an event in a space-time diagram. They may instead be given a modulus in such a diagram, an actual 'length'. This measure is not positive definite (i.e. always delivering a positive interval), as is the measure of Pythagoras in our familiar three dimensions of space. However, the signature adds information, as it allows an elegant classification of space-time regions.

Introducing such a measure converts space-time diagrams into a representation in Euclidian space (i.e. the sheet of paper) of a metric space of a non-Euclidian (actually hyperbolic) character. The representation should not be taken literally, however. From the diagram we can only be sure of intervals along the axes given the coordinates of events, and not of intervals in the 'volume' of space-time. These must be found by calculation using the prescribed metric.

One is reminded here of the painting by the Belgian french painter Magritte that shows a traditional tobacco pipe with the caption 'Ceci n'est pas une pipe' (Figure 4.1). Indeed it is not, it is a picture of a pipe that does not possess all of a pipe's properties that we may wish to know. For the same reason we must interpret the representation of metric space-time on a Euclidian space-time diagram with care.

We will pattern four-vectors in general after the position four-vector. Thus we will attribute to any four-vector the same transformation properties between inertial frames as

Practical Relativity: From First Principles to the Theory of Gravity Richard N. Henriksen

Figure 4.1 *Magritte shows us a picture of a pipe that is not the pipe itself. The pipe has three dimensions, weight, size, smell and texture, among other things. The two-dimensional drawing cannot represent all of these. The same is true for Minkowski space as represented on a space-time diagram. The 'sphere' in space-time is represented as a hyperbola on the diagram. Mixed space and time distances are not represented correctly on a space-time diagram.* Source: © *ADAGP, Paris and DACS, London 2010 (See Plate 6.)*

exist for the position vector, and we will calculate their moduli by the same rule. A slight generalization of the discussions in Chapter 1 will give us a description of vectors and tensors in terms of their contravariant and covariant components in space-time.

Our ultimate objective is to re-write Newtonian mechanics given the replacement of the Galilean transformations (1.78) by the Lorentz transformations (e.g Equation (2.27)). This can be done without four-vectors and metric space-time, although it is relatively tedious. Nevertheless we will devote some space to this approach particularly for constraint forces, inertial forces, and for the electromagnetic force. This demonstrates the continuity with classical mechanics, and allows a ready applicability to practical problems. It also serves to emphasize again that the Minkowski construction of inertial space-time is not inevitable, however inspirational and convenient we find it to be below.

4.2 Metric Space-Time

The column or contravariant form of the position vector in a space-time diagram is given in Equation (1.1). Initially we will use Cartesian coordinates x^i so that $\{q^i\} \equiv \{x^i\}$. We recall that $\{x^i\}_{i=1,...,3}$ refers to the three Cartesian spatial coordinates while the $\{q^i\}_{i=1,...,3}$ are the three generalized spatial coordinates. Previously this notation designated an event. How do we give this event a modulus or 'length' relative to the origin event O?

The modulus will necessarily be a scalar (number) if it is to be an objective property of the space, rather than to vary according to different inertial observers who 'regard' these two events. We accept, as we found in Chapter 2 following the positivist argument, that the Lorentz transformations relate events between any two inertial frames. Under these transformations, the scalar available that depends only on differences in position is that of Equation (2.23). This suggests that we take the squared modulus s^2 of the space-time position vector to be (we continue to use units whereby $c = 1$)

$$s^2 = k\left(t^2 - x^i x_i\right) \equiv k\left(t^2 - \left(x^2 + y^2 + z^2\right)\right) \equiv k\left(t^2 - \mathbf{r}^2\right). \tag{4.1}$$

Here $k = +1$ when $t^2 > x^i x_i$ and $k = -1$ when $x^i x_i > t^2$. With this factor k we may use a real position vector, despite the lack of positive definiteness of the modulus. This modulus is the equivalent in space-time of the theorem of Pythagoras in Euclidian space, except that it is not so much a theorem as a defining axiom of Minkowski space.

When $k = +1$ the modulus is dominated by the time coordinate, be it later or earlier than O. Such positions are taken up sequentially by all known particles whose motion takes them through O (even signals; with the exception of those having the speed of light *in vacuo*). The position vectors of such events relative to O are then said to be 'time-like' as they may be traversed by particles or signals propagating in time. This is the domain of causality, since events before O may influence O and those after may be affected by O.

In the second case ($k = -1$) the modulus is dominated by the squared spatial position $\mathbf{r}^2$. Such events cannot be causally connected to O as no signal or particle may join the two events. They would have to propagate superluminally to do so. These events may be causally connected to other members of the O world (friends of the origin O), but these friends are causally disconnected from the event O. The position vectors relative to O with this modulus are said to be 'space-like'. They are simply 'elsewhere' as far as causes and effects of the event O are concerned.

One should note some additional peculiarities of this definition of 'distance' in space-time. It creates a matter-independent[1] metric space-time that we refer to as 'Minkowski space' for brevity. However, it is grossly non-Euclidian. Along a light ray, for example, there is no distance ($s = 0$), no matter when or where the event. In this sense a pure electromagnetic wave or photon seems to pervade space-time. Of course a wave front traverses finite space in a finite time, but the non-positive-definite character of our measure or modulus allows the net space-time distance to be null. It is as though space-time were 'pinched' into the origin along the light ray directions in space-time.

Although we often speak of rays in one spatial dimension or light 'cones' in two spatial dimensions, we should remember that these are only sections of a three-dimensional light sphere emitted from the origin. Showing the four-dimensional world structure of this sphere is beyond our imagination (but not that of topologists!) at present, but space-time has zero distance on the boundary of this world structure according to Equation (4.1).

Figure 4.2 illustrates the previous discussion in one spatial dimension. The set of events 'after O' (according to our 'external foliation' defined below) that include the world line of O (the t axis) and that lie within the outgoing light rays are the time-like future of O. They may be influenced by the origin event since a physical signal may travel along the position vector from O to each of them. The set of events 'before' O that include the t axis and that lie within the incoming light-rays are the time-like past of the event O. The event O may be influenced by them because a

[1] Particles are 'in' space-time rather than defining it or disturbing it, despite Mach and Bishop Berkeley.

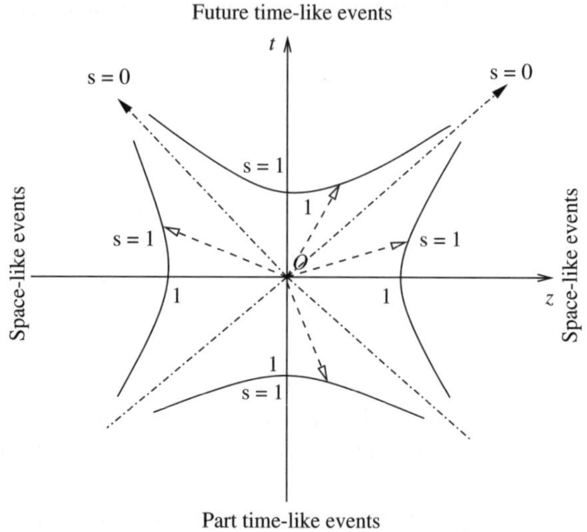

Figure 4.2 *The figure shows light rays as dash-dotted lines through the origin observer O. Another spatial dimension may be represented by allowing the figure to be axi-symmetric. The light rays are then the generators of light cones. According to the Minkowski metric of space-time (4.1) these cones separate the world of O into future time-like events, past time-like events and space-like events. The time-like events can be connected to O by signals so as to influence O when 'earlier' and to be affected by O when 'later'. These are all events within the light cones, including those on the t axis. The space-like events cannot be connected to O by any signal. These include all events outside the light cones including those on the z axis. The solid curves in the figure indicate the shape of a 'unit sphere' in space-time (actually a unit hyperboloid in two spatial dimensions), as distorted by a representation in Euclidian space. Dashed lines ending in open arrows are all of the same space-time 'length'. The light rays have zero 'length'*

signal may travel from each of these events backwards along the position vector to the origin.

The space-like events are those events that encompass the space defined by O observers with synchronized clocks (the positive and negative z axis) and that lie 'outside' one outgoing and one incoming light ray on each half of the space. They are causally disconnected from O. In Chapter 1 we discussed this isolation in terms of an event on Mars and an observer on the Earth. To a lesser extent, global telecommunication also makes such acausality evident.

We also show on the figure the locus of a space-time distance of unit 'length'. This locus would be a circle about the origin in Euclidian space, but we see that a 'sphere' in Minkowski space is represented on a Euclidian plane by two rectangular hyperbolae, each of two sheets. One hyperbola lies in the space-like regions and one lies in the time-like regions.

One of the most peculiar features of assigning a metric structure to space-time is that it becomes a fixed structure made from all historical and future events. The world of O and friends is this fixed web-like structure of all situated events. One may indeed ask if anything intrinsic to Minkowski space reflects our subjective sense of passing time? Strictly the answer appears to be no. Instead we must impose externally a foliation of Minkowski space that consists of parallel hyperplanes marching along the t axis. The separation of these planes is set by the resolution of our clocks. Clocks are regular periodic motions as we have seen previously, and they can be represented in Minkowski space. Each tick of a light clock, for example, would mark a 'new' hyperplane, or in fact an 'old' hyperplane, if we extrapolate the ticks to all relevant earlier times. In this sense subjective time is really 'that which is produced by clocks'. However, our consciousness does not use such precise clocks. There seems to be 'blurring' of the subjective instant due to biological processes. Moreover, a direction of time is sensed due to the local action of the laws of thermodynamics, notably the second law of increasing entropy. The extension of the instant is seen very clearly when listening to a musical composition. The persistence allows us to sense a small segment of the music and 'hear the tune'. A direction is not always present in the music if there are no words (whose intelligibility only works when pronounced in one direction), but if a long orchestral piece were not disciplined by a conductor then disorder would appear. This would be due to the slightly different sense of the music by different members of the orchestra. Time as we sense it goes in that direction of increasing dissonance or chaos. The reverse occurs only by the action of strong external agents.

The above is a useful way of inserting the passage of time into our considerations of Minkowski space. We may speak of earlier or later events according to our foliation. However, the world of Minkowski space is not re-created in every instant. It always is, as a complete set of events regarded as points in space-time. This is because the space-time exists independently of the matter in it, which is uniquely subject to the laws of life and thermodynamics.

We can recognize in this description of space-time yet another peculiarity. There are as many worlds or webs of events as there are inertial observers, namely an infinity. Moreover, the hyperplane foliations are not parallel in different worlds, and neither are the time axes. This is the content of the Lorentz transformations. Subjective times must therefore transform accordingly, as we have seen in our discussion of the twin 'paradox'.

Just as in ordinary space (Chapter 1), it is mathematically more convenient to use the differential form of the metric. This becomes

$$ds^2 = k(dt^2 - d\mathbf{r}^2) \equiv k(dt^2 - dx^i dx_i), \tag{4.2}$$

where the last form assumes Cartesian coordinates. Our classification of time-like and space-like position vectors, relative to a fixed observer O, generalizes to the relative position vectors between any two events in space-time. If $dt > |d\mathbf{r}|$ the relative vector is time-like since its modulus ds is positive ($k = +1$), and the two events may be in causal connection. Otherwise the relative vector is space-like ($k = -1$) and the events cannot be connected causally. The slopes (dt/dz) of time-like intervals in a space-time diagram are greater than 45°, while those of space-like intervals are less than 45°.

The modulus of the relative vector ds is the scalar displacement between any two events in space-time. In particular for a moving observer it is ($k = +1$), the infinitesimal displacement along the world line. Its most important property is that *it is invariant between inertial observers*. This follows directly by application of the Lorentz transformations as in Equation (2.23), and it is the reason that it defines a workable metric. For an instantaneous inertial observer co-moving with a particle, $d\mathbf{r}' = 0$. Hence we conclude that $ds = dt'$, the moving particle's 'proper' time.

This latter observation introduces a profound way of analyzing the twin 'paradox'. It will ultimately allow us to derive the mechanical 'action' of a free particle in Minkowski space. Thus, Equation (4.2) shows (recall that $k = +1$) that $\int_1^2 ds$ is always *greatest* between events 1 and 2, *if they can be connected by a path in space-time that sets* $d\mathbf{r} = 0$. This implies that the longest path between events 1 and 2 that lie on the world line of an inertial observer O, is the world line itself. That is, the longest path between causally connected events in a space-time diagram is the straight line joining them that is parallel to the t axis (we hold $c = 1$). In Euclidian space it would be the shortest path, so this is an amusing contrast. Space-like intervals are largest when $dt = 0$.

This longest distance is the interval of coordinate time in the O world (the proper time interval for O and friends) $\int_1^2 dt$. The elapsed proper time $\int_1^2 ds$ on any 'curved' path in space-time (hence accelerated) that connects events 1 and 2 is necessarily less than the coordinate time interval. This is the geometric statement of the twin 'paradox', which is not in fact a 'paradox'. It is rather another distortion of Minkowski space as represented on the Euclidian plane.

We return to our discussion of the measure of space-time and construct a compact notation for ds^2 in Cartesian spatial coordinates. To this end we define a generalization of the Dirac delta function to space-time that we label $\underline{\underline{\eta}}$. This quantity is defined as the diagonal 4×4 matrix

$$\underline{\underline{\eta}} = k \begin{pmatrix} 1, & 0, & 0, & 0 \\ 0, & -1, & 0, & \\ 0, & 0, & -1, & 0 \\ 0, & 0, & 0, & -1 \end{pmatrix}. \tag{4.3}$$

We refer to the matrix components as $\eta^a{}_b$ where each of 'a','b' may take on the values from 0 to 3. The contravariant index 'a' labels the rows and the covariant index 'b' labels the columns as usual. Using this definition of a metric matrix we may write Equation (4.2) as

$$ds^2 = dx_a \eta^a{}_b dx^b, \tag{4.4}$$

where the implied sums indicate matrix multiplication. We set $x^0 \equiv t$ (it will be ct dimensionally), so that dx^a is the ath component of our familiar column position vector. The corresponding row vector or covariant position vector is dx_a.

Although we readily visualize vectors as either column or row matrices and the metric as a 4×4 matrix (3×3 in Chapter 1), the matrix approach loses its facility with quantities that possess more than two indices. For this reason we introduce the notion of tensors.

Tensors are defined essentially by their transformation properties under a change of coordinates. They may have any number of indices that label the ensemble of their

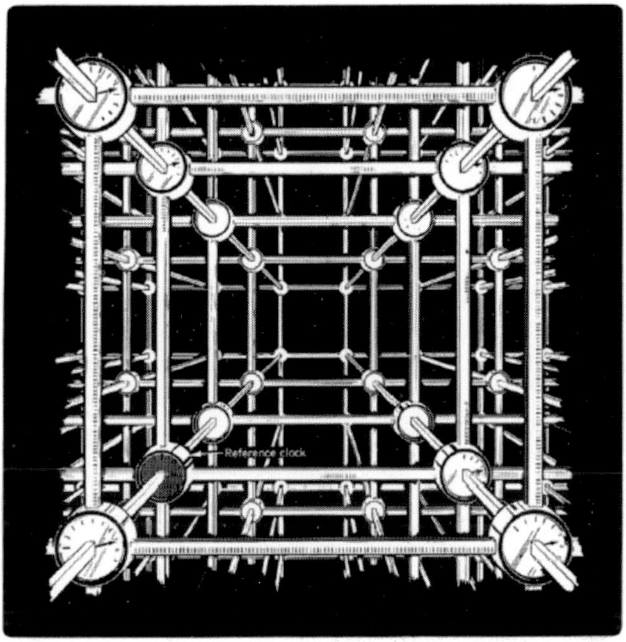

Plate 1 *After a rigid spatial frame of reference is established locally by measurement and synchronization, it might appear as shown in this cartoon. Each ruler indicates a unit of distance and any point on the grid is located with three numbers giving the three independent spatial steps relative to the reference point. The fourth number is the coordinate time, which is the same over the grid. The reference point is shown as having the reference clock with which all of the other clocks are synchronized. Extended to infinity, the grid is the instantaneous world of the reference observer O and friends. It is their inertial frame of reference. Source: Reproduced with permission from Taylor & Wheeler, Spacetime Physics (1966) W.H. Freeman & Company.*

Plate 2 *This beautiful picture of the disc of our galaxy shows it to be an arc in the sky. In ancient times it might have been regarded as the projection of a straight line on the stellar sphere of the sky. In fact it is real. We live on a rotating disc of stars and dust and gas. The interesting point is that the terrestrial coordinate system is obviously rotated with respect to the natural system of the disc. Globally our terrestrial coordinates are spherical polar with the axis defined by the rotation axis. Galaxy coordinates would be oriented relative to the rotation axis of the galaxy. Astronomers must convert frequently between the two systems, using methods outlined in the text. Source: Reproduced with permission from cielosdelteide.com. Copyright 2010 Daniel López.*

Plate 3 *The Earth rotates! Well at least these star trails circling the pole star 'Polaris', indicate that either the distant stars rotate around us or that we rotate. However measurements of 'inertial forces' on Earth show that they are present, according to the assumption of terrestrial rotation. The stars indicate an inertial frame on average that astronomers refer to as a 'local standard of rest'. On the scale of the visible stars it becomes apparent that this inertial reference is not exact and one must regress to larger scales. The regression continues until the mean Universe itself is reached. Source: Reproduced with permission from Dr F.-J. (Josch) Hambsch, http://www.astronomie.be/hambsch/namibia06/startrails1.htm.*

Plate 4 *This is a dramatic image of a railway station in Milan. The parallel tracks appear to converge as the distance between them subtends a smaller and smaller angle, but in fact in Euclidian space they never cross. This is not so on a curved surface. Much can be gained by loitering in railway stations. A train moving steadily and slowly along the track will give the illusion of a bystander moving in the opposite sense. The location of the spatial origins on the train and in the station are arbitrary. Source: Reproduced with permission from www.flickr.com/photos/paolomargari/2550814754. Copyright 2008 Paolo Margari.*

Plate 5 *A few cubes are set in a row (bottom). A second row of cubes on top moves along the first row from left to right at 90% of the speed of light. All cubes, whether moving or at rest, have the same orientation: the face with three dots is in front while that with four dots is on the trailing side. The fact that we can see the trailing sides of the moving cubes is a consequence of the finiteness of the speed of light. Source: Reproduced with permission from www.spacetimetravel.org/galerie/galerie.html. Copyright 2002 Ute Kraus. Universität Hildesheim.*

Plate 6 *Magritte shows us a picture of a pipe that is not the pipe itself. The pipe has three dimensions, weight, size, smell and texture, among other things. The two-dimensional drawing cannot represent all of these. The same is true for Minkowski space as represented on a space-time diagram. The 'sphere' in space-time is represented as a hyperbola on the diagram. Mixed space and time distances are not represented correctly on a space-time diagram. Source: © ADAGP, Paris and DACS, London 2010.*

S130E012141

Plate 7 *Here is a much better inertial frame than the Earth itself. The freely falling space station removes the gravitational field of the Earth (close to the centre of mass). It is not inertial with respect to the distant Universe, but these are small effects. Source: Reproduced by permission of NASA.*

Plate 8 *This picture of the cluster of galaxies (CL0024+1654: courtesy of Hubble Space Telescope) demonstrates the phenomenon of gravitational lensing. The mass of the foreground cluster of galaxies is bending and magnifying the light from background galaxies. These appear as lenticular, distorted objects of colours that differ from those of the cluster galaxies. They are intrinsically blue galaxies. Individual galaxies may also act as gravitational lenses. In this way nearby massive objects act as gravitational telescopes for the more distant Universe. Source: Reproduced by permission of NASA, ESA, and H. Ford (Johns Hopkins University).*

components. Each index runs from 0 to 3 and labels a vector 'dimension' of the tensor. Dimensions that transform against the base vectors are contravariant and are written up, while those transforming with the base vectors are covariant and are written down. In general the components change their values from event to event in space-time to form 'tensor fields'.

These definitions are much the same as in Chapter 1 (e.g. Equations (1.29) and (1.30)). In that chapter we became familiar with the idea of a metric matrix $\underline{\underline{\mathbf{g}}}$ in generalized coordinates, which we wrote in either covariant form g_{ij} or in the inverse contravariant form g^{ij}. This is a compact notation that implies consideration of all of the components as the indices take on their possible values. It is an example of a 'rank two' covariant tensor, but we usually refer to $\underline{\underline{\mathbf{g}}}$ simply as the metric 'tensor'. Although general tensors may have non-trivial mixed covariant and contravariant components, for the metric tensor these are trivially δ^i_j since by definition $g^{ik}g_{kj} = \delta^i_j$.

In space-time described by Cartesian components, the metric tensor may be written η_{ab} for which the components are given in Equation (4.3). Then by adopting the tensor notation Equation (4.4) becomes

$$ds^2 = \eta_{ab}dx^a dx^b, \tag{4.5}$$

where one simply computes the double sum. The invariance of ds^2 is assured so long as η_{ab} transforms with the (coordinated, unnormalized) base vectors $e^a_{(b)} \equiv \partial x^a/\partial q^b$. For then

$$ds^2 = \eta_{ab}dx^a dx^b = \eta_{ab}\frac{\partial x^a}{\partial q^c}\frac{\partial x^b}{\partial q^d}dq^c\,dq^d \equiv g_{cd}dq^c dq^d. \tag{4.6}$$

This defines the Minkowski space metric tensor in generalized (or curvilinear) coordinates by the rules of covariant index transformation.

If the transformed coordinates q^a are due to a pure Lorentz boost in standard configuration, the spatial coordinates will be a non-rotated Cartesian set. Hence the new metric tensor $g_{cd} \equiv \eta'_{cd}$ (see Problem). In that case $\partial q^a/\partial x^c \equiv L^a{}_c$ (see Equation (2.27)). However, the inverse boost, where one reverses the sign of u, occurs in Equation (4.6). It may also be written in this mixed notation as (first index names the row as usual) $L_a{}^c = \partial x^c/\partial q^a$, and thus

$$\eta'_{cd} = L_c{}^a L_d{}^b \eta_{ab} \equiv \eta_{cd}. \tag{4.7}$$

The transformation must be worked out as a double summation rather than as a matrix operation, but one obtains that $\eta'_{ab} = \eta_{ab}$.

Such coordinates we will call 'Galilean' coordinates. They are the simplest parameterization of an inertial space. If in addition to the boost to primed coordinates there is a spatial rotation to double primed coordinates, then we will have (again from the double summation)

$$g''_{cd} = \mathcal{L}_c{}^a \mathcal{L}_d{}^b \eta_{ab}. \tag{4.8}$$

Here $\mathcal{L}_c^a$ is the inverse (reverse the sign of u) of $\mathcal{L}^a{}_c$ (see Equation (2.40)). Despite the tedious summation (see Problem) one finds again that $g''_{cd} = \eta''_{cd}$. We call such coordinates 'rotated Galilean' coordinates.

Problems

4.1 Extract the covariant metric transformation rule implied in Equation (4.6). Use this together with the Lorentz transformations between Galilean coordinates in standard configuration (that is $q^a = x'^a$), to show that $g_{cd} = \eta_{cd}$.

4.2 A slightly more tedious but ultimately satisfying calculation is to show that Equation (4.8) yields η''_{cd}. The $\mathcal{L}_c{}^a = \mathcal{L}^a{}_c(-u)$, and the latter quantities are given in Equation (2.40). The algebra is the greatest difficulty.

4.3 Use Equation (4.11) to show that Equation (4.13) is the correct form of the Minkowski metric in generalized coordinates. Find in particular the g_{ij} in spherical polar coordinates $\{r, \theta, \phi\}$.

In 'special' relativity, that is, the relativity of inertial frames, only boosts require transforming the time. The most convenient transformation procedure is normally to effect the boost in Galilean coordinates in standard configuration. Subsequently one can transform to the desired, rotated Galilean coordinates using the methods of Chapter 1 that led to Equation (2.40). Since both the time part of the Minkowski metric and the spatial part are invariant under spatial rotations of Galilean axes, it is evident that the metric will have the same form in rotated Galilean coordinates (but see Problem 4.3).

Thus under transformations between Galilean coordinates, we shall always have $g_{ab} = \eta_{ab}$ as summarized in Equation (4.7). The inverse of η_{ab} is written η^{ab}. This is identical to η_{ab} since it is easily verified that

$$\eta^{ac}\eta_{cb} = \delta^a_b. \tag{4.9}$$

Hence the identity tensor δ^a_b is the 'mixed' (one covariant and one contravariant index) form of the Galilean metric.

We have seen in Chapter 1, however, that general spatial coordinates include curvilinear orthogonal coordinates, and even non-orthogonal coordinates, in addition to rotated Cartesian coordinates. Consequently in general we must write

$$ds^2 = g_{ab}dx^a dx^b, \tag{4.10}$$

where

$$g_{ab} = \frac{\partial x^c}{\partial q^a}\frac{\partial x^d}{\partial^b}\eta_{cd}. \tag{4.11}$$

This is very similar to the expression for the spatial metric (1.30) in Chapter 1, except that in Minkowski space η_{cd} replaces δ_{cd} and we are working with four vectors. Between two sets of generalized coordinates $\{q'^a\}$ and $\{q^a\}$, the above expression written for g'_{ab} would have q replacing x, q' replacing q and g_{cd} replacing η_{cd}. This yields

$$g'_{ab} = \frac{\partial q^c}{\partial q'^a}\frac{\partial q^d}{\partial'^b}g_{cd}. \tag{4.12}$$

In a given inertial frame the coordinate time is independent of the spatial coordinate and so Equation (4.11) becomes in matrix form

$$g_{ab} = \begin{pmatrix} 1, & 0, & 0, & 0 \\ 0 & & & \\ 0 & & g_{ij} & \\ 0 & & & \end{pmatrix}, \tag{4.13}$$

where $g_{ij} = \eta_{k\ell}(\partial x^k/\partial q^i)(\partial x^\ell/\partial q^j)$ is the spatial metric in generalized coordinates. This is exactly the same as the expression in terms of unnormalized base vectors that we used in Chapter 1, namely $g_{ij} = \mathbf{e}_i \cdot \mathbf{e}_j$.

We emphasize that Equation (4.13) is not valid under a transformation that involves the time in the transformation of spatial coordinates, unless that transformation is a boost to another inertial frame. By definition such time-dependent coordinates are non-inertial. In general they are introduced by observers who are accelerated with respect to an inertial observer.

Example 4.1

Suppose that we want to write the space-time metric in a given inertial world using spherical polar coordinates $\{r,\theta,\phi\}$. A direct calculation from Equation (4.11) using the well known functions $x^a(r,\theta,\phi)$ is left to the Problems. However, we know the g_{ij} from Equation (1.31) in Chapter 1. This allows us to write immediately

$$ds^2 = k\left(dt^2 - \left(dr^2 + r^2 d\theta^2 + r^2\sin^2\theta d\phi^2\right)\right). \tag{4.14}$$

It is just as easy to write this in cylindrical coordinates, or plane polar coordinates, or in fact any set of generalized coordinates using Equation (1.30). In any set of Galilean coordinates we have Equation (4.2).

We write an inverse of g_{ab} as g^{ab} where according to the contravariant indices we expect the contra-base-vector transformation

$$g^{ab} = \frac{\partial q^a}{\partial x^c}\frac{\partial q^b}{\partial x^d}\eta^{cd}. \tag{4.15}$$

Once again, between two sets of generalized coordinates x would be replaced by q' and η^{cd} by g'^{cd}. In a Problem it may be shown that this expression does indeed define an inverse to the metric tensor in Equation (4.11) such that $g^{ac}g_{cb} = \delta^a_b$.

Problems

4.4 Show directly that $g^{ac}g_{cb} = \delta^a_b$ from Equation (4.15).

4.5 Show directly by expressing all vectors in covariant form that $A'_c = g'_{ca}A'^a$ transforms from A_d according to Equation (4.19). You may need to transform A^b into A'^a and then write A^b by raising the index on its covariant form A_d.

4.6 Show that the transformations of Equations (4.18) and (4.19) become respectively

$$A'^a = \frac{\partial q'^a}{\partial q^b} A^b, \tag{4.16}$$

and

$$A'_c = \frac{\partial q^d}{\partial q'^c} A_d \tag{4.17}$$

between two sets of generalized coordinates $\{q'^a\}$ and $\{q^a\}$.

Any four-vector in space-time, A^a, will by definition transform like the position vector according to

$$A'^a = \frac{\partial q^a}{\partial x^b} A^b, \tag{4.18}$$

or in the corresponding covariant form (see Problem)

$$A'_c = \frac{\partial x^d}{\partial q^c} A_d. \tag{4.19}$$

Its modulus is calculated just as for the position vector from

$$A = g_{ab}A^a A^b = g^{cd}A_c A_d = A_d A^d, \tag{4.20}$$

where the proof follows from the definitions. In Galilean coordinates, $g_{ab} = \eta_{ab}$. Problem 4.6 gives the corresponding transformations between two sets of generalized coordinates.

A consequence of the non-positive-definite metric used in the modulus is that a non-zero four-vector may have a zero modulus. This requires only that $(A^0)^2 = (\mathbf{A})^2$, where $(\mathbf{A})^2 \equiv g_{ij}A^i A^j$, and the vector is said to be a 'null vector'. We have seen an example of this already in the position vector, which is null along the light rays.

Another concept that carries over into space-time is the 'orthogonality' of two four-vectors. This is the case for vectors A^a and B^b if

$$g_{ab}A^a B^b = A^a B_a = 0. \tag{4.21}$$

In Galilean coordinates this requirement simplifies, since $g_{ab} = \eta_{ab}$. It requires that $A^0B^0 = \mathbf{A} \cdot \mathbf{B}$, where the usual dot product (Chapter 1) of three vectors is indicated. Null vectors are in this sense orthogonal to themselves.

The cross product of two four-vectors can also be defined using a generalization to space-time of the epsilon symbol introduced in Chapter 1, Equation (1.108). We shall discuss this only as needed, however, since it produces a two-index tensor whose meaning is less intuitive than that of the scalar product.

4.2.1 Two Metric Derivations of the Lorentz Transformation

We have seen two essential derivations of the Lorentz transformations. One follows the path blazed by Lorentz and Poincaré, which deduces the transformations as those that leave Maxwell's equations invariant under the change of inertial observer. The second derivation was based on positivism or the supremacy of measurement in physical theory, and was the path taken by Einstein. The invariance of the speed of light together with the principle of relativity was required in either approach. However, the second path revealed the universality of the transformations if the speed of light is regarded as a maximum signal speed.

With the introduction of the concept of a metric space-time, however, several other derivations of the transformations between inertial observers present themselves. Each of these reveals a slightly different insight and we present two of them in this section. We adopt in each case the metric description of space-time imposed by the Minkowski metric (Equation (4.5)), as a conceptual and calculational convenience.

In the first approach we will require the modulus of a four-vector to be invariant under the transformation between inertial frames in standard configuration. We know that this invariance holds for the position vector itself (which defines the metric as in Equation 2.23) if the Lorentz transformations are used. Here we will deduce the Lorentz transformations from the condition that the modulus of any four-vector should be invariant between inertial observers. By analogy with the invariance under rotation of the moduli of three vectors in Euclidian space (see Chapter 1), we may regard a Lorentz 'boost' as a 'rotation in space-time' even though a single boost effects no rotation in space. We have seen, however, that a multiple boost does (recall Thomas precession).

The argument is rather simple and familiar. Considering any four-vector A^a, we let $A'^a = L^a{}_c A^c$ and require the invariance of the modulus as $\eta_{ab}A'^aA'^b = \eta_{cd}A^cA^d$. From this we deduce (see Problem)

$$\eta_{ab}L^a{}_cL^b{}_d = \eta_{cd}. \tag{4.22}$$

With the assumption of standard configuration we know that $L^1{}_c = \delta^1_c$ and $L^2{}_c = \delta^2_c$ so we need only concern ourselves with the zero and three components. We find easily (see Problem) that the components $c = 0 = d$, $c = 3 = d$, and $c = 0$, $d = 3$ give respectively

$$\begin{aligned}(L^0{}_0)^2 - (L^3{}_0)^2 &= 1,\\ (L^0{}_3)^2 - (L^3{}_3)^2 &= -1\\ L^0{}_3L^0{}_0 - L^3{}_3L^3{}_0 &= 0.\end{aligned} \tag{4.23}$$

The fourth equation follows as before by requiring the Galilean transformation (1.78) at low velocities. This implies $L^3{}_0/L^3{}_3 = -u$, and $L^3{}_3 > 0$. Then proceeding as usual by excluding time reversal between inertial observers, one obtains the Lorentz transformations in standard boost form (see Problem). This is not a surprising result, but it demonstrates a necessary coherence between Lorentz transformations and the modulus or 'measure' of four-vectors. The modulus or measure of a four-vector is invariant under boosts between inertial observers, which may therefore be regarded as rotations in space-time.

Problems

4.7 Deduce Equation (4.22) in the text using the invariant modulus of an arbitrary four-vector.

4.8 Deduce the relations (4.23) and use them to deduce the Lorentz transformations in the form of Equation (2.27).

Another derivation uses the idea that inertial observers 'construct' their respective worlds by synchronizing clocks. In this way a surface of constant t in each inertial frame defines the three-space 'hyper-surface'. It is even better described as a 'simultaneity' to emphasize the constructed nature, and because it orthogonalizes the time axis relative to the three space.

We begin again with the Minkowski metric as the true measure of space-time for an inertial observer O. We transform to coordinates $(t, \overline{z})$ that would be the coordinates of an inertial observer O' moving with relative velocity u in standard configuration, if the classical 'Galilean' transformation applies. That is, we substitute $dz = d\overline{z} + udt$ (leaving time unchanged) into the Minkowski metric $ds^2 = dt^2 - dz^2$ to obtain

$$ds^2 = \frac{dt^2}{\gamma^2} - 2udtd\overline{z} - d\overline{z}^2. \tag{4.24}$$

As usual we have set $c = 1$ and we have suppressed the two perpendicular space components that do not enter into the transformation.

This last result is clearly unsatisfactory as the metric of the new inertial world, since it is not invariant under $dt \leftarrow -dt$. This would mean that distance between space-time events would depend on the direction in which the interval was traversed! This might be the case if time were inhomogeneous, but we do not expect this for inertial observers. More basically, the form of the metric is not preserved for O', which contradicts the principle of relativity.

We can improve the result by completing the square and collecting terms in $d\overline{z}$ (we are 'diagonalizing' the metric which renders the time axis and space orthogonal) so that we obtain

$$ds^2 = \left(\frac{dt}{\gamma} - \gamma u d\overline{z}\right)^2 - \gamma^2 d\overline{z}^2. \tag{4.25}$$

This has the Minkowski form in the new inertial frame (O') if we insist that $dz' = \gamma d\overline{z}$ and $dt' = (dt - \gamma^2 u d\overline{z})/\gamma$. Remembering the initial ansätz, $dz = d\overline{z} + udt$, the expression for dz' becomes

$$dz' = \gamma(dz - udt). \tag{4.26}$$

We recognize this as the differential form of the spatial Lorentz transformation. Moreover, substituting $d\overline{z} = dz'/\gamma = dz - udt$ into our presumed form for dt' yields the Lorentz time transformation

$$dt' = \gamma(dt - udz). \tag{4.27}$$

Consequently we see that by assuming the Minkowski metric to reflect the 'real' metric structure of space-time, the Lorentz transformations follow. We have shown this by requiring the metric structure to be the same for any inertial observer. However, this raises a question that is essential to the 'modern' (post-Minkowski, post-Einstein) view of space-time. If space-time has this absolute structure, why should we be restricted to describing it by inertial observers? In fact we can allow the metric to take on a different form for non-inertial observers, or more generally in non-inertial coordinates, so long as there is a one-to-one mapping with an inertial system. The scalar nature of the interval when the metric is transformed as a covariant tensor allows the Minkowski structure to be preserved in arbitrary coordinates.

We should pause here to distinguish once again 'observers' from 'coordinates'. Inertial observers are well-defined once an archetype is defined, as we discussed in Chapter 1. Each inertial observer constructs a view of the world with the help of 'friends' at relative rest. In some cases, as in our discussion of the rotating disc, such a set of observers also exists in a non-inertial frame. However, we can imagine writing the metric of space-time in terms of arbitrary coordinates (such as $\overline{z}$ above). These might not correspond to the coordinates of events as measured by any physical observers. All that is required is that there be a continuous, onc-to-onc, functional rclation between inertial coordinates of events and events in these arbitrary coordinates. Such a relation the mathematicians call an 'isomorphism'.

We may now return to the rotating disc of Chapter 3. If we write the space-time metric for the observer O in plane-polar coordinates we have (see Equation (4.13) and related discussion)

$$ds^2 = dt^2 - dr^2 - r^2 d\phi^2. \tag{4.28}$$

To transform to coordinates that describe a point on the rotating disc, we need only use t, ϕ', classically where $\phi = \phi' + \Omega t$. But this is formally just the previous transformation if we identity $d\overline{z} = rd\phi'$, $dz = rd\phi$ and $\Omega r = u$. Consequently the diagonalization will yield again

$$rd\phi' = \gamma(rd\phi - (\Omega r)t), \tag{4.29}$$

$$dt' = \gamma(dt - \Omega r^2 d\phi). \tag{4.30}$$

The first of these equations implies as before ($dt = 0$, an inertial simultaneity) that the circumference is $2\pi\gamma r$ on the disc. However, the second equation also confirms that the simultaneity $dt = 0$ is not a simultaneity for disc observers. Around the whole circumference we obtain the cut in disc time equal to $2\pi\gamma\Omega r^2$, as was found by elementary arguments in Chapter 3.

We shall see in the next section that the description of the absolute structure of inertial space-time in arbitrary coordinates is a powerful tool. When coupled with the diagonalization procedure, it allows the recognition of the 'proper time' for an observer who may be at rest in these coordinates. There are coordinates in which no physical observer can be at rest (such as a grid of inward- and outward-going light rays), but in which nevertheless the invariant metric interval is preserved. But this leads too far afield at the moment.

4.3 Four-Vector Dynamics

Some four-vectors arise naturally in mechanics, such as the four-velocity, v^a, and the four-acceleration, a^a. These are defined in terms of the position four-vector in Galilean coordinates and the invariant displacement element ds as

$$v^a \equiv \frac{dx^a}{ds} = \frac{d}{ds}\begin{pmatrix} t \\ \mathbf{r} \end{pmatrix}, \tag{4.31}$$

and

$$a^a \equiv \frac{dv^a}{ds}. \tag{4.32}$$

These are clearly vectors, because they transform as the archetypal contravariant position vector due to the invariant or scalar nature of ds. For any inertial observer O measuring a moving point O', we know that $ds = dt' = dt/\gamma(v)$ in terms of the coordinate time interval dt. Consequently the four-velocity and the four-acceleration become (by direct calculation in the O world, see Problem)

$$v^a = \begin{pmatrix} \gamma(v) \\ \gamma(v)\mathbf{v} \end{pmatrix} \tag{4.33}$$

$$a^a = \begin{pmatrix} \gamma(v)^4\mathbf{a}\cdot\mathbf{v} \\ \gamma(v)^2\mathbf{a} + \gamma(v)^4(\mathbf{a}\cdot\mathbf{v})\mathbf{v} \end{pmatrix}. \tag{4.34}$$

Here we use the Galilean three-acceleration in the O world $\mathbf{a} \equiv d\mathbf{v}/dt$, and the Galilean $\mathbf{v} \equiv d\mathbf{r}/dt$. These vectors are just the contravariant spatial components of the respective four-vectors.

The four-velocity vector is a unit vector in space-time, since if we calculate its modulus in Galilean coordinates using Equation (4.33) we find

$$\eta_{ab}v^a v^b = \gamma^2 - \gamma^2 v^2 = 1. \tag{4.35}$$

In a space-time diagram of the world of a given inertial observer, this unit vector is always tangent to the world line of a particle, pointing in the direction of increasing time.

An important relation between these two dynamic four-vectors is that they are mutually orthogonal. We see this formally in Galilean coordinates from

$$\eta_{ab} v^a a^b \equiv \eta_{ab} v^a \frac{dv^b}{ds} \equiv v_b \frac{dv^b}{ds} = \frac{1}{2}\frac{d(v_b v^b)}{ds} = 0, \tag{4.36}$$

where we use finally the velocity unit modulus. A useful exercise, however, is to verify this by a direct calculation from Equations (4.33) and (4.34) in Galilean coordinates (see Problem).

Problem

4.9 Show directly from the explicit forms of v^a (Equation (4.33)) and of a^b (Equation (4.34)) for an inertial observer, that in Galilean coordinates, $v^b a_b = 0$.

Now it is apparent that any relation between four-vectors in one inertial frame will have the same form in any other inertial frame. This is because each vector transforms into the corresponding vector in the new inertial frame, in a universal manner that depends only on the change in coordinates, and hence observer. This will also be true for tensor relations in general because of similar universal transformation properties. It is for this reason that the principle of relativity requires physical 'laws' to be expressed in tensor or vector form. The laws *must* have the same form for every inertial observer, according to this principle. There are quantities that have to be expressed in other than tensor form. We shall meet a notorious one (the 'Christoffel' symbol) during the discussion in generalized coordinates of the motion of a force-free particle. Nevertheless they must also transform so as to maintain the equality of inertial observers, even if not the equality of all possible observers.

This realization allows us to obtain our first hint regarding the nature of Newton's law in space-time. One might expect for a particle of mass m that this law becomes the four-vector relation

$$K^b = ma^b = \frac{d}{ds}(mv^b), \tag{4.37}$$

where K^b indicates some four-force. The four-force has to be found independently as either a fundamental force or as a force of constraint, just as in the classical treatment. When the acceleration is known we may use this result to calculate the constraining force, also just as in the classical case. By taking the low-velocity limit of Equation (4.34) and substituting into Equation (4.37) we deduce that

$$K^b \to \begin{pmatrix} m\mathbf{a}\cdot\mathbf{v} \\ m\mathbf{a} \end{pmatrix} = \begin{pmatrix} \dfrac{d\mathcal{E}}{dt} \\ \dfrac{d\mathbf{p}}{dt} \end{pmatrix}. \tag{4.38}$$

Here $\mathcal{E}$ is the classical energy and $\mathbf{p}$ is the classical momentum. Thus it appears that, in space-time, Newton's law will combine conservation of energy and conservation of momentum. This is perhaps not surprising if one remembers the conjugate relation between time and energy in Hamiltonian mechanics. We shall have to confirm the correct generalization of energy and momentum to relativistic velocities, but it is clear from the classical limit that the four-vector

$$p^a = mv^a \tag{4.39}$$

is a candidate for containing both. We therefore refer to this four-vector as the 'energy-momentum vector' or simply as the 'momentum four-vector'. The form it takes in a particular inertial frame follows from Equation (4.33). The three-vector formed by its spatial components is

$$\mathbf{p} \equiv m\gamma(v)\mathbf{v}. \tag{4.40}$$

The force description remains mysterious beyond its four-vector nature at this point. However, because of the orthogonality of v^b and a^b, it will have to satisfy $v_b K^b = 0$ if the rest mass of a particle is to be constant along its path. That is, from Equation (4.37) it then follows that $dm/ds = 0$. Such particles may be thought of as 'fundamental'. Electromagnetism, acting on fundamental particles such as electrons, fits this scheme elegantly as we shall see. A full theory of gravity takes us considerably beyond the Minkowski geometry of space-time, although it also conserves the rest mass for neutral (in geodesic motion; see later chapters) fundamental particles.

At high enough interaction energies, atoms are not such fundamental particles, nor are atomic nuclei, since both may be decomposed into substructure. This is also the case for many other sub-atomic particles. Thus when considering such objects in collision, the rest mass may be regarded as variable.

It is important to become familiar with the various operations on four-vectors. In particular we should be able, by applying a boost to v^a and a^b, to re-derive our transformations (3.16), (3.24), (3.25). We demonstrate this procedure in the examples.

Example 4.2

We consider in this example the velocity transformation. For a boost in standard configuration we will have $v'^a = L^a{}_b v^b$, where the boost matrix is that of Equation (2.27). Taking the 'zeroth' component we obtain

$$v'^0 = \Gamma(u)\gamma(v)(1 - \mathbf{u}\cdot\mathbf{v}). \tag{4.41}$$

We use $\Gamma(u)$ as the Lorentz factor of the transformation, $\gamma(v)$ as the Lorentz factor of the object for O, and $\gamma'(v')$ the Lorentz factor of the object for O'. Recall that in standard configuration $v^3\widehat{\mathbf{e}}_3 \parallel \mathbf{u}$, which allows us to write uv^3 as the dot product. But $v'^0 = \gamma'$, and so we obtain the transformation of the particle Lorentz factor from Equation (4.41) as

$$\gamma'(v') = \Gamma(u)\gamma(v)(1 - \mathbf{u}\cdot\mathbf{v}). \tag{4.42}$$

The transverse components of the velocity four-vector do not transform so that $v'^i = v^i$ for $i = 1, 2$. But each of these is equal to $\gamma v(i)$ in their respective frames of reference so that $v'(i) = (\gamma/\gamma')v(i)$. If we combine these two components into the transverse velocity vector $\mathbf{v}_\perp$ in each frame, then with Equation (4.42) we have

$$\mathbf{v}'_\perp = \frac{\mathbf{v}_\perp}{\Gamma(1-\mathbf{u}\cdot\mathbf{v})}, \tag{4.43}$$

which is the transverse part of Equation (3.16) (replacing $\gamma(u)$ by $\Gamma(u)$).

The three (i.e. parallel) velocity component transforms as $v'^3 = \Gamma\gamma(v^3 - u)$ in the boost. This becomes $\gamma'\mathbf{v}'_\parallel = \Gamma\gamma(\mathbf{v}_\parallel - \mathbf{u})$ on multiplying by the base vector in the 3 direction and reverting to three vectors. Finally, then, with Equation (4.42) we obtain

$$\mathbf{v}'_\parallel = \frac{\mathbf{v}_\parallel}{1-\mathbf{u}\cdot\mathbf{v}}, \tag{4.44}$$

which is the second part of Equation (3.16).

Example 4.3

In this example we show that a boost acting on the four-acceleration (4.34) yields the transformations (3.24) and (3.25). The calculation is more tedious than that for the velocity, so we shall leave gaps to be filled in as problems.

1. From the zero component of the boost transformation together with the zeroth component of Equation (4.34) in each frame and Equation (4.42), show that

$$\mathbf{a}'\cdot\mathbf{v}' = \frac{1}{\Gamma^3(1-\mathbf{u}\cdot\mathbf{v})^3}\left(\mathbf{a}\cdot\mathbf{v} - \frac{\mathbf{a}\cdot\mathbf{u}}{\gamma^2(1-\mathbf{u}\cdot\mathbf{v})}\right). \tag{4.45}$$

2. The transverse contravariant components are the same in each frame. Hence $\gamma'^2\mathbf{a}'_\perp = \gamma^4(\mathbf{a}\cdot\mathbf{v})\mathbf{v}_\perp + \gamma^2\mathbf{a}_\perp - \gamma'^4(\mathbf{a}'\cdot\mathbf{v}')\mathbf{v}'_\perp$. Then by substituting for $\mathbf{a}'\cdot\mathbf{v}'$ from the result in part (1) and using Equations (4.42) and (4.43), one obtains

$$\mathbf{a}'_\perp = \frac{\mathbf{a}_\perp(1-\mathbf{u}\cdot\mathbf{v}) + (\mathbf{a}\cdot\mathbf{u})\mathbf{v}_\perp}{\Gamma^3(1-\mathbf{u}\cdot\mathbf{v})^3}, \tag{4.46}$$

which is the desired result (Equation (3.25) where $\gamma(u) \equiv \Gamma(u)$).

3. The three component of the boost gives $a'^3 = -\Gamma u a^0 + \Gamma a^3$. Writing each component according to Equation (4.34) in their respective reference frames and writing z components as parallel vectors yields $\gamma'^2\mathbf{a}'_\parallel = \Gamma\gamma^4(\mathbf{a}\cdot\mathbf{v})(\mathbf{v}_\parallel - \mathbf{u}) + \gamma^2\Gamma\mathbf{a}_\parallel - \gamma'^4(\mathbf{a}'\cdot\mathbf{v}')\mathbf{v}'_\parallel$. One uses again $\mathbf{a}'\cdot\mathbf{v}'$ from the result in part (1) together with Equation (4.44) to obtain

$$\gamma'^2\mathbf{a}'_\parallel = \frac{\Gamma\gamma^2}{1-\mathbf{u}\cdot\mathbf{v}}(\mathbf{a}_\parallel(1-\mathbf{u}\cdot\mathbf{v}) + (\mathbf{a}\cdot\mathbf{u})(\mathbf{v}_\parallel - \mathbf{u})). \tag{4.47}$$

Finally, using Equation (4.42) and realizing that all vectors inside the brackets of this last expression can be taken parallel to $\mathbf{u}$, we obtain

$$\mathbf{a}'_\parallel = \frac{\mathbf{a}_\parallel}{\Gamma^3(1-\mathbf{u}\cdot\mathbf{v})^3}, \tag{4.48}$$

which is Equation (3.24).

The preceding examples serve to give us confidence in the inertial form of the dynamic four-vectors introduced in this section, as well as in their transformations between inertial frames. They will prove to be practical in subsequent sections and chapters.

4.3.1 Lagrangian Dynamics Without Fundamental Forces

Even without a theory of fundamental forces, particle dynamics in an inertial frame is a physical theory. It consists of the inertial concept itself, plus a description of the motion of a particle that is free to move either in three-dimensional space, or in a subspace of it. The constraints that dictate the subspace motion are normally imposed mathematically. This is because an *a priori* description of a constraining force that is both convenient and correctly Lorentz invariant is rare. Nevertheless, once the motion is determined, one can use $K^b = ma^b$ to determine the effective constraining force, just as in the Newtonian limit.

A general method of defining a physical theory is to write an action $\mathcal{S}$. The variation of this quantity with respect to the key variables of the theory, when set equal to zero to give an extremum, yields the dynamic equations. In classical mechanics the action takes the form found in Hamilton's principle, where it is the integration of a Lagrangian L over Newtonian (that is coordinate) time. Newtonian dynamics under conservative forces is derived by setting the Lagrangian $L = T - V$, where T and V are respectively the kinetic energy and potential energy of the system of particles. These quantities are expressed in terms of generalized coordinates and their time derivatives. These coordinates and coordinate velocities are the key variables of the theory, when it is regarded as existing in phase space (i.e. the combined six-space of position and momentum). There is no way to deduce this ansätz for the action other than to demonstrate its agreement with experiment.

We are therefore confronted with making a choice for the action and for the Lagrangian that is appropriate for a relativistic particle. We choose one particle, since a system of non-interacting particles will be described as a sum over single particles. The possibilities are strictly limited. The action must be a scalar under boosts by the principle of relativity, since otherwise its extremum could have no physical importance. Moreover it must be peculiar to the particle in question. This leaves us with the constant inertial particle mass m, the invariant distance element ds along the particle world-line, and the four-velocity scalar $v_a v^a$, with which to form the action. The velocity modulus is equal to one as we have seen, but we can ignore this until after the variation by treating this normalization as a 'non-holonomic' (that is non-integrable) constraint.

Our candidate for the action of a 'free' (at least in some subspace) relativistic particle is thus

$$\mathcal{S} = mc \int_1^2 F\left(\frac{v_a v^a}{2c^2}\right) ds, \tag{4.49}$$

where F is a dimensionless scalar function and the factor 2 is inserted for convenience. As is usual while varying the action, the endpoints of the motion are not varied. We retain conventional units temporarily for dimensional guidance, while combining mc and ds to have the correct dimensions of action (angular momentum).

The function F is quite arbitrary, but we will use the simplest choice that agrees with 'experiment', that is essentially with the Newtonian limit. Moreover, although for a four-vector treatment of a fundamental force (such as electromagnetic) it turns out to be useful to choose a non-trivial form of F, we do not expect it to add new physics beyond ds itself for a free particle. This is because of the identity $v_a v^a = 1$, which can be recognized as the definition of ds when written in the form $dx_a dx^a = ds^2$. We therefore set $F = \pm 1$ in Equation (4.49) as the simplest choice for the moment.

However, although we require the equations of motion to follow only as providing an extremum of the action ($\delta \mathcal{S} = 0$), it is frequently convenient to have these yield a true minimum of the action. We know that a particle at rest in an inertial frame 'remains at rest'. Thus its world line lies along the t axis, and hence we should require this dynamic path to be the one that mimimizes the action. But from the definition of the metric ds in space-time, this world line is the *longest* path between two points on the t axis. Hence to render the action a minimum, we should take $F = -1$. This makes practical use of the 'twin paradox' effect, which we have seen is an essential feature of space-time and not at all a 'paradox'.

These arguments allow us to write an action for a free particle that is suitable for any inertial observer as

$$\mathcal{S} = -mc \int_1^2 ds = -mc^2 \int_1^2 dt \sqrt{1 + \frac{g_{ij}\dot{q}^i \dot{q}^j}{c^2}}. \tag{4.50}$$

The last expression follows by factoring cdt out of the definition of ds (e.g. Equation (4.13)), so that here $\dot{q} = dq/dt$.

We have chosen the coordinate time t to parameterize the world line in the last form of the expression, in order to maintain the distinction between space and time that exists in classical mechanics. We can write $d\mathbf{r}^2 = -g_{ij} d\dot{q}^i d\dot{q}^j$, since this last expression is equivalent to $-\eta_{ij}\dot{x}^i \dot{x}^j$ in Galilean coordinates. This identifies the free particle Lagrangian finally as

$$L = -mc^2 \sqrt{1 + g_{ij}\dot{q}^i \dot{q}^j} \equiv -mc^2 \sqrt{1 - \mathbf{v}^2} \equiv -mc^2/\gamma, \tag{4.51}$$

where in general we define

$$\gamma = \frac{1}{\sqrt{1 + g_{ij}\dot{q}^i \dot{q}^j}}. \tag{4.52}$$

The action is now in standard classical form, and if varied according to the calculus of variations, will yield the dynamic path. Thus the equations of motion of a free relativistic particle follow as the Euler-Lagrange equations in their autonomous form (we will treat constraint forces later)

$$\frac{d}{dt}\left(\frac{\partial L}{\partial \dot{q}^k}\right) - \frac{\partial L}{\partial q^k} = 0, \quad k = 1 \ldots 3. \tag{4.53}$$

In Galilean coordinates the Lagrangian becomes explicitly (we continue with $c = 1$)

$$L = -m\sqrt{1 - \dot{x}^k \dot{x}^k} \equiv -m/\gamma(v). \tag{4.54}$$

The Lagrange equations for a free particle imply in these coordinates that the 'canonical three-momentum' $\partial L/\partial \dot{x}^k$ is an integral of the motion, that is explicitly

$$m\gamma\mathbf{v} = constant \equiv \mathbf{p}. \tag{4.55}$$

This equation identifies the spatial components of the four-momentum vector as the momentum, 'conjugate' to the Cartesian position vector. This inertial Cartesian form is the most physically transparent, as it obviously reduces to the Newtonian momentum at low velocities. The use of generalized coordinates allows the application of various geometrical constraints, however, as will be illustrated in the Problems and examples.

In generalized coordinates q^k, the Lagrange equations are more complicated. This is due to g_{ij} being dependent on the spatial coordinates in general. Under coordinate transformations that are independent of time or depend on it linearly, the g_{ij} will be independent of t. In this case the equations may be derived formally (see Problem), and they reveal the existence of 'inertial forces'. We shall discuss subsequently the most general formal result for all coordinate transformations in Minkowski space-time, but we remark that it is almost always easier to use the Lagrangian formulation directly.

Problems

4.10 Starting from the Lagrangian in generalized coordinates in the first form of Equation (4.51), show that the Euler-Lagrange Equations (4.53) become (dummy indices have arbitrary names)

$$\frac{d^2q^j}{ds^2} = -\Gamma^j_{i\ell}\frac{dq^i}{ds}\frac{dq^\ell}{ds}, \tag{4.56}$$

where

$$\Gamma^j_{i\ell} \equiv \frac{g^{jk}}{2}\left(\frac{\partial g_{ki}}{\partial q^\ell} + \frac{\partial g_{k\ell}}{\partial q^i} - \frac{\partial g_{i\ell}}{\partial q^k}\right). \tag{4.57}$$

Recall that the operator $d/ds = \gamma(d/dt)$.

This Γ symbol (defined here for three-space) is called a 'Christoffel symbol' (of the second kind). It is not a tensor, since it vanishes in Galilean coordinates but not in general coordinates. Thus the apparent accelerations that this symbol produces in the Lagrange Equations (4.56) are coordinate dependent, and exist only in non-inertial frames. The apparent accelerations are called 'inertial' and the corresponding forces are 'inertial forces'.

4.11 Show that Equation (4.59) follows from the definition (4.58) and the appropriate form of the Lagrangian. The proof is easiest in Galilean coordinates, but you may want to try generalized coordinates for a useful exercise.

We are also able to calculate the second grand scalar of Newtonian mechanics, namely the Hamiltonian $H(q^k, p^k)$. This is an integral of the motion for a conservative system (zero force is conservative) that is free of explicit time dependence, such as might be produced by time-dependent constraints. Using the standard Legendre transformation (e.g. [1]) to transform from $L(q^k, \dot{q}^k)$ to $H(q^k, p^k)$ we have the Hamiltonian in the form

$$H(q^k, p^k) \equiv \dot{q}^k \frac{\partial L}{\partial \dot{q}^k} - L. \tag{4.58}$$

If we are interested only in the numerical value of this integral, rather than in its functional form, it is generally denoted h and referred to as the 'Jacobi Integral'.

We may use either Equations (4.51) and (4.52) or Equation (4.54) to calculate (see Problem) that (on restoring conventional units for guidance)

$$H = \gamma mc^2, \tag{4.59}$$

where in general γ is given in Equation (4.52). Classically this integral of the motion is the energy, whenever the Lagrangian is quadratic in the generalized velocities. Since this is the definitely the case for inertial observers relativistically, we expect it to represent the energy here also. In fact, as the only other conserved quantity besides the momentum $\mathbf{p}$, we may define it to be the energy. Hence

$$\mathcal{E} = \gamma mc^2. \tag{4.60}$$

This latter equation may be substituted into the four-momentum Equation (4.39) to write

$$p^a = \begin{pmatrix} \dfrac{\mathcal{E}}{c^2} \\ \dfrac{\mathbf{p}}{c} \end{pmatrix}, \tag{4.61}$$

and after calculating the modulus $\eta_{ab} p^a p^b$ there follows

$$\mathcal{E}^2 - c^2\mathbf{p}^2 = m^2c^4. \tag{4.62}$$

The remarkable implication of Equation (4.62) is that not $\mathcal{E}$, but rather $\mathcal{E} - mc^2$ tends to the classical kinetic energy at small velocities. This is because $\mathcal{E} = mc^2\sqrt{1 + \mathbf{p}^2/(mc)^2}$, and in the low-velocity limit $\mathbf{p}^2/(mc)^2 \to \mathbf{v}^2/c^2$. Consequently $\mathcal{E} \to mc^2 + mv^2/2$. It seems, therefore, that adopting $\mathcal{E}$ as the generalization of the classical energy is appropriate, with the exception of this constant mc^2.

This constant is referred to as the 'rest mass' energy since it is present in an inertial frame with the particle at rest. In the form

$$\mathcal{E}_o = mc^2, \tag{4.63}$$

it is the famous Einstein relation between rest mass energy $\mathcal{E}_o$ and rest mass. Equation (4.60) is the more general relation for a particle moving relative to an inertial observer. That value has the property of becoming infinite as the speed of light is approached, consistent with the impossibility of doing so. However, even at $v/c = 0.99$ the energy has become only about $7mc^2$, which is hardly infinite.

The rest energy relation suggests that an object that is composed of more fundamental particles will release or gain energy as its inertial mass decreases or increases respectively. Of course, we now know this to be true in nuclear transmutations and explosions. If one could convert an entire kilogram of mass into energy, one would release c^2 joules or 9×10^{16} J. This should be compared with roughly 7.5×10^6 J released by exploding one kilogram of TNT (trinitrotoluene). Even if the efficiency of mass to energy conversion is only 7×10^{-3} (the mass fraction converted into energy when four protons fuse to a helium nucleus), one sees that operating on one kilogram of mass produces energy equivalent to 84 kilotons of TNT.

As a simple example of the description of force-free motion in curvilinear coordinates, consider cylindrical coordinates $\{q^j\} = \{r, \phi, z\}$. An easy tensor transformation using the covariant form (4.11) gives (see Problem) $g_{ij}dq^i dq^j = -dr^2 - r^2 d\phi^2 - dz^2$. Consequently the Lagrangian follows from Equation (4.51) as ($c = 1$)

$$L = -m\sqrt{1 - (\dot{r}^2 + r^2\dot{\phi}^2 + \dot{z}^2)} \equiv -m/\gamma, \tag{4.64}$$

whence follows the Lagrange equations for the q^j from Equations (4.53). One finds (recalling $d/dt = (1/\gamma)d/ds$)

$$\begin{aligned} \frac{d^2r}{ds^2} &= \left(\frac{d\phi}{ds}\right)^2 r, \\ \frac{d(m\gamma\dot{\phi}r^2)}{dt} &= \frac{d\left(m\frac{d\phi}{ds}r^2\right)}{ds} = 0, \\ \frac{d(m\gamma\dot{z})}{dt} &= \frac{d\left(m\frac{dz}{ds}\right)}{ds} = 0. \end{aligned} \tag{4.65}$$

The first equation reveals the inertial 'centrifugal force' associated with rotation in ϕ while the second and third equations express the conservation of angular momentum and of z momentum respectively. If each equation is put into a form compatible with the general Lagrange equation in three dimensions (Equation (4.56)), the corresponding Christoffel symbols in cylindrical polar coordinates may be read off (see Problem). This is generally an easier procedure than calculating them directly from the definition (Equation (4.57)).

Problems

4.12 Starting with the expression of the Cartesian coordinates in terms of cylindrical polar coordinates, use the covariant tensor transformation law (Equation (4.11), which is Equation (1.30) of Chapter 1 if η_{ij} replaces the (implicit) δ_{ij}), to deduce the g_{ij} used in the text for cylindricals.

4.13 Write the Lagrange equations for the separate d^2q^j/ds^2 in cylindrical polar coordinates, and then by comparison with Equation (4.56) deduce the non-zero Christoffel symbols. (Answer: $\Gamma^1_{22} = -r$, $\Gamma^2_{12} = \Gamma^2_{21} = 1/r$, all others are zero.)

4.14 Show, using the general definition of γ (Equation (4.52)), that if d/ds is used in place of $(1/\gamma)(d/dt)$ the general expression for γ becomes

$$\gamma = \sqrt{1 - g_{ij}\frac{dq^i}{ds}\frac{dq^j}{ds}}. \tag{4.66}$$

4.15 Use the Lagrange equations in cylindrical coordinates (e.g. see text) to solve for the force-free motion of a particle for which $dr/ds = 0$ at $r = b$. Note that Equation (4.37) shows that γ is a constant. Construct a spatial diagram that shows that the particle moves in a straight line with b the distance of closest approach to the origin. You should be able to verify from the geometry that the equations give correctly the dependence of $\dot{\phi}$ or r. Verify also that γ, and hence the energy, is a constant.

Equation (4.65) together with the second of the Lagrange equations and an initial condition allow us to solve for the free motion of the particle in cylindrical coordinates. However, we know from the solution in Cartesian coordinates that the motion is in a straight line with constant energy. It is of interest nevertheless to see how this appears in cylindrical coordinates (see Problem).

To find more interesting problems we must apply constraints. For holonomic (i.e. integrable) constraints in the form $f_\alpha(\{q^j\}, t) = 0$, where α runs over the number of constraints, the Lagrange equations become [1]

$$\frac{d}{dt}\left(\frac{\partial L}{\partial \dot{q}^k}\right) - \frac{\partial L}{\partial q^k} = -\Sigma_\alpha\left(\lambda_\alpha \frac{\partial f_\alpha}{\partial q^k}\right), \quad k = 1\ldots 3, \tag{4.67}$$

where the $\{\lambda_\alpha\}$ are the Lagrange undetermined multipliers and the summation is over all constraints. With the sign chosen on the right-hand side, each term in α will yield the back reaction force on the element associated with the constraint.

An interesting set of constrained problems arises when a particle is constrained to move on a subspace that may itself be moving. The subspace may be a curved surface or line. In cosmology, which unfolds in space-time, the subspace may be the three-space. This will not concern us for a while yet, however. We give a set of problems of this sort, but we begin with an example of a bead, sliding without friction on a long, rigid, straight wire that is rotating about an axis perpendicular to itself through one end [2].

The only subtlety to observe in such problems is that there are two approaches. One can write the Lagrangian of the particle according to Equation (4.51) and then apply the constraints afterwards. This will not maintain the energy as an integral of the motion if the constraints are doing work on the particle. As an alternative one might choose to eliminate the degree of freedom associated with the constraint when formulating the Lagrangian. This has the effect of imposing unspecified forces on the particle that do no work in the subspace. In that case the Jacobi integral h continues to exist for a time-independent subspace Lagrangian, but it is not the energy in the inertial space. It contains the influence of the apparent forces in the subspace.

Example 4.4

We take up the problem cited in the text of a long straight wire rotating with angular velocity Ω about a perpendicular axis through one end. We take that end as the origin of plane-polar coordinates (r, ϕ). A particle of mass m slides without friction on the wire. For simplicity we will start the particle at the origin with zero radial velocity $\dot{r}$. This and related problems can approximate the small pitch-angle (the angle the velocity makes with the magnetic field line) motion of a relativistic charged particle along a magnetic field line, in a rotating reference frame in which the electric field is zero [2].

The Lagrangian in these coordinates is (from Equation (4.51))

$$L = -mc^2\sqrt{1 - (\dot{r}^2 + \dot{\phi}^2 r^2)/c^2}. \tag{4.68}$$

But we have a time-dependent ('rheonomous') holonomic constraint as $f \equiv \phi - \Omega t = 0$ if the origins $t = 0$ and $\phi = 0$ are taken together. The radial Lagrange equation is unaffected by this constraint according to Equation (4.67), and it becomes (setting $\gamma^2 \equiv 1/(1 - (\dot{r}^2 + \dot{\phi}^2 r^2)/c^2)$)

$$\frac{d(\gamma \dot{r})}{dt} = \gamma \dot{\phi}^2 r. \tag{4.69}$$

The azimuthal Lagrange equation is (see Equation (4.67))

$$\frac{d(m\gamma \dot{\phi} r^2)}{dt} = -\lambda. \tag{4.70}$$

After writing these equations we apply the constraint in the form $\dot{\phi} = \Omega$.

This constraint treats the wire as a classical rigid body, but such an object is impossible in practice because of the finite and limiting speed of light. In such problems one imagines that the wire has been set into motion very gradually, so that each section has had time to attain the steady motion through the normal action of elastic waves. However, no part of the structure can ever exceed the speed of light.

We see from the azimuthal equation that the constraint force $-\lambda(\partial f/\partial \phi) = -\lambda$ is equal to the inertial 'torque' on the particle. This regards the inertial mass as having increased to γm, as also occurs in the particle energy. Hence λ itself is the back reaction on the wire. This reaction would have to be accepted progressively by the wire, which implies an elastic signal speed greater than the particle radial velocity. One must always remember that these problems are only approximations to more complicated relativistic physics, and one should not draw general conclusions.

The radial Lagrange equation can be written as an equation for the radial four-velocity component $u = \gamma \dot{r}$ by multiplying by γ, using the constraint, and changing the variable from t to r. This gives $udu = \gamma^2 \Omega^2 r dr$. However, Equation (4.66) allows us to write $\gamma^2 = 1 + (u^2 + \gamma^2 \Omega^2 r^2)/c^2$, which becomes

$$\gamma^2 = \frac{(1 + u^2/c^2)}{(1 - \Omega^2 r^2/c^2)}. \tag{4.71}$$

Substituting this back into the equation for u gives a separable differential equation whose solution is

$$1 + u^2/c^2 = \frac{1}{1 - \Omega^2 r^2} \equiv (\gamma_\phi)^2. \tag{4.72}$$

Thus by the previous expression for γ^2 we find that the particle energy varies as

$$\gamma^2 = (\gamma_\phi)^4. \tag{4.73}$$

Substituting this into Equation (4.71) together with $u = \gamma \dot{r} \equiv \gamma v_r$ yields after some algebra

$$v_r^2 = \Omega^2 r^2 (1 - \Omega^2 r^2/c^2) \equiv \dot{r}^2. \tag{4.74}$$

The final integration for $r(t)$ could be effected from this equation. The interesting result, however, is in this expression for the radial velocity. It starts from zero and returns to zero at $r = c/\Omega$. It has a maximum at $r = c/(\sqrt{2}\Omega)$ given by $v_r = c/2$. This already exceeds the speed of any known elastic wave, although perhaps not of electromagnetic waves in conducting plasmas.

The maximum in the radial velocity may be somewhat of a surprise. It is quite understandable in terms of the limiting velocity c. As the azimuthal velocity is constrained to increase inexorably towards c at $r = c/\Omega$, it follows that after an initial increase the radial velocity must drop to zero in order to maintain the total particle velocity less than c.

Now that we have done this problem the hard way, we can better enjoy the more direct solution. We reduce the Lagrangian to the allowed subspace by setting $\dot{\phi} = \Omega$. This gives $L = -mc^2\sqrt{1 - (\dot{r}^2 + \Omega^2 r^2)/c^2}$, which is the Lagrangian in the radial subspace. There are no constraints that do work in this space so that $h = \dot{r}(\partial L/\partial \dot{r}) - L$ is an integral of the motion. A direct calculation shows that $h/(mc^2) = \gamma/(\gamma_\phi)^2$, and our initial conditions set $h = mc^2$. Hence one obtains the same solution for γ as above. Subsequently the procedure to obtain the radial velocity from Equation (4.71) is the same and yields the previous result.

Problems

4.16 Figure 4.3 shows a rotating 'rigid' wire in the form of an Archimedes spiral $r = k\phi'$ with $k > 0$. The spiral is rotating clockwise as indicated.

(a) Write the holonomic constraint in terms of inertial plane-polar coordinates (r, ϕ).

(b) Use the constraint to write the particle Lagrangian in a radial subspace. Find the Jacobi integral h in this subspace. Note its value if $\dot{r} = 0$ at $r = 0$.

(c) From the Jacobi integral (or otherwise) show that the radial motion satisfies the equation

$$\frac{\dot{r}^2}{c^2}\left(1 + \frac{r^2}{k^2} + \left(\frac{\Omega r^2}{kc}\right)^2\right) - 2\left(\frac{\Omega^2 r^2}{c^2}\right)\left(\frac{\Omega r^2}{kc}\right)\frac{\dot{r}}{c} - \left(\frac{\Omega^2 r^2}{c^2}\right)\left(1 - \frac{\Omega^2 r^2}{c^2}\right) = 0. \tag{4.75}$$

Note that as $k \to \infty$ the result of the example in the text is recovered.

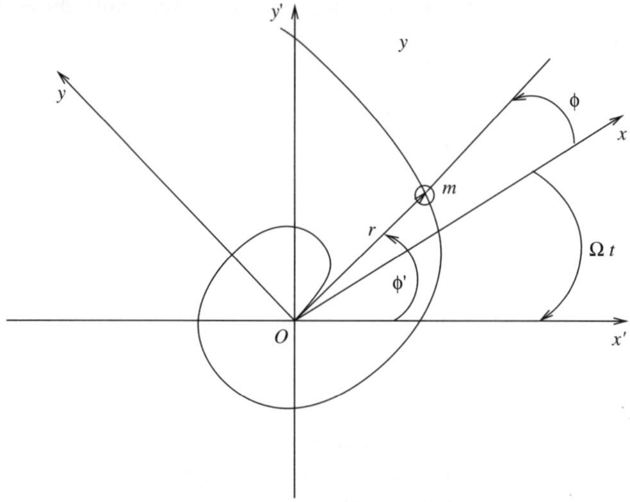

Figure 4.3 The figure shows a wire in the form of a Archimedes spiral $r = k\phi'$ in a rotating (primed) system of coordinates. A bead of rest mass m slides on the wire as shown. The primed frame rotates in a clockwise direction relative to an inertial frame (unprimed) with the angular velocity Ω. The x axes of the two systems coincide at $t = 0$

(d) Write an expression for $\dot{r} = v_r$ at $r = c/\Omega$ using the previous expression and show that it has a maximum of $c/\sqrt{2}$ for $k = \sqrt{2}(c/\Omega)$.

4.17 Consider a wire in the form of a dipolar field line so that in spherical polar coordinates $r = r_e \sin^2\theta$. The polar axis is in the plane of the wire and tangent to it at the origin. Hence r_e is the equatorial radius of the wire. The wire is in rotation about the polar axis according to $\dot{\phi} = \Omega$. Write the Lagrangian for a particle of mass m that is sliding without friction on the wire in the θ subspace. Hence show that the Jacobi constant is $h = \gamma mc^2(1 - (\Omega^2 r_e^2/c^2)\sin^6\theta$. Use this integral to study the motion of the particle in θ if (i) $\Omega r_e/c < 1$; and (ii) $\Omega r_e/c > 1$. Take $\dot{\theta} = 0$ at $\theta = 0$.

Until this point we have mainly used the formulation of relativistic mechanics that most resembles classical mechanics; that is, we use coordinate time as a parameter of the particle trajectory, and so continue the classical 'three plus one split' of space-time. This becomes more cumbersome (although still practical) when forces are introduced in the next chapter, so we conclude this section with a proper space-time discussion of force-free motion. This is not necessary for practical purposes, but it is an introduction to the formalism of general relativity without engaging the theory of gravity directly.

We begin by recalling the form of the action that we inferred at the beginning of this subsection (Equation (4.49)). We choose now $F = v_a v^a/2$ and write $(c = 1)$

$$\mathcal{S} = -m \int_1^2 \frac{v_a v^a}{2} ds. \tag{4.76}$$

To obtain a simple result from this form we do not use generalized coordinates, but rather only Galilean coordinates. This is because changing from the covariant to the contravariant form of the velocity (or vice versa) introduces the metric tensor. In non-Galilean coordinates the metric tensor is a function of the coordinates, and this dependence would introduce considerable complication (as we shall see below).

However, in the variation of the action we must remember that there is a constraint, namely $v_a v^a = 1$. This may be written as the 'semi-holonomic constraint' [1]:

$$df = v_a dx^a - ds = 0, \tag{4.77}$$

which form is familiar in rolling-without-slipping problems in classical mechanics. Under a virtual variation (a change at fixed s indicated by the operator δ) this is simply $\delta f = v_a \delta x^a = 0$. To effect the variation we incorporate the constraint as

$$\delta \mathcal{S} - \int_1^2 \lambda \delta f ds = 0. \tag{4.78}$$

Moreover in Galilean coordinates (since only then $\delta(v_a v^a) = 2 v_a \delta v^a$ during the variation of $\mathcal{S}$) and remembering that $\delta v^a = (d/ds)\delta x^a$ this implies

$$\frac{d(m v_a)}{ds} = \lambda v_a. \tag{4.79}$$

This is after integrating by parts in the varied action $\mathcal{S}$, and remembering to keep the endpoints fixed. We note that had we kept $F(v_a v^a/2)$ an arbitrary function, the only difference in this last expression would be a factor F' (the symbol $()'$ indicates derivative with respect to the argument) on the left. This does not change the subsequent argument.

Now we apply the constraint in its non-varied form by multiplying this last equation by v^a. The right-hand side is simply λ, while the left-hand side is zero by the orthogonality of four-acceleration and four-velocity (which amounts to differentiating the constraint) and the invariance of m. Consequently $\lambda = 0$, and so in Galilean coordinates

$$p^a = constant. \tag{4.80}$$

Therefore a free particle in Galilean coordinates has four-momentum as an integral of the motion. We know the four-momentum to include both the energy and the three-momentum. The combination contained in the modulus will also be an integral. This can be quite practical in the discussion of collisions between force-free particles.

We turn now to find an expression in space-time for the motion of a free particle, which is similar to that found in three-space in Equation (4.56). It suffers from the same calculational impracticality that we noted for Equation (4.56), but it has the merit of being correct for a generalized metric of any sort. This allows the use of Einstein's theory of gravity.

We use the action in its simplest form (Equation (4.50)), but with $ds = \sqrt{g_{ab} dq^a dq^b}$ and g_{ab} is a function of the generalized coordinates $\{q^k\}$. A straightforward virtual variation of the coordinates yields as the variation of the action ($c = 1$)

$$-m\delta \int_1^2 ds = -m \int_1^2 \delta(\sqrt{g_{ab} dq^a dq^b}) \tag{4.81}$$
$$= -m \int_1^2 ds \frac{1}{2}\left(\frac{\partial g_{ab}}{\partial q^c}\delta q^c \frac{dq^a}{ds}\frac{dq^b}{ds} + g_{ab}\frac{d\delta q^a}{ds}\frac{dq^b}{ds} + g_{ab}\frac{dq^a}{ds}\frac{d\delta q^b}{ds}\right).$$

We now integrate the last two terms by parts while keeping the endpoints fixed, and then change the dummy indices so that δq^c factors each term. Since δq^c is arbitrary, this yields

$$\frac{1}{2}\left(\frac{d\left(g_{cb}\frac{dq^b}{ds}\right)}{ds} + \frac{d\left(g_{ac}\frac{dq^a}{ds}\right)}{ds} - \frac{\partial g_{ab}}{\partial q^c}\frac{dq^a}{ds}\frac{dq^b}{ds}\right) = 0. \tag{4.82}$$

This expands to become (again rearranging the dummy indices)

$$g_{ca}\frac{d^2 q^a}{ds^2} + \frac{1}{2}\left(\frac{\partial g_{ca}}{\partial q^b} + \frac{\partial g_{cb}}{\partial q^a} - \frac{\partial g_{ab}}{\partial q^c}\right)\frac{dq^a}{ds}\frac{dq^b}{ds} = 0. \tag{4.83}$$

Multiplying this last equation by g^{dc} gives the result as

$$\frac{d^2 q^d}{ds^2} + \Gamma^d_{ab}\frac{dq^a}{ds}\frac{dq^b}{ds} = 0, \tag{4.84}$$

where the Christoffel symbol of the second kind in four-space is

$$\Gamma^d_{ab} \equiv \frac{g^{dc}}{2}\left(\frac{\partial g_{ca}}{\partial q^b} + \frac{\partial g_{cb}}{\partial q^a} - \frac{\partial g_{ab}}{\partial q^c}\right). \tag{4.85}$$

The earlier form (Equation (4.83)) has the advantage of employing only the covariant form of the metric. If we write the Christoffel symbol of the first kind as

$$\Gamma_{c,ab} \equiv \frac{1}{2}\left(\frac{\partial g_{ca}}{\partial q^b} + \frac{\partial g_{cb}}{\partial q^a} - \frac{\partial g_{ab}}{\partial q^c}\right), \tag{4.86}$$

then the equation of motion is simply

$$g_{ca}\frac{d^2 q^a}{ds^2} + \Gamma_{c,ab}\frac{dq^a}{ds}\frac{dq^b}{ds} = 0. \tag{4.87}$$

These equations give us a general means of computing the motion of a force-free particle (including the motion in a constrained subspace) in general coordinates. In the absence of machine assistance, the calculation is more tedious and indirect than the three plus one Lagrangian approach presented above. However, many programs allow such calculations presently. The Christoffel symbols are identically zero in Galilean coordinates so that $d^2x^d/ds^2 = 0$ and the particle is free or unaccelerated for inertial observers. The Christoffel symbols allow us in principle to compute 'apparent' or 'inertial' forces, which appear in non-inertial coordinates.

Although computationally awkward, this formulation has the advantage of allowing the metric to be a function of all four coordinates, so that we may work coherently in four-space when using general coordinate transformations. However, in Minkowski space-time under only boost transformations and spatial rotations, this is not necessary. In that case $g_{oo} = 1$ while $g_{oi} = 0$. This reduces Equation (4.84) to the previous (4.56).

However, in the geometry of Gauss and Riemann, spaces of any dimension are characterized by their metrics, which may be quite general functions of all coordinates. The formalism corresponds to ours if the most general non-inertial coordinates are used to describe space-time. In geometry ds is again the actual distance in the space, and the process by which we have derived Equation (4.84) amounts to finding the shortest path between two points in the space. Such paths are 'geodesics', and Equation (4.84) is the 'geodesic equation'. We see therefore that free particles follow the geodesics as described in generalized coordinates.

Should the particle be a photon, one must use a parameterization different from s as this is zero along the path. This may be a parameter in space or time. Such paths are referred to as 'null geodesics' and they are often most readily found from the explicit condition $ds = 0$.

Generalized transformations from Minkowski space can also result in interesting space-time metrics. Let us briefly return to our rotating disc. We choose to use non-inertial coordinates (r, ϕ') where $\phi = \phi' + \Omega t$. These are rotating plane-polar coordinates and we may consider ϕ' to be measured by disc observers. We know that if we synchronize clocks on the disc by diagonalizing the metric then we obtain the local tangential inertial frame. These cannot be joined together around the disc without a cut in the disc time. However, we do not insist on inertial observers here, and we may regard the use of rotating plane-polar coordinates as simply a generalized transformation.

The metric (Equation (4.28)) becomes in the new coordinates by direct substitution

$$ds^2 = dt^2(1 - \Omega^2 r^2) - 2\Omega r^2 d\phi' dt - dr^2 - r^2(d\phi')^2, \tag{4.88}$$

so that $g_{oo} = 1 - \Omega^2 r^2$, $g_{11} = -1$, $g_{22} = -r^2$ and $g_{02} = g_{20} = -\Omega r^2$. We leave as a Problem the derivation from Equation (4.87) of the equations of motion in the form (after one integration)

$$\frac{d^2 r}{ds^2} = r\left(\frac{d\phi'}{ds} + \Omega\frac{dt}{ds}\right)^2 \equiv r\left(\frac{d\phi}{ds}\right)^2, \tag{4.89}$$

$$L = r^2\left(\frac{d\phi'}{ds} + \Omega\frac{dt}{ds}\right) \equiv r^2\frac{d\phi}{ds}, \tag{4.90}$$

where L is an integral of the motion. To complete the set of equations one must append the metric in the form (or use the zero component of (4.83))

$$1 = \left(\frac{dt}{ds}\right)^2 - r^2\left(\frac{d\phi'}{ds} + \Omega\frac{dt}{ds}\right)^2 - \left(\frac{dr}{ds}\right)^2. \tag{4.91}$$

These equations describe the trajectory of a force-free particle in rotating coordinates (r, ϕ'). We have indicated the presence of the inertial ϕ only to emphasize the simplicity of inertial coordinates. The discussion is continued in the Problems.

Problems

4.18 **(a)** Derive the equations of motion of a particle sliding without friction on a disc using rotating plane-polar coordinates. Equations (4.83) are most convenient. The metric coefficients are summarized in the text and only the index values $c = 1, 2$ are required for what is given there.

(b) Consider the geodesic equation when the *index* $c = 0$. In order to be compatible with the preceding equations, what quantity must be conserved? You may want to recall the classical Jacobi constant in a rotating frame.

(c) Show that $dr(s)/ds$ and hence dt/ds may be found explicitly from the equations of motion.

4.19 **(a)** For pure radial motion on the disc ($d\phi' = 0$), show that the motion reduces to that of Example 4.4 using equations (a) and (c) of Problem (4.18) when $dr/ds = 0$ at $r = 0$.

(b) By using the metric in rotating coordinates directly, show that $(dt/ds)^2 = (1 + (dr/ds)^2)/(1 - \Omega^2 r^2)$. Together with equation (a) of Problem 4.18 in the form $d^2r/ds^2 = (\Omega^2 r(1 + (dr/ds)^2)/(1 - \Omega^2 r^2)$ (which may be integrated directly), solve again for $\dot{r}^2$. Show that it is the same as that for a bead sliding on a rotating straight rigid wire as discussed in a previous example. Initial conditions are again $dr/ds = 0$ at $r = 0$.

4.3.2 Collisions Between Free Particles

A 'collision' between point particles is necessarily an idealization. For there to be a finite cross-section there must be a region associated with each particle where some force repels any incoming object. In such a region the interacting particles are not free. However, if there are no long range forces, the particles are free both before the collision and after the collision. During the collision the colliding particles form a single system whose momentum is conserved in the absence of external forces (Equation (4.37)). Thus, when regarding the relation of the scattered state to the incoming state, we may regard the total four-momentum of all the interacting particles to be the same. The central tracks of Figure 4.4 show the non-relativistic collision of an incoming fast α-particle with a nearly stationary nucleus of helium (α-particle plus two electrons). The non-relativistic limit of Equation (4.94) can be used to predict the observed scattering at right angles.

After a collision between real particles such as nuclei or sub-atomic particles, there is not necessarily the same set of particles as existed before the collision. This depends on the 'energy' (e.g. defined in the zero momentum frame - see below for definition). Such collisions are *inelastic* and the momentum sum may be over different particles before

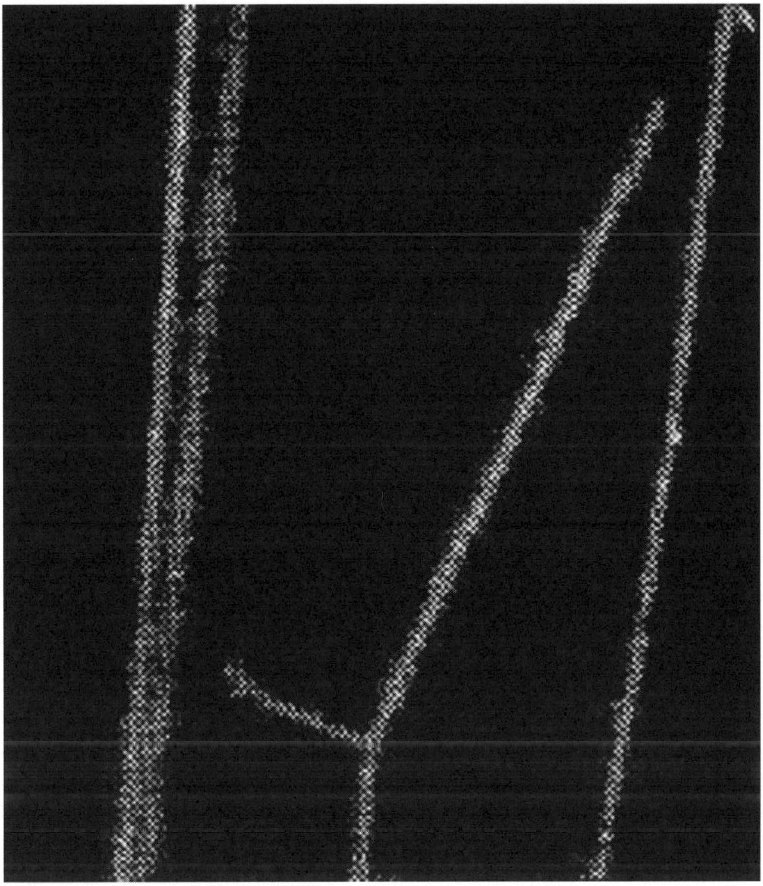

Figure 4.4 *This is a cloud chamber photograph of an α particle (an energetic nucleus of helium) colliding with a relatively stationary helium atom. The mass of the α particle and that of the helium atom are essentially equal. The collision is not relativistic. The collision has occurred in the plane of the photograph and the post collision tracks of the two nuclei are at right angles to one another. Nevertheless momentum is conserved.* Source: *Reproduced by permission of the Royal Society. Plate 2 from N. Feather (1933) Collisions of alpha-particles with fluorine nuclei.* Proceedings of the Royal Society A, ***141****, 194–209*

and after the collision. At lower energies purely *elastic* collisions are relevant and these are defined as those in which all particles maintain their 'identities' (i.e. properties, essentially the mass for a neutral point particle) throughout the collision.

We may write the momentum sum in Galilean coordinates, remembering Equation (4.62), as (the index n counts the interacting particles)

$$\begin{aligned} \Sigma_n p^a &= \Sigma_n \begin{pmatrix} \mathcal{E}/c \\ \mathbf{p} \end{pmatrix} \\ &\equiv P^a. \end{aligned} \tag{4.92}$$

Here the sum is over all the particles and $\mathbf{p}$ is the relativistic three-momentum (4.40). In general coordinates this would be replaced by dq^i/ds, but we will not have occasion to use these. For brevity we will refer to the time component of P^a as U/c and the spatial components as $\mathbf{P}$. Each of these is equal to the sums indicated in Equation (4.92).

Equation (4.92) is not independent of reference frame, and indeed there may be some doubt as to whether transforming it as a four-vector corresponds to the sum of the p^a in different frames of reference. This is because a fixed instant in one frame, say the frame of O, for all of the measurement events of the $\mathbf{p}^a$ will correspond to different instants in another frame of reference O'. However, O can choose to measure the p^a along a 'surface' of constant t' by measuring different particles at the appropriate different times. This sum will be the same as over a surface of constant t by conservation of the p^a. Even if a collision occurs, the total momentum remains conserved. But the sum over the surface $t' = constant$ is just what is achieved by summing the transformed p'^a. Consequently the sum has the same physical significance in each frame when transformed as a four-vector.

Much of the art in calculating relativistic collisions lies in a judicious choice of inertial frame. One common choice is the 'zero momentum frame' (denoted by subscript 0 where necessary, and by the letters ZP in the text). This is possible because P^a is a time-like vector, and like any time-like vector, the spatial part may be rendered zero by choice of inertial frame. The simple proof follows.

In an arbitrary frame we take the z axis to lie along the direction of $\mathbf{P}$. Then the transverse components are zero. The parallel component transforms in another frame moving in standard configuration with relative velocity u_0 to $P_0 = \gamma(u_0)(P - u_0 U/c^2)$ (P is the z component). To set this equal to zero we take $u_0/c = cP/U$, which is possible if the ratio is less than one. This is always true for a time-like vector since by definition the time component is always greater than the space component. A space-like vector may always be reduced to a pure space-like vector by a similar argument. The zero momentum frame is the relativistic analogue of the classical centre of mass frame $u_{cm} = \Sigma mv/\Sigma m$, as may be seen by taking the low-velocity limit of u_0 to obtain u_{cm}.

For a two-particle collision another useful reference frame is one in which one of the particles is at rest. Suppose that this is particle 1 and the other is particle 2. Then using covariant components $p_a(1) = (m_1 c, \mathbf{0})$ and $p_a(2) = (\gamma(u_{21})m_2 c, -\gamma(u_{21})m_2\mathbf{u}_{21})$, where u_{21} is the velocity of m_2 relative to m_1. The quantity $p(1)^a p(2)_a = c^2 m_1 m_2 \gamma(u_{21})$ is a Lorentz invariant.

For an elastic collision, $p(1)^a + p(2)^a$ will be the same before and after the collision. Let A^+ denote a quantity A after the collision. Then 'squaring' (i.e. the scalar product with itself) this sum before and after the collision yields

$$m_1^2c^2 + m_2^2c^2 + 2p(1)^a p(2)_a = m_1^2c^2 + m_2c^2 + 2p^+(1)^a p^+(2)_a. \tag{4.93}$$

Hence the Lorentz invariants are equal for both input and scattered states, namely

$$p^+(1)^a p^+(2)_a = p(1)^a p(2)_a. \tag{4.94}$$

If we calculate these invariants in the frame in which particle 1 is at rest before and after the collision, then we see from the form of the invariant in these frames given above that $u'_{21} = u_{21}$. That is, the relative speed of two particles is unchanged during an elastic collision. This does not require the directions to be the same, however.

Another quantity that is not frame-dependent is the modulus of P^a, namely

$$\frac{U^2}{c^2} - \mathbf{P}^2 = \frac{U_0^2}{c^2}, \tag{4.95}$$

where U_0 is the energy in the ZP frame. For one particle this reduces to Equation (4.62). This quantity also has the property of being unchanged by a collision so that

$$\frac{(U^+)^2}{c^2} - (\mathbf{P}^+)^2 = \frac{U^2}{c^2} - \mathbf{P}^2 = \frac{U_0^2}{c^2}. \tag{4.96}$$

In the single particle version of this Equation (4.62), we see that if a particle were to have zero rest mass then $\mathcal{E} = cp$ where $p \equiv |\mathbf{p}|$. Classically there are no such particles since mass is the only mark of existence. Neither are there particles of negative mass, which would certainly behave in a very strange way (and incidentally destroy the weak principle of equivalence). Quantum mechanically, however, there are particles of zero rest mass, namely 'photons' or light quanta. Until relatively recently neutrinos were considered to be another example, but this is now no longer the case (see [3] and references therein).

A photon has the quantum of energy $\mathcal{E}_\nu = h\nu$ where h is Planck's constant of action. Experimentally it behaves as though it had momentum $h\mathbf{k}$, where $\mathbf{k} \equiv (\nu/c)\mathbf{n}$ and $\mathbf{n}$ is the direction of motion. Consequently the photon four-momentum becomes

$$p_\nu^a = h\begin{pmatrix} k \\ \mathbf{k} \end{pmatrix}. \tag{4.97}$$

Here $k \equiv |\mathbf{k}| = \nu/c$ and it is sometimes convenient to write $p_\nu^a = hk^a$ with an obvious definition of the null four-vector, k^a.

We are now equipped to study collisions of greater or lesser complexity and we shall content ourselves with a few simple examples. Consider first two particles of equal mass that enter into an elastic collision. In the zero momentum frame O' one particle (say particle 1) may be regarded as travelling in the negative z' direction towards the origin with speed $v' = v$ while the other particle (say particle 2) travels in the positive z' direction towards the origin with speed $v' = v$. After the collision in O' they must have the same relative speed in O'. Moreover their actual z velocities must be again v and $-v$ in order to conserve four-momentum. For true point particles only a head-on collision is possible. Thus the expected solution has particle 1 rebounding along the positive z' axis with speed v, while particle 2 rebounds along the negative z' axis also with speed v. That is all there is to be said in this frame.

In a reference frame O in which particle 1 is at rest initially, the ZP frame of O' moves along the positive z axis with speed $u = v$. Particle 2 moves with speed v along the z' axis relative to the world of O', and hence the initial incident velocity of particle 2 in the frame of particle 1 is given by the parallel velocity transformation (Equation 2.46) as $w \equiv 2v/(1 + v^2/c^2)$. After the collision by the same formulae and argument, particle 2 is at rest and particle 1 travels along the z axis with the speed w (the incident speed of particle 2).

However, we do not know or care about the detailed physics in the collision zone. There is a more general possibility permitted by the conservation laws for the asymptotic scattered state. The post collision in the frame of O' may take place at an angle θ' to the z' axis for particle 1 and at an angle $\pi - \theta'$ to the z' axis for particle 2, always in the same plane. According to the velocity aberration formula of Equation (3.20) we have, in the frame of O (i.e. that of particle 1 before the collision), for the angle that particle 1 makes with the z axis (note that $u = v$ and $v' = v$)

$$\tan\theta_1 = \frac{\sin\theta'}{\gamma(v)(\cos\theta' + 1)}, \tag{4.98}$$

while that for particle 2 is ($u = v$, $v' = v$ and $\pi - \theta' \leftarrow \theta'$)

$$\tan\theta_2 = -\frac{\sin\theta'}{\gamma(v)(\cos\theta' - 1)}. \tag{4.99}$$

Hence also by taking the product, there is the tidy constraint on the scattered angles namely $\tan\theta_1 \tan\theta_2 = 1/\gamma(v)^2$. It is clear that for very energetic collisions (large $\gamma(v)$), the collision reduces to the head-on example. Otherwise there is an array of possible scattering angles, each weighted by the solid angle available to θ', that can be used to calculate a cross-section for scattering.

An inelastic collision does not yield a set of scattered products equal to the set of input products (at least the masses have changed due to the excitation of internal states). Such collisions are the material of high-energy physics where the details of the collision zone are of prime importance. The forces involved produce an array of fundamental particles, whose detection and characterization allow the forces to be studied. The particles are the quanta of the forces.

The present analysis allows us to impose energy-momentum constraints on what products are possible. Suppose that U_0 in Equation (4.95) is the total energy in the ZP frame of the products. Then in an inertial frame that 'creates' the input particles, say the Laboratory frame, we have

$$U^2 = U_0^2 + c^2\mathbf{P}^2. \tag{4.100}$$

One can minimize this requirement in two ways. In the first instance we can arrange that the Laboratory frame be the zero-momentum frame of the input particles. This sets $\mathbf{P}^2 = 0$. This is the technique used in collider accelerators such as the Large Hadron Collider (LHC) at CERN. In that machine, protons are collided against protons exactly as described in the preceding elastic example. However, the energies delivered to the collision (at the time of writing about 1.2 TeV,[2] but soon to be capable of 7 TeV and ultimately ≥ 14 TeV) are many hundreds of times larger than the rest-mass energy of the proton. Thus protons may be transmuted into a vast array of products by a highly inelastic process.

The second way to minimize $U = U_0$ is to minimize the total energy of the products. Their rest masses cannot be reduced, but their kinetic energies can be. If then the particles

[2] One TeV is 10^{12} eV, or about 544 proton rest masses.

are actually at rest in the ZP frame, we have the laboratory frame 'threshold energy' equal to the sum of the product rest masses.

Collisions can become complex as the number of particles involved increases. We will not explore these complications here, but rather content ourselves with one or two simple but important applications. Some are given in the Problem set, but we consider two problems involving photons in the examples below.

Example 4.5

Consider the historically important problem of Compton Scattering. A photon is incident along the z axis on an electron at rest with mass m_e. The collision is inelastic since some energy is taken from the photon by the scattered electron. This changes its direction in general as well as its frequency. In fact, four-momentum requires that

$$p_e^a + p_\nu^a = (p_e^+)^a + (p_\nu^+)^a. \tag{4.101}$$

Compton's experiment was sensitive to the frequency and the angle of the scattered photon. We can isolate these quantities first by making use of the fact that $p_e^a(p_e)_a = m_e^2c^2 = (p_e^+)^a(p_e^+)_a$, since both moduli can be evaluated in the rest frame of the electron where $(p_e)_a = (m_e c, \mathbf{0})$. In addition we take the interesting quantity $(p_\nu^+)^a$ to the other side of the conservation equation and 'square' the equation (dot product of each side with itself [4]) to find (the photon four-momentum is null)

$$p_\nu^a(p_e)_a - (p_\nu^+)_a(p_\nu^a + p_e^a) = 0. \tag{4.102}$$

Hence by performing the indicated calculations using Equation (4.97) one finds

$$k^+ \quad k = \frac{h}{m_e c}k^+k(1 - \cos\theta), \tag{4.103}$$

where θ is the angle that the scattered photon makes with the incident direction. In terms of the photon frequency this becomes the Compton law

$$\nu^+ - \nu = -\frac{h}{m_e c^2}\nu^+\nu(1 - \cos\theta). \tag{4.104}$$

This may also be written in terms of the wavelength of the photon as

$$\lambda^+ - \lambda = \frac{h}{m_e c}(1 - \cos\theta). \tag{4.105}$$

The quantity $h/(m_e c) \equiv \lambda_c$ is the Compton wavelength ($\approx 2.426 \times 10^{-12}$ m) of the electron. This law is well verified experimentally. Equation (4.104) shows that the fractional frequency change (difference divided by the new frequency) is proportional to $h\nu/(m_e c^2)$. The electron rest mass energy is about 511 keV, which is in the energy range of gamma photons. Photons with higher energy than this are capable of producing a different kind of inelastic collision, namely electron-positron pair creation. The pure Compton effect is best seen in X-rays, having photon energies of a few keV.

For 'inverse Compton scattering' (wherein the electron gives energy to the photon) we consider a very relativistic electron moving in the negative z direction where it encounters the same photon as before moving in the positive z direction. One might have thought that the result of such a collision could be obtained by transforming Compton scattering to a frame O' that is moving towards the origin with the electron speed. However, this does not allow for the Döppler shift of the photon relative to the oncoming electron. Instead we again use Equation (4.102) to find (see Problem) for the energies of the initial and final photons

$$\frac{\mathcal{E}\mathcal{E}^+}{c^2}(1-\cos\theta) = \gamma_e m_e \mathcal{E}(1+v_e/c) - \gamma_e m_e \mathcal{E}^+(1+\cos\theta(v_e/c)). \tag{4.106}$$

This can be rearranged into the form

$$\frac{\mathcal{E}^+}{\mathcal{E}} = \frac{2}{1+\frac{v_e}{c}\cos\theta + \frac{\mathcal{E}}{\gamma_e m_e c^2}(1-\cos\theta)}. \tag{4.107}$$

Hence with the initial photon energy $\mathcal{E}$ much less than the electron energy $\gamma_e m_e c^2$, the right-hand side becomes $\approx 2/(1+(v_e/c)\cos\theta)$. Since $v_e/c = \sqrt{1-1/\gamma_e^2} \approx 1 - 1/(2\gamma_e^2)$, we see that the head-on collision in which $\cos\theta = -1$ (i.e. the photon rebounds along the negative z axis) transfers the maximum energy to the photon. In fact this becomes

$$\frac{\mathcal{E}^+}{\mathcal{E}} \approx 4\gamma_e^2. \tag{4.108}$$

This effect is remarkably present in astronomical sources. Relativistic electrons are common as perceived at the Earth in cosmic rays (see [5]) and low-energy photons are also abundant in many sources. Moreover, microwave photons, relics from the beginning of cosmological time, fill the Universe. In many radio sources electrons with $\gamma_e = 10^4$ are present, and these can inverse Compton scatter on 10^9 Hz radio photons to produce 4×10^{17} Hz X-rays.

We conclude this section and this chapter with a selection of Problems that are of practical importance. We turn in the next chapter to a survey of the real forces that one may incorporate into Lorentzian relativity.

Problems

4.20 Show that Equation (4.107) follows from Equation (4.102) by inserting the appropriate four-momenta.

4.21 Show that a free electron and a free positron cannot annihilate into a single freely propagating photon. Note that one can work in the ZP frame of the initial electron and positron. A positron is the anti-electron and has the same mass.

4.22 Relative to some inertial frame, a particle is seen to have four-momentum p^a. An observer O' is also moving relative to this frame with the four-velocity u^a. Show that the energy $\mathcal{E}'$ of the particle as measured by O' is given by the invariant $p^a u_a$.

4.23 Ultrarelativistic neutrinos are emitted by a supernova explosion at a distance D from Earth. Assume that they are all emitted at the same coordinate time (Earth and the supernova are assumed to be at rest in an inertial frame), but with varying energies $\mathcal{E}_\nu$. Derive a relation between $\mathcal{E}_\nu$ and the coordinate time when the neutrinos of this energy are detected at the orbit of the Earth (of radius zero compared with D). Suppose that the neutrinos have a mass of $m_\nu = 10\,eV$ and an energy of $\mathcal{E}_\nu = 10\,MeV$ and that $D = 10^4$ light-years. What is the arrival coordinate time if they were launched at $t = 0$? What is the difference in arrival time between a neutrino of mass $20\,eV$ and that of mass $10\,eV$.

References

1. Goldstein, H., Poole, C. and Safko, J. (2002) *Classical Mechanics*, Addison-Wesley, San Francisco.
2. Henriksen, R.N., and Rayburn, D.R. (1971) *Monthly Notices of the Royal Astronomical Society*, **152**, 323.
3. SNO Collaboration (2008) *Physical Review Letters*, **101**, 111301.
4. Rindler, W. (2006) *Relativity*, Oxford University Press, New York.
5. Irwin, J.A. (2007) *Astrophysics: Decoding the Cosmos*, John Wiley & Sons Ltd., Chichester, p. 156.

5

Electromagnetic Theory in Space-Time

Let there be light!

Genesis, Revised Standard Bible

5.1 Prologue

In the previous chapter we introduced a four-vector relation between a hypothetical four-force and the four-acceleration, namely Equation (4.37). In that form a prescribed four-acceleration allows a corresponding four-force to be calculated, which force is taken to be provided by the constraints that produce the given acceleration. This corresponds to a usage in normal Newtonian physics that arises often in engineering applications. The interpretation more frequently used in fundamental physics may be written as an apparently trivial modification

$$\frac{dp^a}{ds} = K^a. \tag{5.1}$$

However, this modified equation implies a physical theory that yields the appropriate four-vector force in any given circumstance. By substituting it into this version of the equation of motion, the four-acceleration of a 'test particle' (i.e. one that does not disturb the force prescription) may be calculated at each point in space-time. Subsequently the particle's four-momentum trajectory may be found by integration from initial conditions.

We recall also that the force prescription should be such that $u_a K^a = 0$ in order that the inertial mass of a particle be conserved. Until this point we have only found this to be true for inertial forces as expressed in Equation (4.84). This is because $\Gamma^d_{ab} u_d u^a u^b = 0$ in any coordinates. This follows by differentiating $g_{ca} u^c u^a = 1$ partially with respect to q^b while recalling that $u_c u^c = 1$, whence $u_c \partial u^c / \partial q^b = 0$ (see Problem).

Practical Relativity: From First Principles to the Theory of Gravity Richard N. Henriksen

Problem

5.1 Following the suggestions given in the text, show that $\Gamma^d_{ab}u_d u^a u^b = 0$ so that the inertial mass is constant. Equation (4.84) should be written in terms of the four-velocity $u^a = dq^a/ds$, and it helps to recognize when various terms in an expression are the same due to summation.

In this chapter and the next we introduce two forces that fit this prescription. The electromagnetic force is the subject of this chapter, and the gravitational force is the subject of the next chapter. In the first case the relativistic applications are eminently practical, since charged fundamental particles are readily made to approach the speed of light both in the laboratory and in the cosmos. We do not include the radiation reaction force, however, so we must remain in the regime where this is negligible (see e.g. [1]).

The strong gravitational forces necessary to require major deviations from Newtonian gravity are not encountered in our terrestrial experience. Such effects are encountered cosmically, in gravitationally collapsed objects and indeed in the Universe itself. First-order corrections to Newtonian gravity are readily apparent on Earth, however. These are manifest in global positioning systems, in anomalous planetary orbits, and in the distorted view of the cosmos produced by foreground gravitational lenses. We shall therefore present some rudiments of gravitational theory in the next chapter.

5.1.1 Electromagnetic Four-Potential

The theory of the electromagnetic field is a vector-field theory that is specified by Clerk-Maxwell's equations. It is coupled to matter through 'charge', e, which like inertial mass is taken to be a scalar invariant. The electromagnetic force acts on elementary charges through the Lorentz prescription, which we shall have to modify slightly in the relativistic limit.

We will have to decide on a system of units that includes both mechanical and electromagnetic quantities. The SI (Système Internationale) units are generally recommended. However, these have the very real disadvantage, from the perspective of notational simplicity, of attributing different units to the electric field **E** and the magnetic induction field **B**. Moreover the SI system is most useful when condensed media are present. The magnetic and electric properties of such media require that a distinction be made between the net magnetic field **H** and the magnetic induction (flux density) **B**, as well as between the electric displacement **D** and the electric field **E**.

In this chapter we will not be considering the presence of condensed media, since these are rarely moving relativistically. Moreover we choose to use a system where **E** and **B** have the same units. These units are called 'Gaussian' and they use the 'cgs' (centimetre, gram, second) mechanical system of units. Clerk-Maxwell's equations in an inertial frame in these units are

$$\nabla \cdot \mathbf{B} = 0, \qquad \nabla \cdot \mathbf{E} = 4\pi\rho_e, \tag{5.2}$$

$$\nabla \wedge \mathbf{E} = -\frac{1}{c}\frac{\partial \mathbf{B}}{\partial t} \qquad \nabla \wedge \mathbf{B} = \frac{1}{c}\frac{\partial \mathbf{E}}{\partial t} + \frac{4\pi}{c}\mathbf{j}. \tag{5.3}$$

In these equations **j** is the three-current due to the motion of elementary charges. When only test charges are being considered, this may be set equal to zero together with the charge density ρ_e. With these substitutions the equations describe the electromagnetic field ('light') *in vacuo*.

Relative to SI units, charges are measured in statcoulombs, where $c/10$ statcoulombs is one coulomb. The magnetic induction is measured in gauss where 10^4 gauss is one Tesla. The electric field is measured in statvolts per cm, where $c/10^6$ volts per metre is one statvolt per cm. One statvolt is therefore $c/10^8$ volts. In this system, **E** and **B** have the same dimensions, but not the same units. We observe that setting $c = 1$ in the vacuum equations yields a theory free of physical constants in the absence of boundaries.

We turn now to the identification of the electromagnetic four-vector. It is shown in standard electromagnetic texts [1] that in Galilean coordinates ($c = 1$)

$$\eta^{cd} \frac{\partial^2 A^b}{\partial x^c \partial x^d} = 4\pi j^b, \tag{5.4}$$

where the column vectors are

$$A^b = \begin{pmatrix} \Phi \\ \mathbf{A} \end{pmatrix}, \quad j^b = \begin{pmatrix} \rho_e \\ \mathbf{j} \end{pmatrix}. \tag{5.5}$$

The inverse of η_{cd} is written η^{cd} and is identical to η_{cd}.

The operator

$$\eta^{cd} \frac{\partial^2}{\partial x^c \partial x^d} \equiv \frac{\partial^2}{\partial t^2} - \nabla^2, \tag{5.6}$$

is the 'wave operator' in Galilean coordinates. We have seen that it is an invariant under boosts and rotations (i.e. linear transformations), although it is not under curvilinear transformations. To write the operator in general coordinates we shall have to recall the notion of 'true derivative' introduced at the end of Chapter 1 for curvilinear spatial coordinates. This concept must be extended to space-time.

The 'potentials', consisting of the scalar function Φ and the three-vector function **A**, result from integrating the first of Equations (5.3) and satisfying automatically the first of Equations (5.2) so that

$$\mathbf{E} = -\nabla\Phi - \frac{\partial \mathbf{A}}{\partial t} \tag{5.7}$$

$$\mathbf{B} = \nabla \wedge \mathbf{A}. \tag{5.8}$$

Finally, to obtain Equation (5.4), we must apply the Lorentz gauge condition to the potentials. This condition becomes in Galilean coordinates

$$\eta_{ab} \frac{\partial A^a}{\partial x^b} = 0. \tag{5.9}$$

Equation (5.4) written *in vacuo* (yielding the speed of 'light' *in vacuo*) requires that the column vector transforms linearly under transformations between Galilean coordinates, since the operator is an invariant and it computes zero which is a scalar. We are free to take this linear transformation to be that of a four-vector between Galilean coordinates, provided that in the end this assumption agrees with experiment. This would also require that the column vector j^b is a four-vector.

However, it is instructive to argue in the other sense, by first determining that j^b is physically a four-vector. To this end we introduce several tools. The first useful concept is the generalization of the permutation (epsilon) symbol (Chapter 1) to space-time in the form

$$\epsilon^{abcd}, \tag{5.10}$$

which is defined to be $+1$ for an even permutation of $\{0, 1, 2, 3\}$, -1 for an odd permutation, and zero if any two indices are the same. Note that the permutations interchange under the interchange of any two indices so that the sign changes also under this operation.

We want this object to transform as a tensor under Galilean transformations, and yet it should have the same numerical character in all inertial frames. This is possible because under such transformations

$$\epsilon'^{\ abcd} = \frac{\partial x'^a}{\partial x^e}\frac{\partial x'^b}{\partial x^f}\frac{\partial x'^c}{\partial x^g}\frac{\partial x'^d}{\partial x^h}\epsilon^{efgh}. \tag{5.11}$$

The expression on the right of this last equation is just $\epsilon^{abcd} det(\underline{\underline{\mathbf{S}}}\,\underline{\underline{\mathbf{L}}})$ by the definition of a determinant. We have written the transformation between Galilean coordinates $\{x'\}$ and $\{x\}$ as a boost $\underline{\underline{\mathbf{L}}}$ followed by a spatial rotation $\underline{\underline{\mathbf{S}}}$, so we might have written $\underline{\underline{\mathcal{L}}}$. However $det(\underline{\underline{\mathbf{S}}}\,\underline{\underline{\mathbf{L}}}) = det(\underline{\underline{\mathbf{S}}})det(\underline{\underline{\mathbf{L}}})$ which is equal to $+1$ under rotations without reflection ($det(\underline{\underline{\mathbf{S}}}) = +1$), since in addition $det(\underline{\underline{\mathbf{L}}}) = +1$ (see Equation (2.27)). Consequently the permutation symbol can be regarded as a tensor under transformations of Galilean coordinates and

$$\epsilon'^{\ abcd} = \epsilon^{abcd}. \tag{5.12}$$

The indices are to be lowered and raised with the Galilean metric $\underline{\underline{\eta}}$, so that $\epsilon_{abcd} = -\epsilon^{abcd}$.

Our second useful result now follows readily. The element of space-time volume in Galilean coordinates d^4x is itself an invariant. Thus (expand the second line in standard configuration)

$$\begin{aligned} d^4x' &\equiv dx'^{\ 0}dx'^{\ 1}dx'^{\ 2}dx'^{\ 3} \\ &= \frac{\partial x'^{\ 0}}{\partial x^a}\frac{\partial x'^{\ 1}}{\partial x^b}\frac{\partial x'^{\ 2}}{\partial x^c}\frac{\partial x'^{\ 3}}{\partial x^d}\epsilon^{abcd}dx^0dx^1dx^2dx^3 \\ &= \epsilon^{0123}det(\underline{\underline{\mathbf{S}}}\,\underline{\underline{\mathbf{L}}})d^4x = d^4x. \end{aligned} \tag{5.13}$$

The desired transformation behaviour of j^a now follows from the scalar nature of charge together with the scalar four-volume d^4x. The number of charges in a spatial three-volume d^3x is $\rho_e d^3x$ in any inertial frame. Since d^4x is invariant, this implies that ρ_e should transform like the time component of a four-vector. Similarly we may count the number of charges crossing the surface normal to the three axis in a given time interval as $j^3dx^1dx^2dt$ in any inertial frame. For this to be invariant j^3 must transform as the 'three' spatial component of a four-vector. The argument may be repeated for the other two components of the charge current. This proves that the column vector j^b is in fact a four-vector under Galilean transformations, and hence so is the column vector A^b.

We have thus arrived at the four-vector that describes a given electromagnetic field. Its relation to its sources is given by Clerk-Maxwell's equations, but we shall find later a more succinct expression of these. Let us turn now to the dynamics of a charge in a given electromagnetic field A^b.

We seek the four-vector action for a particle of charge e and mass m that is coupled to a given electromagnetic field. We know the action for a free particle (Equation (4.76)) and we expect it to be present as we let the electromagnetic field (effectively A^b) go to zero. Thus the part of the action that couples the particle to the electromagnetic field must be added to the action of a free particle. Almost the only scalar available that couples the field and the charged particle is $e\mathbf{v}_b A^b$, where the charge e is the coupling constant and its presence guarantees zero coupling for zero charge. One might ask why not $ea_b A^b$, where a_b is the four-acceleration? Ultimately the choice is such as to agree with experiment. However, the scalar involving the charge's three-velocity linearly has the merit of guaranteeing that the three-vector $\mathbf{A}$ will not contribute to the Hamiltonian (see below), as it should not. Experimentally we recall, the magnetic field cannot work on a charge.

We are therefore led to propose the action for a charge in an electromagnetic field in the four-vector form

$$\mathcal{S} = -\int_1^2 ds \left(m\frac{\mathbf{v}_b \mathbf{v}^b}{2} + e\mathbf{v}_b A^b \right), \tag{5.14}$$

where the minus sign is placed before the coupling term in order to have a potential term like $-e\Phi$ in the three-Lagrangian (see below). Dimensionally with $c = 1$, the product eA^b has the dimensions of mass so that the dimensions in Equation (5.14) are coherent. For the subsequent variation of this action to be simple, we will work as before in Galilean coordinates. Moreover there is the same non-holonomic constraint

$$\mathbf{v}_a dx^a - ds = 0 \tag{5.15}$$

to apply.

We will divide our attention in the following subsections between the classical three plus one treatment of the dynamics and the fully four-vector approach. The classical approach allows all of the techniques of advanced classical mechanics to be used, and is therefore eminently practical.

5.2 Lagrangian Dynamics of an Electromagnetic Charge

We pass from the four-vector action (5.14) to the normal three-Lagrangian by writing $ds = dt/\gamma(\mathbf{v})$ along the particle path. In addition we apply the constraint $\mathbf{v}_a \mathbf{v}^a = 1$ immediately, and also dispense with the factor $1/2$ that is only present in order to avoid varying twice the free-particle four-action. Thus we deduce the action in the classical form of a time integral over the three-Lagrangian, as (expanding Equation (5.14))

$$\mathcal{S} = \int_1^2 dt \left(-m\sqrt{1 - \mathbf{v}^2} - e\Phi + e\mathbf{v} \cdot \mathbf{A} \right) \equiv \int_1^2 L dt. \tag{5.16}$$

The three-vectors in this equation are constituted from the physical components, which are computed as in Chapter 1 by taking the scalar product with the ortho-normal base vectors. Conventional units may be restored by inserting c so as to make L have the dimensions of energy. Recall that $e\Phi$ and $e\mathbf{A}$ do have this dimension in conventional units.

Our first task is to calculate the Lagrange equations of motion. We work in Galilean coordinates so that the canonical momentum $\mathbf{P} \equiv \partial L/\partial \mathbf{v}$ becomes ($c = 1$)

$$\mathbf{P} = m\gamma\mathbf{v} + e\mathbf{A} \equiv \mathbf{p} + e\mathbf{A}, \tag{5.17}$$

and ∇L follows as

$$\nabla L = -e\nabla\Phi + e\nabla(\mathbf{v}\cdot\mathbf{A}). \tag{5.18}$$

Using a standard vector calculus identity to expand $\nabla(\mathbf{v}\cdot\mathbf{A})$ in this last equation, and recalling that $d\mathbf{A}/dt = \partial\mathbf{A}/\partial t + (\mathbf{v}\cdot\nabla)\mathbf{A}$, the Lagrange equations give the three-vector form of the equations of motion (see Problem) as

$$\frac{d\mathbf{p}}{dt} = e(\mathbf{E} + \mathbf{v}\wedge\mathbf{B}). \tag{5.19}$$

This vector form can be resolved along any set of base vectors for convenience.

We recognize the familiar Lorentz force on a charged particle in this last expression. The only difference from the non-relativistic expression is the gamma factor that modifies the scalar inertial mass in the particle momentum $\mathbf{p}$. This correspondence might have been expected since the theory of the electromagnetic field is the archetypal relativistic theory. This force and this equation of motion can be used to give the standard results concerning the motion of a charged particle, although occasionally it is more useful to work directly from the Lagrangian. We shall demonstrate such cases in the Examples and the Problems.

The form of the Lorentz force is valid in any inertial frame, where $\mathbf{v}$ is the particle velocity in that frame. This implies that all vectors in the expression must transform between inertial frames so as to maintain the same form. The equation of motion can be used to find how the electric and magnetic fields transform, given what we know about velocity and acceleration transformations. We delay this discussion, but only after such an analysis can the student's question 'Whose velocity?' be answered definitively.

The Hamiltonian will yield the energy of the particle from $H = \mathbf{v}\cdot\mathbf{P} - L$ which gives

$$H = m\mathbf{v}^2\gamma + m/\gamma + e\Phi = m\gamma + e\Phi \equiv \mathcal{E}. \tag{5.20}$$

Although this is numerically equal to the energy, it does not have the canonical functional form $H(\mathbf{r},\mathbf{P})$. This canonical form is necessary in order to use Hamilton's equations or the Hamilton-Jacobi equation to solve dynamic problems (see e.g. [2]).

We recall that the canonical equations take the following form in Galilean coordinates

$$\begin{aligned}\frac{d\mathbf{P}}{dt} &= -\nabla H(\mathbf{r},\mathbf{P},t)\\ \frac{d\mathbf{r}}{dt} &= \frac{\partial H(\mathbf{r},\mathbf{P},t)}{\partial\mathbf{P}},\end{aligned} \tag{5.21}$$

where we have used an obvious notation for the gradient in Galilean momentum space.

The Hamilton-Jacobi method also needs the canonical form as it requires solving the Hamilton-Jacobi equation for 'Hamilton's principal function' $\mathcal{S}$ as

$$H(\mathbf{r}, \mathbf{P}, t) + \frac{\partial \mathcal{S}}{\partial t} = 0, \tag{5.22}$$

where we must substitute for the canonical momentum

$$\mathbf{P} = \nabla \mathcal{S} \tag{5.23}$$

to complete the partial differential equation. Its solution yields $\mathcal{S}(\mathbf{r}, \boldsymbol{\alpha}, t)$ if $\boldsymbol{\alpha}$ comprises the three constants of integration (a single particle). The solution for the dynamics is completed by using the auxiliary condition

$$\boldsymbol{\beta} = \frac{\partial \mathcal{S}}{\partial \boldsymbol{\alpha}}. \tag{5.24}$$

The components of the vector $\boldsymbol{\beta}$ are also constants. The vectors $\boldsymbol{\alpha}$ and $\boldsymbol{\beta}$ may be regarded as defining initial conditions, or in fact any complete set of constants of integration.

To obtain the Hamiltonian in canonical form we employ Equations (5.17) and (5.20) to write

$$(\mathbf{P} - e\mathbf{A})^2 = m^2\gamma^2\mathbf{v}^2$$
$$(H - e\Phi)^2 = \gamma^2 m^2.$$

We substitute $\mathbf{v}^2 = 1 - 1/\gamma^2$ in the first of these equations, and use the result to eliminate $m^2\gamma^2$ in the second equation. In this way we find the canonical form (restoring c temporarily to obtain familiar conventional units)

$$H = e\Phi + \sqrt{m^2c^4 + \left(\mathbf{P} - \frac{e\mathbf{A}}{c}\right)^2}. \tag{5.25}$$

In this series of equations the square of a three-vector is a short expression for the scalar product with itself. It is now a straightforward exercise to deduce the Lorentz force and equation of motion from Hamilton's equations (see Problem).

Problems

5.2 Following the procedure outlined in the text, show that the Lagrange equations in Galilean coordinates yield the three-vector equation of motion with the Lorentz force (5.19).

5.3 Use Hamilton's canonical equations together with the definitions of $\mathbf{P}$ and $\mathbf{p}$ and the vector calculus expansion of $\nabla(\mathbf{P} - e\mathbf{A})^2$, to show that they yield Equation (5.19).

If we take the scalar product of Equation (5.19) with $\mathbf{v}$ and multiply both sides by $\gamma(\mathrm{v})$ we obtain $(m/2)d(\gamma^2\mathbf{v}^2)/dt = e\gamma\mathbf{v}\cdot\mathbf{E}$. Then by using $\mathbf{v}^2 = 1 - 1/\gamma^2$ there follows the energy equation

$$\frac{d(m\gamma)}{dt} = e\mathbf{v}\cdot\mathbf{E}. \tag{5.26}$$

In the four-vector version of the equation of motion we can expect this result to follow from the time component.

We are now able to look at typical charged particle motion in *given* electromagnetic fields. These will be studied in the Problems and Examples.

Example 5.1

We look first at a relativistic particle moving in a uniform and constant magnetic field $\mathbf{B}$ that may be regarded as pointing in the direction of the z axis. We treat this problem by both the vector and scalar methods starting with the equation of motion (5.19). Figure 5.1 indicates the qualitative behaviour.

(i) *Vector Solution*

Equation (5.19) tells us that the momentum parallel to $\mathbf{B}$, namely $\mathbf{p}_{\|}$ is constant. We can thus choose an inertial frame that is moving along the z axis such that $\mathbf{v}_{\|}$ is zero, although we must remember that in an arbitrary frame this translational motion is possible. We shall see later that this transformation does not change the magnetic field. Equation (5.26) tells us moreover that the particle energy, that is γ, is constant and hence so is $|\mathbf{v}_{\perp}|$. Consequently from the perpendicular components of Equation (5.19)

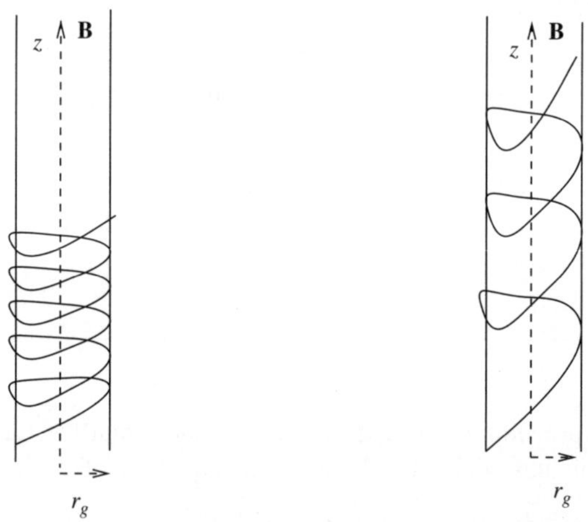

Figure 5.1 *The sketch shows the motion of a charged particle in an inertial frame containing only a uniform magnetic field. On the left there is only a small velocity parallel to the magnetic field so the pitch angle is large. On the right the parallel velocity is larger and the pitch angle is smaller. The radius of gyration* $\mathrm{r_g}$ *defines a cylinder on which the particle trajectory is wound*

we obtain (replacing $c = 1$ by conventional units)

$$\frac{d\mathbf{v}_\perp}{dt} = \boldsymbol{\omega}_G \wedge \mathbf{v}_\perp, \tag{5.27}$$

where the vector gyro-angular frequency is $\boldsymbol{\omega}_G \equiv -(e\mathbf{B})/(m\gamma c)$. This equation implies, according to Equation (1.118), that the particle velocity is constant in axes rotating with the gyro-angular velocity. That is, the perpendicular velocity is a constant vector in pure rotation in this particular inertial frame. In a general inertial frame we recall that the motion will be helical with a pitch angle $\tan\psi = v_\parallel/v_\perp$.

To find the particle position vector $\mathbf{r}_\perp$, we use $\mathbf{v}_\perp = d\mathbf{r}_\perp/dt$ to integrate Equation (5.27) to give

$$\frac{d\mathbf{r}_\perp}{dt} = \boldsymbol{\omega}_G \wedge \mathbf{r}_\perp, \tag{5.28}$$

where the vector constant of integration is dropped in order to guarantee that $|\mathbf{r}_\perp|$ is also a vector in pure rotation. Taking magnitudes in this last equation one finds

$$r_\perp^2 = \left(\frac{v_\perp}{\omega_G}\right)^2. \tag{5.29}$$

These results imply a gyro-frequency for an electron of $(2.8 \times 10^6 B(gauss)/\gamma)$ cycles per second, in the right-hand sense about the magnetic field if $e < 0$. The gyro-radius is conveniently written using the definition of ω_G as

$$r_\perp = \frac{cp_\perp}{eB} \approx \frac{\mathcal{E}_\perp}{eB}. \tag{5.30}$$

The approximation is for a very relativistic particle. Numerically this becomes $\mathcal{E}_\perp(eV)/(300B(gauss))$ cm. An ultra-high-energy cosmic ray (UHCR) with energy 10^{20}eV can traverse the galaxy undeviated since the mean galactic field of the galaxy is of the order of 10^{-6} gauss.

(ii) *Lagrangian Solution*

A uniform and constant magnetic field may be described by the vector potential $\mathbf{A} = (\mathbf{B} \wedge \mathbf{r})/2$, as may be confirmed by calculating the curl of this expression using the appropriate vector-calculus identity. The magnetic field may be supposed to lie once again along the z axis. Using cylindrical polar coordinates ($\{r, \phi, z\}$) with this vector potential, the Lagrangian (5.16) becomes ($(\dot{}) \equiv d/dt()$)

$$L = -m(\sqrt{1 - \dot{r}^2 - r^2\dot{\phi}^2 - \dot{z}^2}) + \frac{eBr^2\dot{\phi}}{2}. \tag{5.31}$$

There is no dependence on t in this expression, so that $\mathcal{E} = m\gamma$ is an integral of the motion. Moreover, there is no dependence on z so that $p_z = m\gamma\dot{z}$ is also an integral of the motion. We again take $\dot{z} = 0$ by a choice of reference frame. The remaining Lagrange equations in r and ϕ yield respectively

$$\begin{aligned} \frac{d(m\gamma\dot{r})}{dt} - m\gamma\dot{\phi}^2 r - eBr\dot{\phi} &= 0, \\ (m\gamma\dot{\phi}r^2) &= \ell, \end{aligned} \tag{5.32}$$

where ℓ is an integral of the motion (angular momentum) corresponding to the absence of ϕ in the Lagrangian. Using the energy integral $\dot{r}^2 + \dot{\phi}^2 r^2 = 1 - 1/\gamma^2$, and combining this with the angular momentum integral, yields (restoring c temporarily)

$$\dot{r}^2 + \frac{\ell^2}{m^2 c^2 \gamma^2 r^2} = 1 - \frac{1}{\gamma^2}. \tag{5.33}$$

This equation has an inner turning point but no outer turning point. It describes the nearly straight-line motion of a particle past the origin when the particle energy is so high that the field does not deflect the particle before it leaves the magnetized region. The trapped solutions are always appropriate for an infinite magnetized region (cf. the vector solution) and these require $r = constant$. With $r = constant = r_\perp$ we find immediately from the radial equation that $\dot{\phi} = -eB/(m\gamma c)$, where $\dot{\phi}$ is positive in the right-hand sense relative to the z axis. Using the angular momentum equation with $\ell = r_\perp p_\perp$ one obtains once more Equation (5.30).

The Hamilton-Jacobi method is excessively awkward for this rather degenerate problem, so we will apply this method in a different context below.

Problem

5.4 Consider a uniform and constant electric field **E** parallel to a uniform and constant magnetic field **B**. Both fields may be taken to be parallel to the z axis. A particle of charge e and inertial mass m starts from rest on the z axis and nearly at the space-time origin (but $r \neq 0$) of an inertial system.

(a) Find the integrals of the motion and so show that $\gamma = 1 + (eE/mc^2)z$ and $P_\phi = m\gamma\dot{\phi}r^2 + eBr^2/2c$, where P_ϕ is the canonical azimuthal momentum. We look for solutions that are trapped in cylindrical radius so that $r = constant = r_\perp$.

(b) Show that the particle gyrates with the angular frequency $\dot{\phi} = \omega_G$ on a cylinder of radius $r_\perp^2 = (-2cP_\phi/eB)$.

(c) Integrate the z equation of motion to find $\gamma^2 = 1 + ((eE/mc)t)^2$ and hence $z = (\gamma - 1)(mc^2/eE)$.

(d) Show that the coordinate time t and the proper time of the particle t' are related through $(eE/mc)t = \sinh(eE/mc)t'$. Hence, by recalling the definition of ω_G show finally that $z = (mc^2/eE)(\cosh(eE/mc)t' - 1)$ and note the trajectory.

A well-known problem because of its special character (see subsection on the four-vector electromagnetic theory in this chapter) is dealt with in the next example. It may be efficiently treated using either one of the vector or Lagrangian approaches, but we will leave these to the Problems. We proceed with the Hamilton-Jacobi formulation as an example of the power of this method when only particle trajectories are required.

Example 5.2

We consider a uniform and constant electric field E directed along the x axis of an inertial system, and a uniform and constant magnetic field B directed along the y axis. The special character of this Problem exists when $E = B$, so we are imagining a kind of 'frozen' plane wave, which in reality can only be a rather local approximation. To simplify the algebra we start the particle at the space-time origin with zero velocity.

The electrostatic potential can be taken as $\Phi = -Bx$ and the vector potential is similarly $\mathbf{A} = -Bx\widehat{\mathbf{z}}$. Consequently the Hamiltonian in canonical form is $H = -eBx + \sqrt{m^2 + (\mathbf{P} + eBx\widehat{\mathbf{z}})^2}$. One sees that $\partial H/\partial t = 0$ and so the energy is an integral of the motion. This gives $\mathcal{E} \equiv \gamma m - eBx = m$, because of our initial conditions, and so

$$\gamma = 1 + \frac{eB}{m}x. \tag{5.34}$$

Moreover $\partial H/\partial y = 0$, which requires $P_y = p_y = constant$ by the canonical equations. We shall set this constant equal to zero in our initial conditions, which amounts to a choice of reference frame. Finally, $\partial H/\partial z = 0$, which implies $P_z = p_z - eBx = constant$. However, the motion will take place in part in the z direction, so we must not set this constant equal to zero. Similarly, we must not set $\mathcal{E} = m$ which is its initial value, before finding the solution. This is because these constants appear in the solution for $\mathcal{S}$ and must be free to be varied until the solution is found.

The Hamilton-Jacobi equation for this problem is

$$-eBx + \sqrt{m^2 + (\nabla\mathcal{S} + eBx\widehat{\mathbf{z}})^2} + \frac{\partial \mathcal{S}}{\partial t} = 0. \tag{5.35}$$

We assume separation of variables in the form $\mathcal{S} = -\mathcal{E}t + P_z z + W_x(x)$ since $\partial\mathcal{S}/\partial t = \mathcal{E}$ and $P_z = \partial\mathcal{S}/\partial z$. Substituting the assumed form of the solution and rearranging we obtain

$$\left(\frac{dW_x}{dx}\right)^2 = \mathcal{E}^2 - P_z^2 - m^2 + 2eB(\mathcal{E} - P_z)x, \tag{5.36}$$

and so the solution for the action is ($\mathcal{E}$ is both here and in the Lagrangian approach the *total* energy)

$$\mathcal{S} = -\mathcal{E}t + P_z z \pm \int^x dx\sqrt{\mathcal{E}^2 - P_z^2 - m^2 + 2eB(\mathcal{E} - P_z)x}. \tag{5.37}$$

We would obtain $x(t)$ and hence $\gamma(t)$ and $p_z(t)$ by writing explicitly a constant $\beta_t = \partial\mathcal{S}/\partial\mathcal{E}$. But this is too much information for the purpose of finding the particle orbit in space. For this we calculate another constant $\beta_z = \partial\mathcal{S}/\partial P_z$ which yields the orbit directly as

$$\beta_z = z \mp \int^x dx \frac{(P_z + eBx)}{\sqrt{\mathcal{E}^2 - P_z^2 - m^2 + 2eB(\mathcal{E} - P_z)x}}. \tag{5.38}$$

Now we may apply our initial conditions and set $P_z = 0$, $\beta_z = 0$ (starting at the origin) and $\mathcal{E} = m$, after which one finds the orbit from the preceding equation as (for a positive charge and on restoring c dimensionally)

$$z = \sqrt{\frac{2eB}{9mc^2}}x^{3/2}. \tag{5.39}$$

Thus $x/z \to 0$ as $x \to \infty$. Hence the charge moves ultimately wholly in the direction given by $\mathbf{E} \wedge \mathbf{B}$, namely the z direction. We may see this in terms of the velocity by noting that $p_x = dW_x/dx = \sqrt{2eBmx}$ while $p_z = eBx$. Dividing by γ from Equation (5.34), we see that $v_z \to 1$ while $v_x \to 0$ for large x.

This problem is usefully solved by vector methods using Equations (5.19) and (5.26), which procedure we sketch in the following Problem, leaving gaps for the reader.

Problems

5.5 Solve the problem discussed in Example (5.2) using Equations (5.19) and (5.26). The secret is to proceed systematically so we will sketch one approach. The initial conditions are as in the example and we reduce the problem to two dimensions by taking $p_y = 0$ in our chosen inertial frame, since it is constant in any frame (show). The magnitudes E and B are again equal. We set $c = 1$.

- **(a)** Combine the energy equation and the z equation of motion to find (here the energy is the *particle* energy and is not constant) $\mathcal{E} = m + p_z$ and $p_z = eBx$.
- **(b)** Recall $p_a p^a = m^2$ and combine with previous results to find $\mathcal{E} = m + p_x^2/(2m)$ and $p_z = p_x^2/(2m)$.
- **(c)** Use the results above and the x equation of motion to show that $(eB/m)t = p_x/m + (1/6)(p_x/m)^3$. Multiplying the equation of motion by $\mathcal{E}$ gives a simple integration. One can now find agreement with the conclusions in the example.
- **(d)** By noting that $\mathbf{v} \equiv \mathbf{p}/\mathcal{E}$ and changing the independent variable from t to p_x, show that the orbit parameterized by p_x is $z = p_x^3/(6eB)$ and $x = p_x^2/(2eBm)$ (c must be put back by dimensional analysis to recover conventional units). The orbit is the same as that found by the Hamilton-Jacobi method.

5.6 Solve the same Problem using Lagrangian methods. This gives the easiest solution if the trick of changing the variable to p_x in $\gamma m(dz/dt) = p_z$ is used. Remember that the energy here is the total energy and is an integral of the motion.

We conclude this section with two examples that are of some practical interest in themselves, but they may also demonstrate methods of examining 'wakefield' particle acceleration in plasmas (e.g. [3] plus many earlier references).

We consider a given low-frequency electromagnetic plane wave that is incident on a charged test particle. We have considered this previously in the section on Compton scattering, but in that context the energy is high enough that individual photons are studied. At low energies or frequencies, it is the phase of the wave that is important. Our purpose is to study the induced motion of the charge as a function of the phase of the wave. If we were to calculate the radiation from the accelerated charge, this would also be a scattering problem. But we ignore this as being beyond the scope of this chapter. In any case it should be negligible dynamically unless the electron reaches very high energies.

Let the plane wave move in the direction $\widehat{\mathbf{k}}$, which for convenience we identify with the z axis. Thus $\widehat{\mathbf{k}} = \widehat{\mathbf{z}}$ where convenient. The wave fields lie therefore in the $x - y$ plane. In a Coulomb gauge the electrostatic potential of the plane wave (far from its sources) is zero, while the electric and magnetic wave fields follow directly from the vector potential $\mathbf{A}$ as

$$\mathbf{E} = -\frac{\partial \mathbf{A}}{\partial t}, \tag{5.40}$$

$$\mathbf{B} = \widehat{\mathbf{k}} \wedge \mathbf{E}. \tag{5.41}$$

The wave vector potential is a function only of the phase $\xi = ct - z$, and we shall set $c = 1$ throughout the argument for convenience.

It is useful to use the Hamilton-Jacobi method since this allows us to find the particle orbit directly. But we emphasize that all three approaches discussed in this section have their merits. We illustrate two and leave the vector approach to a Problem.

Using Equations (5.25), (5.22) and (5.23), the Hamilton-Jacobi equation for this Problem may be written, after squaring and rearranging, as

$$\left(\frac{\partial \mathcal{S}}{\partial t}\right)^2 - \left(\frac{\partial \mathcal{S}}{\partial z}\right)^2 = m^2 + (\nabla_\perp \mathcal{S} - e\mathbf{A}(\xi))^2. \tag{5.42}$$

We have set $\nabla_\perp = \widehat{\mathbf{x}}\partial_x + \widehat{\mathbf{y}}\partial_y$ for brevity, in this last expression. However, there is no dependence on either x or y in the Hamiltonian so that the canonical momenta are constants. We take the constants to be respectively α_x and α_y so that

$$P_x \equiv \frac{\partial \mathcal{S}}{\partial x} = p_x + eA_x(\xi) = \alpha_x, \tag{5.43}$$

$$P_y \equiv \frac{\partial \mathcal{S}}{\partial y} = P_y + eA_y(\xi) = \alpha_y. \tag{5.44}$$

We are justified therefore in looking for a solution in the form $\mathcal{S} = \boldsymbol{\alpha}_\perp \cdot \mathbf{r}_\perp + \mathcal{S}_{tz}(t,z)$ where $()_\perp$ indicates the x, y components of the vector.

The right-hand side of Equation (5.42) is dependent only on ξ, while on the left we have a function of t, z. This suggests that we change variables from t, z to $\eta \equiv t + z$ and ξ. Performing this change and expanding yields

$$4\frac{\partial \mathcal{S}_{tz}}{\partial \eta}\frac{\partial \mathcal{S}_{tz}}{\partial \xi} = m^2 + \boldsymbol{\alpha}_\perp^2 - 2e\boldsymbol{\alpha}_\perp \cdot \mathbf{A} + e^2\mathbf{A}^2. \tag{5.45}$$

For a separated solution of the additive kind in these coordinates (which corresponds to independent degrees of freedom) $\mathcal{S} = \boldsymbol{\alpha}_\perp \cdot \mathbf{r}_\perp + \mathcal{S}_\eta(\eta) + \mathcal{S}_\xi(\xi)$ we must have

$$\frac{d\mathcal{S}_\eta}{d\eta} = -\frac{\alpha_\eta}{2} \tag{5.46}$$

$$-2\alpha_\eta \frac{d\mathcal{S}_\xi}{d\xi} = m^2 + \boldsymbol{\alpha}_\perp^2 - 2e\boldsymbol{\alpha}_\perp \cdot \mathbf{A} + e^2\mathbf{A}^2. \tag{5.47}$$

The arbitrary constant on the right of Equation (5.46) is chosen in the form shown for convenience below. It remains arbitrary.

We may now write the complete solution for the action from the Hamilton-Jacobi equation as

$$\mathcal{S} = -\frac{\alpha_\eta}{2}\eta + \boldsymbol{\alpha}_\perp \cdot \mathbf{r}_\perp + \mathcal{S}_\xi(\xi), \tag{5.48}$$

where, on recalling that additive constants are unimportant, we find from Equation (5.47)

$$\mathcal{S}_\xi(\xi) = -\frac{m^2 + \boldsymbol{\alpha}_\perp^2}{2\alpha_\eta}\xi + \frac{e\boldsymbol{\alpha}_\perp}{\alpha_\eta} \cdot \int_0^\xi \mathbf{A}d\xi - \frac{e^2}{2\alpha_\eta}\int_0^\xi \mathbf{A}^2 d\xi. \tag{5.49}$$

We employ Equation (5.24) to find the orbits in the various planes. One finds the orbit in the $x - y$ plane from $\boldsymbol{\beta}_\perp = \partial\mathcal{S}/\partial\boldsymbol{\alpha}_\perp$, that is

$$\boldsymbol{\beta}_\perp = \mathbf{r}_\perp - \frac{\boldsymbol{\alpha}_\perp}{\alpha_\eta}\xi + \frac{e}{\alpha_\eta}\int_0^\xi \mathbf{A}d\xi \tag{5.50}$$

where the betas are new constants. At $\xi = 0$ these give the location of the particle in the $x - y$ plane, and by choosing this to be the origin, $\boldsymbol{\beta}_\perp = \mathbf{0}$. Moreover we expect the motion in this plane to be periodic in ξ since the applied wave fields are such. We may ensure this by setting the canonical momenta $\boldsymbol{\alpha}_\perp = \mathbf{0}$. Then Equations (5.43) and (5.44) show that the time averages $\langle \mathbf{p}_\perp \rangle = 0$ are indeed zero. Hence

$$\mathbf{r}_\perp = -\frac{e}{\alpha_\eta}\int_0^\xi \mathbf{A}d\xi. \tag{5.51}$$

It may not be immediately obvious that this last statement agrees with Equations (5.43) and (5.44), but recall that $\mathbf{p} = md\mathbf{r}/ds$, where ds is the proper interval for the particle. Moreover $d\xi = dt - dz = dt(1 - v_z)$ and $dt = \gamma ds$. Hence there is a simple relation between $d\xi$ and ds, namely

$$d\xi = ds(\gamma - p_z/m). \tag{5.52}$$

But the energy of the system $\mathcal{E} = -\partial\mathcal{S}/\partial t$ coincides in this case with the particle energy γm, since $\Phi = 0$. Moreover the canonical momentum P_z coincides with p_z. Consequently from Equation (5.48), $\mathcal{E} = \alpha_\eta/2 - d\mathcal{S}_\xi/d\xi$ and $p_z = -\alpha_\eta/2 - d\mathcal{S}_\xi/d\xi$ $(\partial_z = -\partial_\xi)$ whence

$$\mathcal{E} - p_z = \alpha_\eta. \tag{5.53}$$

Writing the particle energy again as γm, this last equation gives $\gamma - p_z/m = \alpha_\eta/m$. Whence with $d\xi$ above we obtain

$$d\xi = ds\frac{\alpha_\eta}{m}. \tag{5.54}$$

This allows the Equations (5.43) and (5.44) to be integrated into the form of Equation (5.51). We have found the transverse orbit for any polarization of the wave as implied in the choice of **A**.

For the z motion we use again Equation (5.24) to write an arbitrary constant $\beta_\eta = \partial\mathcal{S}/\partial\alpha_\eta$. This becomes (with $\boldsymbol{\alpha}_\perp = \mathbf{0}$)

$$\beta_\eta = -\frac{\eta}{2} + \frac{m^2}{2\alpha_\eta^2}\xi + \frac{e^2}{2\alpha_\eta^2}\int_0^\xi \mathbf{A}^2 d\xi. \tag{5.55}$$

If the motion is taken to begin at $t = 0$ and $z = 0$ then $\beta_\eta = 0$, and the resulting equation gives the orbit $\eta(\xi)$. It is more useful as $z(\xi)$ which we may calculate as $z = (\eta - \xi)/2$. We then find the orbit (implicit in z and t) normal to the wave front as

$$z(\xi) = \left(\frac{m^2}{\alpha_\eta^2} - 1\right)\frac{\xi}{2} + \frac{e^2}{2\alpha_\eta^2}\int_0^\xi \mathbf{A}^2 d\xi. \tag{5.56}$$

This really completes the solution for the motion of a charge in a plane wave of arbitrary polarization, but we have not yet found the significance of α_η. With this objective in mind we calculate $p_z = \partial\mathcal{S}/\partial z$ explicitly to find ($\boldsymbol{\alpha}_\perp = 0$)

$$p_z = -\frac{\alpha_\eta}{2} + \frac{m^2}{2\alpha_\eta} + \frac{e^2}{2\alpha_\eta}\mathbf{A}^2. \tag{5.57}$$

This component has both a secular and an oscillatory component. We place ourselves in the inertial frame moving along the z axis in which $\langle p_z\rangle = 0$. Averaging Equation (5.57) requires a particular value of the constant α_η to satisfy the zero average, namely

$$\alpha_\eta^2 = m^2 + e^2\langle\mathbf{A}^2\rangle. \tag{5.58}$$

Substituting this back into Equation (5.57) gives for p_z

$$p_z = \frac{e^2}{2\alpha_\eta}(\mathbf{A}^2 - \langle\mathbf{A}^2\rangle). \tag{5.59}$$

Since $p_z = \alpha_\eta dz/d\xi$ we can integrate this equation directly to get $z(\xi)$, but we already have this in Equation (5.56). Finally, Equation (5.53) gives the energy of the particle at any phase. We leave special cases of particular polarizations to the Example and a Problem.

Example 5.3

Consider the motion of a charge excited by a linearly polarized plane wave. We find its motion in the frame in which $\langle p_z \rangle = 0$ so that we may use the general result of the text. Let $\mathbf{A} = -(E_x/\omega) \sin \omega\xi$, where E_x is the amplitude of the electric field. Then a careful calculation using Equation (5.56) gives the orbit as (restoring conventional units)

$$z = -\frac{e^2 E_x^2 c^3}{8\omega^3 \alpha_\eta^2} \sin 2\omega\xi, \tag{5.60}$$

where

$$\alpha_\eta^2 = m^2 c^2 + \frac{e^2 E_x^2 c^2}{2\omega^2}. \tag{5.61}$$

The shape of the orbit is indicated in Figure 5.2.

In this inertial frame the momentum is (e.g. from Equation (5.59))

$$p_z = \frac{e^2 E_x^2 c}{\omega^2 \sqrt{m^2 c^2 + e^2 E_x^2 c^2/(2\omega^2)}} (\sin^2 \omega\xi - 1/4), \tag{5.62}$$

from which the energy follows as $\mathcal{E} = \alpha_\eta + p_z$. One sees that the maximum energy at phase $\pi/2$ is $\approx eE_x c/\omega$ if relativistic. This is roughly $E_x \lambda$ eV for an electronic charge if the field amplitude is in volts per metre and the wavelength is in metres. For metre wavelengths one requires an amplitude of megavolts/metre to attain relativistic energies, and there is nothing to be gained over an electrostatic field of the same magnitude.

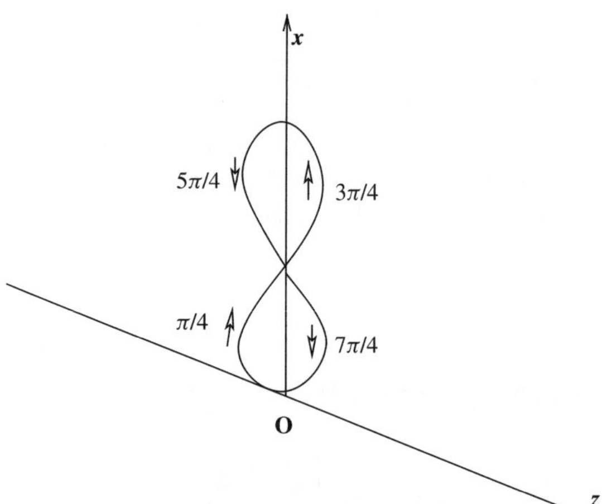

Figure 5.2 The sketch shows the motion in the x − z plane of a charge moving under the influence of a linearly polarized plane wave propagating in the z direction. The inertial frame is that of zero mean motion. The positions at different phases are indicated. The particle is unable to remain in phase, that is 'surf', with the wave. The wave moves at speed c so this is to be expected

Problems

5.7 Find the normal orbit of a charge in an elliptically polarized plane wave in a frame in which $\langle \mathbf{p} \rangle = 0$. You may use the general solution given in the text. If the polarization is described by $\mathbf{A} = E_x \cos(\omega\xi)/\omega\widehat{\mathbf{x}} + E_y \sin\omega\xi/\omega\widehat{\mathbf{y}}$, show in particular that ($\xi = t - z/c$)

$$z = \frac{e^2c^3(E_x^2 - E_y^2)}{8\omega^3\left(m^2c^4 + \frac{e^2c^2(E_x^2+E_y^2)}{2\omega^2}\right)} \sin 2\omega\xi. \qquad (5.63)$$

Discuss the situation for circular polarization. Show, based on Equation (5.57), how α_η determines the mean z displacement of the charge.

5.8 Solve for the motion of a charge in a plane wave using vector methods based on Equations (5.19) and (5.26). You can retain a general polarization.

Example 5.3 shows that electromagnetic plane waves incident on a charge at a given frequency generate motion at the first harmonic frequency. This frequency would be re-radiated by the charge. However, this bare wave-charge interaction does not accelerate the charge more efficiently than an applied electrostatic field. Very strong transient electric fields can be produced in plasmas by creating highly non-linear waves. These may be produced by laser irradiation and are currently described as ‘wake fields’. This process will not concern us here directly, but such experiments have found it useful for beaming purposes to impose a magnetostatic field on the plasma. This magnetic field creates a gyro-frequency in the system of charges and allows the possibility of resonance with an applied electromagnetic wave.

Because of the possibility of resonance, as a final discussion in this section we examine the action of a vacuum electromagnetic wave on a test charge in the presence of a uniform magnetic field. The Hamilton-Jacobi method is ill-suited to this problem, but the Lagrange procedure is straightforward. One must simply remember that in the presence of the wave there is an explicit time dependence in the Lagrangian, so that the energy is not an integral of the motion.

Once again we consider the plane wave to propagate along the z axis and to be described by the vector potential $\mathbf{A}(\xi)$. There is also a uniform and constant magnetic field $\mathbf{B}_s$ directed along the z axis, which may be derived from the vector potential $\mathbf{A}_s = B_s x\widehat{\mathbf{y}}$. The Lagrangian of the particle-field system is therefore

$$L = -m\sqrt{1 - \mathbf{v}^2} + e\mathbf{v}\cdot\mathbf{A} + eB_s x v_y. \qquad (5.64)$$

The canonical momentum in the y direction $\partial L/\partial v_y$ is an integral of the motion, say α_y, and $\alpha_y = p_y + eA_y + eB_s x$. The canonical momentum in the x direction is not constant, but the equation of motion is simply $dP_x/dt = eB_s v_y$. Hence

$P_x = eB_s y + \alpha_x$, where α_x is an integral of the motion, and $P_x = \partial L/\partial v_x = p_x + eA_x$. Collecting these two results yields

$$p_x = eB_s y - eA_x + \alpha_x \tag{5.65}$$

$$p_y = -eB_s x - eA_y + \alpha_y. \tag{5.66}$$

The z canonical momentum is equal to p_z and the equation of motion is

$$\frac{dp_z}{dt} = e\mathbf{v} \cdot \frac{\partial \mathbf{A}}{\partial z} \equiv -e\mathbf{v} \cdot \frac{\partial \mathbf{A}}{\partial \xi}. \tag{5.67}$$

The system energy is again the particle energy and satisfies $d\mathcal{E}/dt = -\partial L/\partial t$ which becomes $d\mathcal{E}/dt = -e\mathbf{v} \cdot (d\mathbf{A}/d\xi)$. Combining this with Equation (5.67) yields, as in the previous Equation (5.53) (energy is not changed by a magnetic field),

$$\mathcal{E} - p_z = \alpha_\eta. \tag{5.68}$$

This means that Equation (5.54) continues to apply. Consequently the z equation of motion becomes

$$\frac{dp_z}{d\xi} = -\frac{e}{\alpha_\eta}\mathbf{p}_\perp \cdot \frac{d\mathbf{A}}{d\xi}. \tag{5.69}$$

We proceed by solving Equations (5.65) and (5.66) directly using $\mathbf{p}_\perp = m d\mathbf{r}_\perp/ds$ and Equation (5.54). We adopt the complex variables $\zeta = x + iy$, $Y = A_x + iA_y$ and $\alpha = \alpha_x + i\alpha_y$ and find the solution as

$$\zeta = C_o e^{i\Omega\xi} - \frac{i\alpha}{eB_s} - e\frac{e^{i\Omega\xi}}{\alpha_\eta}\int_0^\xi Y(\xi)e^{-i\Omega\xi}\,d\xi, \tag{5.70}$$

where the 'gyro-frequency' is $\Omega \equiv -eB_s/\alpha_\eta$ and C_o is an arbitrary constant that we may take real. To resonate with an electron we apply a circularly polarized plane wave $Y(\xi) = (E_o/\omega)e^{i\omega\xi}$. Moreover we are most interested in the resonance condition, so we set $\omega = \Omega$. The solution (5.70) then becomes explicitly

$$x = C_o \cos\Omega\xi + \frac{eE_o}{\Omega\alpha_\eta}\xi\cos\Omega\xi - \frac{\alpha_y}{eB_s} \tag{5.71}$$

$$y = C_o \sin\Omega\xi + \frac{eE_o}{\Omega\alpha_\eta}\xi\sin\Omega\xi + \frac{\alpha_x}{eB_s}. \tag{5.72}$$

A somewhat tedious calculation from Equation (5.69) yields p_z and hence the energy. This is facilitated by expanding Equation (5.69) using Equations (5.65) and (5.66) written in vector form. After using Equation (5.54) we find

$$\frac{dp_z}{d\xi} = \frac{e^2}{2\alpha_\eta}\frac{d\mathbf{A}^2}{d\xi} - \frac{e}{\alpha_\eta}\boldsymbol{\alpha}_\perp \cdot \frac{d\mathbf{A}}{d\xi} + \frac{e^2 B_s}{\alpha_\eta}\left(x\frac{dA_y}{d\xi} - y\frac{dA_x}{d\xi}\right). \tag{5.73}$$

An integration using the solutions (5.71) and (5.72) yields (after restoring c, take ξ to have dimensions of time)

$$p_z = \frac{ce^2\mathbf{A}^2}{2\alpha_\eta} - 2\frac{e\boldsymbol{\alpha}_\perp \cdot \mathbf{A}}{\alpha_\eta} + C_o\frac{e^2B_sE_o}{\alpha_\eta}\xi + \frac{ce^2E_o^2}{2\alpha_\eta}\xi^2. \tag{5.74}$$

We see that there is secular acceleration with phase as the wave passes, with the dominant term going as ξ^2.

From the energy Equation (5.68) we obtain $\mathcal{E}^2 = \gamma^2m^2c^4 = (\alpha_\eta + cp_z)^2$ that is, recalling the square of the four-momentum, $m^2c^4 + p_\perp^2 = \alpha_\eta^2 + 2\alpha_\eta cp_z$. This allows the perpendicular energy to be found. However, with $\mathbf{p} = \mathbf{0}$ initially, $\alpha_\eta = mc^2$. Thus, returning to the total energy and using the dominant term in Equation (5.74), at large phase the energy varies as

$$\mathcal{E} \approx \frac{c^2e^2E_o^2}{2mc^2}\xi^2 + mc^2. \tag{5.75}$$

Taking the ratio of the first term to the second we have the pure number $e^2E_o^2c^2\xi^2/(2m^2c^4)$, which should be larger than 1 for relativistic acceleration. This is numerically $\approx 0.015E_o^2c^2\xi^2$, where E_o is in volts per metre and $c\xi$ is in metres. For metre wavelengths one requires a wave amplitude of the order of ten volts per metre to attain relativistic motion over a distance of several metres. However, the last term in Equation (5.74) must be dominant, which requires by comparison with the first term that $\xi \gg 1/\Omega$. Moreover there would be radiation loss from the gyrating particle due to small pitch-angle gyration that leads to synchrotron radiation [1].

5.2.1 Field Transformations Between Inertial Frames

Equation (5.19) often presents a puzzle to critical students, when it is introduced at an elementary level. The charge velocity that appears is relative to an inertial system of reference and therefore changes simply with the inertial perspective. In order for this force to remain a valid description for any inertial observer, it is clear that the fields must also change with the reference frame so as to maintain the same form. That this is the case is most readily seen in the four-vector treatment of the next section. However, that argument is rather formal, and tends to be confused with the necessity of a metric space-time. Consequently we present in this section a derivation of the three-vector transformations based directly on the necessity of preserving the three-vector force.

We consider two inertial frames O and O' in standard configuration, and we use Galilean coordinates. The O' frame is co-moving with a charged particle, so that the relative velocity $\mathbf{u}$ is also the particle velocity for O observers. We may take the component of the force Equation (5.19) parallel to the charge velocity $\mathbf{u}$ and write $d\mathbf{p}_\parallel/dt = e\mathbf{E}_\parallel$. A direct calculation of the derivative shows that each side of this equation is equal to $m\gamma(u)^3\mathbf{a}_\parallel$, where $\mathbf{a}_\parallel = d\mathbf{u}/dt$. Thus, recalling Equation (3.24) wherein we set $\mathbf{v} = \mathbf{u}$, we see that $\gamma(u)^3\mathbf{a}_\parallel = \mathbf{a}'_\parallel$. Consequently $e\mathbf{E}_\parallel = m\mathbf{a}'_\parallel$.

Equation (5.19) must also apply in the frame of the charge (i.e. for O' observers) so that $d\mathbf{p}'_\parallel/dt = e\mathbf{E}'_\parallel$. In this frame the calculation of the derivative gives $m\mathbf{a}'_\parallel$ since $\mathbf{v}'_\parallel = \mathbf{0}$. Equating the two expressions for $\mathbf{a}'_\parallel$ yields the transformation of the parallel electric field (i.e. to the relative velocity of the boost) as

$$\mathbf{E}'_{\parallel} = \mathbf{E}_{\parallel}. \tag{5.76}$$

This result allows us to see how a constant proper acceleration may be established for a charged particle. A uniform electric field parallel to the motion will by our result produce a constant proper acceleration. This will cause the charge to perform hyperbolic motion as described in Chapter 3. Any realistic motion will not persist over all space-time, which removes difficulties concerning the radiation of the charge [1,4].

Proceeding with the same arrangement for the perpendicular component of the force, we have $d\mathbf{p}_{\perp}/dt = e(\mathbf{E}_{\perp} + \mathbf{u} \wedge \mathbf{B}_{\perp})$. A direct calculation of the derivative of the momentum yields only $m\gamma(u)\mathbf{a}_{\perp}$, since $\mathbf{v}_{\perp} = \mathbf{0}$ by definition. However, Equation (3.25) with $\mathbf{v} = \mathbf{u}$ gives $\mathbf{a}_{\perp} = \mathbf{a}'_{\perp}/\gamma(u)^2$. Consequently $e(\mathbf{E}_{\perp} + \mathbf{u} \wedge \mathbf{B}_{\perp}) = m\mathbf{a}'/\gamma(u)$.

In the frame of the charge, $d\mathbf{p}'_{\perp}/dt = e\mathbf{E}'_{\perp}$, and the derivative of the momentum is $m\mathbf{a}'_{\perp}$. Consequently by equating the two expressions for $\mathbf{a}'_{\perp}$ we find the transformation under a boost as

$$\mathbf{E}'_{\perp} = \gamma(u)(\mathbf{E}_{\perp} + \mathbf{u} \wedge \mathbf{B}_{\perp}). \tag{5.77}$$

These transformations of the electric field may evidently be written together as

$$\mathbf{E}' = (1 - \gamma(u))(\widehat{\mathbf{e}}_u \cdot \mathbf{E})\widehat{\mathbf{e}}_u + \gamma(u)(\mathbf{E} + \mathbf{u} \wedge \mathbf{B}). \tag{5.78}$$

The inverse of these transformations is found as usual by reversing the sign of the relative velocity and interchanging the primes.

We turn now to the magnetic field transformation. The perpendicular component is rather straightforward. We write the inverse of Equation (5.77) and take the cross product with $\mathbf{u}$ to obtain

$$\mathbf{u} \wedge \mathbf{E}_{\perp} = \gamma(u)(\mathbf{u} \wedge \mathbf{E}'_{\perp} + \mathbf{u}^2\mathbf{B}'_{\perp}). \tag{5.79}$$

We can eliminate $\mathbf{E}'_{\perp}$ on the right of this equation using Equation (5.77) to obtain, after rearranging and collecting terms,

$$\gamma(u)\mathbf{u}^2\mathbf{B}'_{\perp} = (1 - \gamma(u)^2)\mathbf{u} \wedge \mathbf{E}_{\perp} + \gamma(u)^2\mathbf{u}^2\mathbf{B}_{\perp}. \tag{5.80}$$

Hence simplifying and dividing by $\gamma(u)\mathbf{u}^2$ we find the transformation of the perpendicular magnetic field as

$$\mathbf{B}'_{\perp} = \gamma(\mathbf{B}_{\perp} - \mathbf{u} \wedge \mathbf{E}_{\perp}). \tag{5.81}$$

To find the parallel magnetic field transformation, we imagine a charge moving instantaneously along the x axis in the frame of O with velocity v_x. A magnetic field $\mathbf{B}_{\parallel}$ points along the z axis, which is the direction of the relative motion $\mathbf{u}$ of O'. This arrangement is sketched in Figure 5.3. There is no electric field in either frame because of this arrangement of the magnetic field (see Equation (5.78)). Then Equation (5.19) tells us that $dp_y/dt = -ev_xB_{\parallel}$ and $dp_x/dt = 0$, where we assume a set of right-handed Cartesian coordinates. The latter constraint tells us (after expansion) that $dv_x/dt = 0$.

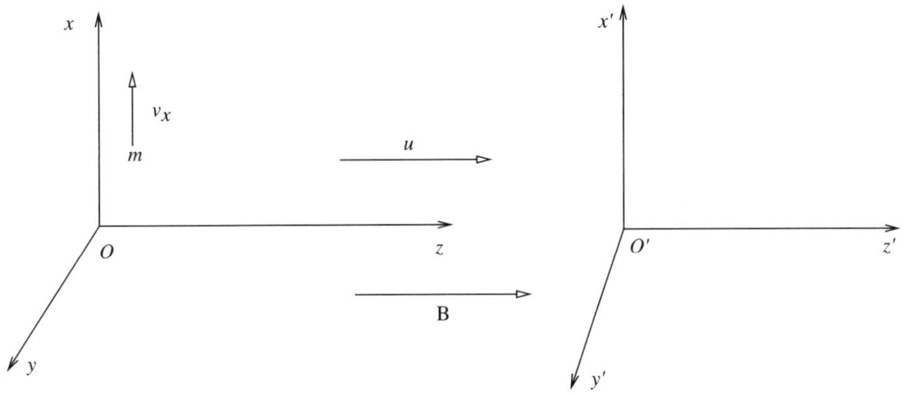

Figure 5.3 *The sketch indicates the two inertial observers O and O′ with relative motion u along the z axis. The particle m is moving purely along the x axis for O, while the uniform magnetic field* **B** *points along the common z axis. The third axis y completes the Cartesian set for each observer*

Hence, because of zero acceleration in the x direction, $dp_y/dt = m\gamma(v_x)a_y$ by direct expansion.

We adopt the notation $\gamma \equiv \gamma(v_x)$ and $\gamma' \equiv \gamma'(v'_x)$ for the Lorentz factors of the charged particle in the frame of O and O' respectively. For clarity we denote the Lorentz factor of the relative motion between the frames by $\Gamma(u)$. We know from Equation (3.25) on setting $\mathbf{u} \cdot \mathbf{a} = 0$ and $\mathbf{u} \cdot \mathbf{v} = 0$ that $a_y = \Gamma(u)^2 a'_y$. Hence from the two expressions for dp_y/dt we find $-ev_x B_\parallel/(m\gamma\Gamma^2) = a'_y$.

In the reference frame of O' we require the same force law. Thus again $dp'_y/dt' = -ev'_x B'_\parallel$ and $dp'_x/dt' = 0$, so that $dp'_y/dt' = m\gamma' a'_y$. That is, we have explicitly that $a'_y = -ev'_x B'_\parallel/(m\gamma')$. But Equation (3.16) tells us that $v'_x = v_x/\Gamma$ when $\mathbf{u} \cdot \mathbf{v} = 0$. After substituting this and equating the last two expressions for a'_y, we obtain

$$B'_\parallel - \frac{\gamma'(v'_x)}{\gamma(v_x)\Gamma(u)} B_\parallel. \tag{5.82}$$

Now either a direct calculation from the definition of the various Lorentz factors (see Problem), or by equating the various expressions for the particle proper time given below, shows that

$$\gamma'(v'_x) = \Gamma(u)\gamma(v_x), \tag{5.83}$$

and hence finally

$$B'_\parallel = B_\parallel. \tag{5.84}$$

The necessary proper time expressions to prove the identity are $ds = dt'/\gamma' = dt/\gamma$. Moreover the coordinate times between events on the world line of the particle are related by $dt' = \Gamma(dt - udz)$. However, the particle moves only in x at this instant, so $dz = 0$ and $dt' = \Gamma dt$. Thus $\gamma' ds = \Gamma\gamma ds$ and the identity follows.

Problem

5.9 Show directly from the definitions of $\gamma' \equiv \gamma'(u^2 + (v_x')^2)$ (show this also from a velocity transformation), $\Gamma(u)$ and $\gamma(v_x)$, that $\gamma' = \gamma\Gamma$.

The two transformations for the magnetic field can be combined vectorially as

$$\mathbf{B}' = (1 - \gamma(u))(\widehat{\mathbf{e}}_u \cdot \mathbf{B})\widehat{\mathbf{e}}_u + \gamma(u)(\mathbf{B} - \mathbf{u} \wedge \mathbf{E}). \tag{5.85}$$

Here we have returned to $\gamma(u)$ for the Lorentz factor of the relative motion. The inverse transformations are found as usual by reversing the sign on $\mathbf{u}$ and interchanging the primes.

We have derived these transformations by considering special (although physical) arrangements and by insisting that the force law should hold for all inertial observers (the Poincaré principle of relativity). We have not shown the inverse, namely that the form of Equation (5.19) is preserved for any inertial observer under any physical arrangement. That requires an involved algebraic manipulation using not only the field transformations, but also those of velocity, acceleration, Lorentz factor and energy. We will leave this proof to the section on four-vectors where it becomes almost trivial. We can, however, expect these field transformations to be general, since the three-vectors by our arguments have been resolved parallel and perpendicular to the relative motion. Such a resolution provides a complete description of each vector.

The most important general conclusion to draw from these transformations is that electric and magnetic fields are intimately related. A charge at rest produces a Coulomb electric field by definition, but no magnetic field. However, Equation (5.81) shows that in another inertial frame, a magnetic field arises. This would be attributed to the moving charge by observers in this frame, and indeed one of Clerk-Maxwell's equations (Ampère's law) reveals the sources of the magnetic field to be moving charges in a stationary field.

The electric field of a charge remains electric in all inertial frames, but Equation (5.77) shows that the additional presence of a magnetic field leads to the electric field (and indeed the magnetic field) becoming a linear combination of these fields for another inertial observer. These considerations suggest that the division of the electromagnetic field into electric and magnetic components is artificial. Neither quantity is invariantly defined. In the next section we see that the electromagnetic field is properly an anti-symmetric tensor of rank two (two indices) in space-time, whose six independent components define the electric and magnetic three-fields. It was the first successful unification in physics.

5.3 Electromagnetism for Arbitrary Inertial Observers

Let us turn now to study the four-vector implications of the action (5.14). The action must be varied subject to the non-holonomic constraint $\mathbf{v}^b \delta x_b = 0$. A straight-forward statement of this procedure in Galilean coordinates is

$$\delta\mathcal{S} = \int_1^2 ds\left(-m\eta_{ab}\delta v^a v^b - e\eta_{ab}A^a\delta v^b - e\eta_{ab}\delta A^a v^b\right), \tag{5.86}$$

and

$$\delta\mathcal{S} - \int_1^2 \lambda v^b \delta x_b = 0, \tag{5.87}$$

where λ is the Lagrange multiplier. We manipulate $\delta\mathcal{S}$ only and recall the constraint term at the end of the calculation. Integrate the first two terms by parts using $\delta v^a = d\delta x^a/ds$ and keep the end points fixed to obtain

$$\delta\mathcal{S} = \int_1^2 ds\left(\delta x_b\left(\frac{dp^b}{ds}\right) + e\eta_{ab}\frac{\partial A^a}{\partial x^d}v^d\delta x^b - e\eta_{ab}\frac{\partial A^a}{\partial x^c}\delta x^c v^b\right), \tag{5.88}$$

where $\delta x_b = \eta_{ab}\delta x^a$ and we have expanded dA^a/ds and δA^a as functions of the coordinates. In the second term 'a','b' and 'd' are dummy indices, serving only to indicate summation from 0 to 3. To obtain a consistent labelling among the terms we interchange 'a' and 'b' and change 'd' to 'c'. In the third term we rename the dummy indices such that *on the original indices only* $a \to c$, $c \to b$ and $b \to d$. Then we have

$$\delta\mathcal{S} = \int_1^2 ds\,\delta x_b\left(\frac{dp^b}{ds} + e\partial^c A^b v_c - e\partial^b A^c v_c\right), \tag{5.89}$$

where $\partial^c A^b \equiv \eta^{cd}\partial_d A^b$.

For brevity in the last expression we introduce the four-gradient (row or co-variant) operator in Galilean coordinates

$$\partial_a \equiv \frac{\partial}{\partial x^a} = (\partial_t, \nabla). \tag{5.90}$$

In Galilean coordinates one obtains the contravariant (or column) form by raising the index as

$$\partial^a \equiv \eta^{ab}\partial_b = \begin{pmatrix}\partial_t \\ -\nabla\end{pmatrix}. \tag{5.91}$$

Thus in particular we used the property $\partial_c A^b v^c \equiv \partial^c A^b v_c$.

With this notation it follows from the last form of the varied action together with the constraint that

$$\delta\mathcal{S} = \int_1^2 ds\,\delta x_b\left(\frac{dp^b}{ds} + ev_c(\partial^c A^b - \partial^b A^c) - \lambda v^b\right) = 0. \tag{5.92}$$

Since δx_b is arbitrary, we set the bracket in the integrand equal to zero to find the Euler-Lagrange equations for the motion of the charge. The resulting equation contains λv^b, but multiplying by v_b (and doing the implicit sum) this term becomes λ while the other two terms are identically zero. Hence $\lambda = 0$. Moreover the quantity

$$F^{bc} \equiv \partial^b A^c - \partial^c A^b, \tag{5.93}$$

is a contravariant tensor of the second rank (two indices) since A^b is a four-vector and the differential operator transforms like a four-vector. We call it the electromagnetic field tensor. It is a quantity made, as we shall see, from both electric and magnetic three-vectors. As such it represents the unified electromagnetic field. Our choice of the sign of the free particle part of Equation (5.14) (which is essentially arbitrary) has led to a definition of the field tensor that is the same as those in standard texts [1,5]. The opposite choice of the sign of the free particle action would give a sign change to the field tensor, which form is used in some texts. This tensor is clearly anti-symmetric which gives it the expected six independent components (those above the zero diagonal elements) corresponding to **E** and **B**.

Our equation of motion for a charge following from the action may now be written as (using Equation (5.92) and restoring c temporarily and changing the indices for clarity)

$$\frac{dp^a}{ds} = e\frac{v_b}{c}F^{ab}. \tag{5.94}$$

We note that multiplying this equation by v_a yields zero on the right because of the antisymmetry of the field tensor. Hence inertial mass is conserved.

By the nature of tensors and vectors which transform linearly, this equation of motion always has the same form in any inertial frame of reference. We proceed to show (after a discussion of the field tensor) that this equation of motion agrees with the Equations (5.19) and (5.26) in any inertial frame, and so guarantees their form for every inertial observer.

From the definition of the field tensor we calculate $F^{0a} = \partial^0 A^a - \partial^a A^0$ so that (remembering $\partial^i = -\partial_i$ and Equations (5.7) and (5.8)) $F^{0i} = \partial_t \mathbf{A} + \partial_i \Phi = -E(i)$. We use the index in a bracket to denote the physical component in orthogonal spatial coordinates, $\widehat{\mathbf{e}}_i \cdot \mathbf{E}$. We also note that $F^{00} = 0$ as it must to be antisymmetric. Similarly $F^{kj} = \partial^k A^j - \partial^j A^k$ obviously produces zero diagonal elements. The non-zero spatial elements are $F^{12} = -B(3)$, $F^{13} = B(2)$ and $F^{23} = -B(1)$. In summary then

$$\begin{aligned} F^{0i} &= -F^{i0} = -E(i), \\ F^{kj} &= -F^{jk} = -\epsilon^{0kj\ell} B(\ell). \end{aligned} \tag{5.95}$$

When performing explicit calculations it is convenient to have computed these values in the matrix representation. This is

$$F^{ab} = \begin{pmatrix} 0, -E_1, -E_2, -E_3 \\ E_1, 0, -B_3, B_2 \\ E_2, B_3, 0, -B_1 \\ E_3, -B_2, B_1, 0 \end{pmatrix}. \tag{5.96}$$

A covariant form F_{ab} and a mixed form F_a^b of the field tensor follow by raising or lowering indices with the Minkowski metric η. We leave the mixed form to the reader, but the covariant form is $F_{ab} = \eta_{ac}\eta_{bd}F^{cd}$. This implies that $F_{0i} = -F^{0i}$ and $F_{ij} = F^{ij}$. So in summary

$$F_{ab} = \begin{pmatrix} 0, E_1, E_2, E_3 \\ -E_1, 0, -B_3, B_2 \\ -E_2, B_3, 0, -B_1 \\ -E_3, -B_2, B_1, 0 \end{pmatrix}. \tag{5.97}$$

It is also useful to express the three vectors in terms of the components of the field tensor. The electric field is just $-F^{0i}$ but the magnetic field is slightly more subtle (the epsilon symbol indices are lowered with η_{ab} and the covariant form is positive for an odd permutation of symbols: see discussion after Equation (5.12). The expression is

$$B(\ell) = \frac{1}{2}\epsilon_{0\ell kj}F^{kj} = -\frac{1}{2}\epsilon^{0\ell kj}F^{kj} \tag{5.98}$$

since on substituting for F^{kj} we calculate that $\frac{1}{2}\epsilon_{0\ell kj}\epsilon^{0kjm}B(m) \equiv B(\ell)$. The last statement follows from the general expression of the permutation symbol in terms of the Kronecker delta as ([5], p.17)

$$\epsilon^{abkj}\epsilon_{cdkj} = -2\left(\delta_c^a\delta_d^b - \delta_d^a\delta_c^b\right). \tag{5.99}$$

The proof follows by trying all possible values of the indices.

Returning now to the equation of motion (5.94), we express the four components in three-vector form. Setting $a = 0$ gives

$$\frac{d(\gamma(v)mc)}{ds} = \frac{e}{c}v_jF^{0j} = \frac{e}{c}v^jE(j), \tag{5.100}$$

and hence again Equation (5.26) as

$$\frac{d(\gamma mc^2)}{dt} = e\mathbf{v}\cdot\mathbf{E}. \tag{5.101}$$

In order to make sense of the signs, recall that $v^a = dx^a/ds = -v_a$ and that $v(j) = v^j$.

Setting $a = j$ gives $dp^j/ds = eF^{jb}v_b = eF^{j0}v_0 + eF^{jk}v_k$. However, $F^{j0} = E(j)$ and $F^{jk} = -\epsilon^{0\ell jk}B(\ell)$. Remembering that $v_k = -v(k)$ we obtain

$$\frac{dp^j}{ds} = e\gamma(v)\left(E(j) + \epsilon^{0jk\ell}v(k)B(\ell)\right), \tag{5.102}$$

and hence Equation (5.19) when $dt = ds/\gamma$ is used.

We have seen in a previous section that the three-vector dynamic formulation allows the immediate application of the techniques of Lagrangian mechanics. The exceptional strength of the four-vector treatment lies in its easy treatment of the coordinate transformation properties of the electromagnetic field. Consider for example the transformation of the three-vectors **E** and **B**. We know from the tensor nature of F^{ab} that between two inertial frames, each using Galilean coordinates, we have

$$F'^{cd} = \frac{\partial x'^c}{\partial x^a}\frac{\partial x'^d}{\partial x^b}F^{ab}. \tag{5.103}$$

Since the partial derivatives for two frames in standard configuration are simply the matrix (2.27), one can write this expression in matrix form as

$$\underline{\underline{\mathbf{F}}}' = \underline{\underline{\mathbf{L}}}\,\underline{\underline{\mathbf{F}}}\,\underline{\underline{\mathbf{L}}}. \tag{5.104}$$

However, it is possible to apply the boost directly from the transformation equation. We leave the complete statement to a Problem and only give some examples. Thus $F'^{0i} = (\partial x'^0/\partial x^a)F^{ab}(\partial x'^i/\partial x^b)$, which on remembering the boost Lorentz transformation (2.21) yields

$$F'^{0i} = \left(\gamma(u)F^{0b} - \gamma(u)uF^{3b}\right)\frac{\partial x'^i}{\partial x^b}. \tag{5.105}$$

For example, $F'^{01} = \gamma(F^{01} - uF^{31}) = \gamma(-E(1) + uB(2))$, hence $E'(1) = \gamma(E(1) - uB(2))$. Similarly $E'(2) = \gamma(E(2) + uB(1))$ and added together we find Equation (5.77). It is found in the Problem that $F'^{03} = F^{03}$, which gives Equation (5.76).

For the magnetic part, recall that $F^{kj} = -\epsilon^{0kj\ell}B(\ell)$. Then for the parallel field we consider F^{12}. The transformation (5.103) gives immediately that $F'^{12} = F^{12}$, which is the result (5.84). For $B(1)$ we consider $F'^{23} = (\partial x'^2/\partial x^a)F^{ab}(\partial x'^3/\partial x^b) = F^{2b}\partial x'^3/\partial x^b$. This becomes $F'^{23} = \gamma(F^{23} - uF^{20})$ or $B'(1) = \gamma(B(1) + uE(2))$. This is the 1 component of Equation (5.81). The other perpendicular component follows similarly.

However, this tensor treatment allows us to discover important Lorentz invariants that are not immediately obvious otherwise. One such is

$$I_1 \equiv F_{ab}F^{ab} = 2F^{0i}F_{0i} + 2F^{jk}F_{jk} = 2(\mathbf{B}^2 - \mathbf{E}^2), \tag{5.106}$$

on multiplying the two matrices element by element. The factor two is not significant, but the sign and magnitude are. We might have envisioned such an invariant under Lorentz transformations since in a light wave $\mathbf{E}^2 = \mathbf{B}^2$, and this condition must be invariant. This makes the null value of this invariant characteristic of electromagnetic radiation. When it is positive the magnetic field dominates, and we can always find an inertial frame in which the field is purely magnetic. The converse is true when the invariant is negative. We demonstrate this below in an example.

We might motivate our second invariant once again by acknowledging the invariant nature of light. The three-vector fields in a light wave satisfy $\mathbf{E}\cdot\mathbf{B} = 0$, and this must be an invariant condition. But how does it follow in general from our field tensor representation?

To motivate the formal approach, let us consider a simple symmetry of the electromagnetic field. It is seen from the vacuum Clerk-Maxwell equations that letting $\mathbf{E} \to \mathbf{B}$ and $\mathbf{B} \to -\mathbf{E}$ leaves the equations invariant. This is an example of a 'dual transformation'.

We can see that this transformation is contained in the definition of a 'dual field tensor' according to

$$\mathcal{F}^{ab} = \frac{1}{2}\epsilon^{abcd} F_{cd}. \tag{5.107}$$

From this definition one calculates (see Problem)

$$\mathcal{F}^{ab} = \begin{pmatrix} 0, -B_1, -B_2, -B_3 \\ B_1, 0, E_3, -E_2 \\ B_2, -E_3, 0, E_1 \\ B_3, E_2, -E_1, 0 \end{pmatrix} \tag{5.108}$$

Comparing this to the matrix F^{ab} shows that the action of the dual transformation on the three-vectors transforms the field tensor F^{ab} to the dual field tensor $\mathcal{F}^{ab}$.

With the dual tensor in hand, a second invariant suggests itself, namely

$$I_2 \equiv 2\mathcal{F}^{cd} F_{cd} \equiv -|det\ \underline{\underline{\mathbf{F}}}|. \tag{5.109}$$

Either a direct calculation of the determinant of F, or a term-by-term multiplication of the dual tensor with F_{cd}, shows that

$$I_2 = 2(\mathbf{E} \cdot \mathbf{B})^2. \tag{5.110}$$

This shows that it is essentially $\mathbf{E} \cdot \mathbf{B}$ that is invariant.

Problems

5.10 Complete the derivation of the three-vector transformations between inertial frames using Equation (5.103). Sample calculations are given in the text.

5.11 Verify the components of the dual tensor as given in Equation (5.108).

5.12 Show that Equation (5.111) implies the explicit Equations (5.112).

5.13 Deduce Equation (5.113) directly from the definition of the field tensor in terms of the four-potential.

The simple duality symmetry is so characteristic of the electromagnetic field equations in a vacuum that it constrains their form. The equations that couple the field to matter must be more independent, since they break the simple symmetry in the presence of charged matter. The pair of Clerk-Maxwell equations that do not change in the presence of matter (Faraday's law and the absence of magnetic monopoles) have no source terms, and they retain a memory of the duality symmetry. This suggests that they might follow from

$$\partial_b \mathcal{F}^{ab} = 0. \tag{5.111}$$

Indeed, a direct calculation from the form of the dual tensor (5.108) yields (see Problem)

$$\nabla \cdot \mathbf{B} = 0$$
$$\nabla \wedge \mathbf{E} = -\frac{\partial \mathbf{B}}{\partial t}. \tag{5.112}$$

Using the formal definition of the dual tensor in Equation (5.107), these equations become $(\epsilon^{abcd}/2)\partial_b F_{cd} = 0$. For each value of 'a', the three other indices can only be an even or an odd permutation of the other three values. The even and odd permutation in the sum will pair up with the same sign because of the antisymmetry of F_{cd}. Thus each choice of the value of 'a' leads to a sum of the even permutations of the other three values. As all values of 'a' are allowed in turn, we conclude that

$$\partial_b F_{cd} + \partial_d F_{bc} + \partial_c F_{db} = 0, \tag{5.113}$$

where all possible values of $\{b, c, d\}$ are permitted. This last expression also follows directly from the definition of the field tensor in terms of the four-potential (see Problem).

Example 5.4

In this example we study archetypal forms of uniform three-vector electromagnetic fields as seen by special inertial observers. These special frames of reference allow charged particle motion to be solved using (as it turns out) the simplest uniform three-vector descriptions. The solution may subsequently be transformed to any relevant inertial frame.

From Equations (5.77) and (5.81), we can infer that there is a reference frame in which a uniform electromagnetic field is purely electric, if in any inertial frame where the field is 'mixed' $E > B$. Similarly there is a frame in which a uniform electromagnetic field is purely magnetic, if $B > E$ in some inertial frame in which the field is 'mixed'. Suppose that in the mixed inertial frame the vectors $\mathbf{E}$ and $\mathbf{B}$ lie in a plane at an angle θ to one another. When $E > B$ we choose a boost $\mathbf{u}$ in a direction perpendicular to the plane formed by the electric and magnetic vectors; that is, $\mathbf{E}_\parallel = \mathbf{B}_\parallel = \mathbf{0}$. Then the boost to the frame in which the entire magnetic field is zero satisfies $\mathbf{B} - \mathbf{u} \wedge \mathbf{E} = \mathbf{0}$. From this it follows by taking the cross product with $\mathbf{E}$ that the desired boost to the electric frame is

$$\mathbf{u} = \frac{\mathbf{E} \wedge \mathbf{B}}{E^2}. \tag{5.114}$$

This has magnitude $(B/E)\sin\theta$, which is less than one. An exactly similar procedure for the magnetic case gives for the boost to the purely magnetic inertial frame

$$\mathbf{u} = \frac{\mathbf{E} \wedge \mathbf{B}}{B^2}, \tag{5.115}$$

which has magnitude $(E/B)\sin\theta < 1$.

As a practical example, we know the motion of a charged particle in a uniform and constant magnetic field. This is a gyration at the frequency found in a previous section of this chapter. We may now conclude that in a mixed field with $E < B$ the motion will be gyration about a centre that drifts with the boost velocity (5.115). Similarly the motion of a charge in a predominantly electric field is subject to the drift (5.114)

We see that the preceding discussion requires that the invariant $I_1 \neq 0$, but permits the invariant I_2 to be zero (i.e. $\theta = \pi/2$). When both invariants are zero we cannot remove either field since together they form an electromagnetic plane wave. So the case that remains is one in which the invariant I_1 may be zero, but not the invariant I_2. An example of this arrangement would have parallel fields with $\mathbf{E} = \mathbf{B}$. The question arises as to whether one can, with $I_2 \neq 0$, always (i.e. independent of the value of I_1) find an inertial frame in which the fields are parallel but not necessarily equal in magnitude?

We begin in a general inertial frame in which the three-vector fields lie in a plane at an angle $\theta \neq \pi/2$ to one another. If there is a primed frame in which the three-vectors are parallel, the component of the boost velocity parallel to this direction is arbitrary (it does not change the fields) and may be taken as zero. The desired boost velocity can also be taken perpendicular to the plane of the three-fields in the original reference frame, since parallel components will remain zero in that frame. Using the inverses of Equations (5.77) and (5.81) we have (requiring the primed fields to be parallel)

$$\mathbf{E} \wedge \mathbf{B} = \gamma(u)^2(\mathbf{E}' - \mathbf{u} \wedge \mathbf{B}') \wedge (\mathbf{B}' + \mathbf{u} \wedge \mathbf{E}') = \gamma^2 \mathbf{u}(\mathbf{E}'^2 + \mathbf{B}'^2), \tag{5.116}$$

or for the desired $\mathbf{u}$

$$\gamma(u)\mathbf{u} = \frac{\mathbf{E} \wedge \mathbf{B}}{\mathbf{E}'^2 + \mathbf{B}'^2}. \tag{5.117}$$

To complete the transformation we use the invariants in the form

$$\begin{aligned} \mathbf{E}'^2 - \mathbf{B}'^2 &= \mathbf{E}^2 - \mathbf{B}^2, \\ E'B' &= EB\cos\theta, \end{aligned} \tag{5.118}$$

from which the new three-field magnitudes may be found if $\theta \neq \pi/2$. The reduction is particularly simple if $I_1 = 0$ so that $E' = B'$.

This example shows that the motion of a charged particle in uniform electromagnetic fields can be reduced to three canonical cases plus the motion in a plane wave.

We have seen that two of Maxwell's equations follow from a homogeneous identity that is satisfied either by the dual field tensor or by the field tensor itself. It is not our intention to explore the many subtleties of formal electromagnetic theory, but for completeness we will discuss the second pair of equations. These are coupled to matter and are not simply a result of symmetry. Rather they require a physical theory. As usual we require an action as a statement of this theory. This action must allow us to vary the electromagnetic field vectors while holding particle properties constant, if we hope to find the field's natural form as a minimum condition of the action.

For the first time we are seeking an action that must encompass an infinite number of degrees of freedom. These represent the value of the fields at every point of space-time. We begin by generalizing the coupling terms in the action (5.14). The free particle term becomes a sum over all discrete particles that are present in the volume of interest, but the second or coupling term must also permit a continuous variation of the electromagnetic potential. Moreover, the field must be coupled to the charge that is present at each point in space. Therefore we need a continuous description of the charged matter. This can be accomplished by introducing a charge density in any inertial frame

$$\rho_e = \Sigma_i\, e_i \delta(\mathbf{r} - \mathbf{r_i}), \tag{5.119}$$

where $\mathbf{r_i}$ is the location of each charge and $\mathbf{r}$ locates an arbitrary point in three-space. The generalized function $\delta(\mathbf{r} - \mathbf{r}_i)$ is the 'Dirac delta function'.

It is $\rho_e d^3x$ that is invariant since it gives the number of scalar charges in a volume, but as we found in the prologue to this chapter, $d^4x = d^3xdt$ is also an invariant. That is, the charge density transforms like coordinate time, and therefore an explicit expression of the four-vector current is

$$j^a = \rho_e \frac{dx^a}{dt} \equiv \rho_e v^a/\gamma. \tag{5.120}$$

In this expression the velocity is physically that of some average four-velocity taken over a small volume of charge, but we need not worry about this average if we assume an infinitesimal d^3x compared with other scales in the Problem.

We may now construct the continuous coupling term in the action by writing

$$-\int e v_b A^b\, ds = -\int \rho_e d^3x\, v_b A^b dt/\gamma = -\int d^4x j_b A^b. \tag{5.121}$$

The last expression succeeds in being sensitive to every point in space-time as well as being a scalar. Classically the integral over the three-volume is the Lagrangian and its integrand is the Lagrangian density.

Thus we may expect the action that includes the degrees of freedom of the field to have the form

$$\mathcal{S} = -\Sigma_m\, m \int ds \frac{v_b v^b}{2} - \int d^4x j_b A^b - \int d^4x\, X, \tag{5.122}$$

where X should be a continuous invariant dependent wholly on the electromagnetic field.

The field 'coordinates' are the A^b, and in order to have second-order Euler-Lagrange equations, X should depend only on their first derivatives. This leaves only I_1 or I_2 as candidates, and since we have used the dual that is associated with I_2 for the first pair of equations, we may settle here on $I_1 = 2(\mathbf{B}^2 - \mathbf{E}^2)$. The minus sign is chosen to make this $\propto (\mathbf{E}^2 - \mathbf{B}^2)$. This means that a wave field in a Coulomb gauge far from sources contributes only positively through the $\partial\mathbf{A}/\partial t$ part of the electric field. Otherwise we could not expect a minimum in the action to yield the field equations for arbitrarily high frequencies.

Apart from the dimensional insertion of c to produce conventional units, there are arbitrary geometric factors inserted in some systems of units. Here we must choose one such system, and in order to continue with identical dimensions for $\mathbf{E}$ and $\mathbf{B}$ we choose the Gaussian system. To obtain these in the standard form, we must take $X = I_1/16\pi$.

Hence the action that includes the electromagnetic field itself is taken to be (restoring conventional units temporarily)

$$\mathcal{S} = -\Sigma_m m \int_1^2 ds \frac{v_b v^b}{2} - \frac{1}{c}\int d^4x j_b A^b - \frac{1}{16\pi}\int F_{ab}F^{ab} d^4x. \tag{5.123}$$

Varying this action with respect to particle coordinates alone treats the last term as constant and will produce the equations of motion for a system of charges in a fixed electromagnetic field. The more charges that are present, the less likely they are to leave the field undisturbed, so this usually reduces to our considerations for a test charge above. We proceed by varying the field coordinates alone, that is the A^b, to find the Euler-Lagrange equations for the field coupled to fixed matter.

This type of variation is very standard in field theory, and the result in this case can be found in many places. However, because of its utility we include a brief derivation here.

A virtual variation of the four-potential in Equation (5.123) gives (remembering that the field tensor depends on the derivatives of the potential)

$$\delta\mathcal{S} = -\int dt \int d^3x \left(\frac{1}{8\pi}F^{ab}\delta F_{ab} + j^b \delta A_b\right). \tag{5.124}$$

However, $\delta F_{ab} = (\partial_a \delta A_b - \partial_b \delta A_a)$ by the commutation properties of the variation and partial coordinate derivatives. Moreover $F^{ab}\delta F_{ab} = F^{ab}(\partial_a \delta A_b - \partial_b \delta A_a) \equiv 2F^{ab}\partial_a \delta A_b$, where the last expression follows by interchanging the dummy indices in the second term of the previous expression and using $F^{ba} = -F^{ab}$. Consequently

$$\delta\mathcal{S} = -\int dt \int d^3x \left(\frac{1}{4\pi}F^{ab}\partial_a \delta A_b + j^b \delta A_b\right). \tag{5.125}$$

We are now able to integrate the first term under the integrals by parts to obtain

$$\delta\mathcal{S} = -\int dt \int d^3x \delta A_b \left(-\frac{1}{4\pi}\partial_a F^{ab} + j^b\right) + \oint d\sigma_a \delta A_b F^{ab}. \tag{5.126}$$

The last integral in this expression is over a closed surface in space-time, and $d\sigma_a = n_a d\sigma$ is the outward normally directed surface element of the surface bounding the system. The four-vector n_a is the normal to the element ($n_a n^a = \pm 1$) and $d\sigma$ is the three-area defined for a given inertial observer and kept invariant [4]. For example, the spatial world of an O observer (that is, a hypersurface which is the spatial three-surface formed by synchronizing clocks of O observers) has $n_a = (1, 0, 0, 0)$ and $d\sigma = dxdydz$. The natural bounding surface for an evolving physical electromagnetic field would be the union of the light-spheres around the various sources. However, for a virtual evolution we might take the time-like hypersphere at spatial infinity to bound any two separated hypersurfaces (i.e. two separate coordinate times) created by O observers. Such a surface would have $n_a = (0, 1, 1, 1)/\sqrt{3}$ and $d\sigma = R^2 \sin\theta d\theta d\phi dt$, where we have used the standard form of the surface element of a sphere of radius R extended over time.

The point of the previous discussion is that the integral over a bounding surface in the varied action (5.126) may be taken between fixed hypersurfaces (times) for O, and over the the time-like hypersphere at infinity. We fix the four-potential ($\delta A_b = 0$) on the two hypersurfaces and on the bounding hypersphere, so that in fact the integral vanishes.

The arbitrariness of δA_b in Equation (5.126) now allows by the usual point by point arguments to infer the field equations in the form (restoring cgs units temporarily)

$$\partial_a F^{ab} = \frac{4\pi}{c} j^b = -\partial_a F^{ba}. \tag{5.127}$$

Remembering Equation (5.95), the zero component of this equation gives $4\pi\rho_e = -\partial_a F^{0a}$ or

$$\nabla \cdot \mathbf{E} = 4\pi\rho_e, \tag{5.128}$$

while from the i component $4\pi j^i/c = \partial_t E^i - \epsilon^{0ji\ell}\partial_j B(\ell)$ or

$$\nabla \wedge \mathbf{B} = \frac{4\pi}{c}\mathbf{j} - \partial_t \mathbf{E}. \tag{5.129}$$

We have therefore succeeded in understanding Clerk-Maxwell's equations as a relativistic tensor field theory. There are many possible developments from this base, but we shall leave these as taking us too far beyond the scope of this book. They can be found in the standard references. The most important reservation about all of the discussion is that we have ignored the radiation reaction force. Accelerated charges radiate energy and momentum and in principle the resultant damping force should be included in our calculations. Fortunately it becomes necessary to do this only at very high energies (e.g. [1]), well beyond the appearance of inertial relativistic effects with which we have been concerned.

The last consideration of this chapter will be to discuss the four-vector presentation of the theory in curvilinear coordinates. This is contained in the next subsection.

5.3.1 Curvilinear Electromagnetic Theory

We know from Chapter one and onwards in this book that, in non-Galilean coordinates, both the metric (that is g_{ab} rather than η_{ab}) and the description of vectors are more complicated. We have become accustomed to covariant, contravariant, and physical (resolved along normalized base vectors) vector components, and these distinctions remain valuable. When we use curvilinear coordinates in inertial frames, it is only the spatial metric coefficients that become non-constant ($g_{00} = 1$ if $c = 1$). However, we have seen, as in the case of a rotating disc, that this can change in non-inertial frames when both g_{00} and g_{0i} in addition to g_{ij} may become non-trivial functions. For this reason and also as an introduction to the metrics encountered in the theory of gravity, we allow in this section the metric coefficients to be general functions of the space-time coordinates.

The first revision we must make is to the invariant volume element in space-time. In Galilean coordinates this is simply d^4x as found in Equation (5.13). Under a transformation to generalized coordinates $\{q^a\}$ we have that $d^4q = Jd^4x$, where J is the Jacobian of the transformation. But at the same time the metric tensor becomes $g_{ab} = (\partial x^c/\partial q^a)(\partial x^d/\partial q^b)\eta_{cd}$. Taking the determinant of this last matrix equation gives

$$g = -1/J^2, \tag{5.130}$$

since $det(\partial x^c/\partial q^a) = 1/J$ and $det(\eta_{cd}) = -1$. Hence the invariant volume element in generalized coordinates is

$$\sqrt{-g}d^4q = d^4x. \tag{5.131}$$

This will be true in any set of generalized coordinates so that between such sets $\{q\}$ and $\{q'\}$ we have $\sqrt{-g}d^4q = \sqrt{-g'}d^4q'$.

In a similar vein the epsilon symbol that was seen to be a Galilean tensor in Equation (5.12) now transforms according to $\epsilon^{abcd} = J\epsilon_G^{abcd}$ as in Equation (5.11) when the $\{x'\}$ are the $\{q\}$. Here ϵ_G^{abcd} is the familiar four-index permutation symbol. Consequently the tensor permutation symbol (therefore invariant in form) in curvilinear coordinates is

$$\epsilon^{abcd} = \frac{\epsilon_G^{abcd}}{\sqrt{-g}} \tag{5.132}$$

Both of these modifications vanish in Galilean coordinates where $\sqrt{-g} = 1$.

Another useful tool when the metric in arbitrary coordinates has a general form, is to recall the process of synchronization. By this process we remove cross terms of the form $dq^j dt$ and so establish a spatial hypersurface. Unlike the situation in Galilean coordinates, this can normally only be done locally; that is, the space-orthogonal time may not be globally consistent on closed loops as we saw in our discussion of the rotating disc. We accomplish this in general by writing the metric g_{ab} in its temporal, spatial and mixed parts as

$$ds^2 = g_{00}dt^2 + 2g_{0j}dq^j dt + g_{ij}dq^i dq^j. \tag{5.133}$$

We synchronize by introducing the locally orthogonal time

$$d\tilde{t} = dt + g_{0j}dq^j/g_{00}. \tag{5.134}$$

This transformation succeeds in producing the hypersurface orthogonal space-time metric as [5]

$$ds^2 = g_{00}d\tilde{t}^2 - \sigma_{ij}dq^i dq^j, \tag{5.135}$$

where the hypersurface metric is itself defined as

$$\sigma_{ij} \equiv \frac{g_{0i}g_{0j}}{g_{00}} - g_{ij}. \tag{5.136}$$

In particular in these local diagonal coordinates we calculate the determinant of $g_{ab} \equiv g$ by $-\epsilon_G^{0bcd}g_{00}\sigma_{1b}\sigma_{2c}\sigma_{3d}$, which implies by definition of the three-determinant

$$g = -g_{00}det(\sigma_{ij}) \equiv -g_{00}\sigma. \tag{5.137}$$

The Jacobian of the transformation to local coordinates from generalized coordinates is one, so that this relation is generally true.

The synchronization procedure defines a three-space (with metric σ_{ij}) in which an observer using the generalized spatial coordinates would usually choose to work. For this reason we will also express the four-vector electromagnetic equations in this space.

The one additional tool that we need is the definition of a true change in a vector component, since in curvilinear coordinates the base vectors are changing from point to point. This implies a change in the components of a vector even if it is in fact parallel transported (constant). The equivalent question of what specifies a constant vector at two different events (i.e. how is a vector parallel-transported in space-time) was addressed for orthogonal curvilinear spatial coordinates in Example (1.4).

We need now to generalize this consideration to arbitrary coordinates in space-time. We can dispense with a long derivation by thinking about the meaning of the geodesic Equation (4.84). This is the expression of straight-line motion in Galilean space-time in arbitrary coordinates. But straight-line motion is one in which the tangent vector u^a to the trajectory is truly constant. Hence Equation (4.84) tells us that to parallel-transport a four-vector in arbitrary coordinates we must apply a change in each vector component equal to

$$\delta u^a = -\Gamma^a_{bc} u^b u^c ds = -\Gamma^a_{bc} u^b dq^c. \tag{5.138}$$

For a general vector A^a, we replace u^a by A^a. Then, following the logic of Example 1.4 we find the true change in a vector over the coordinate interval dq^c as $\nabla A^a = dA^a - \delta A^a$. Neither dA^a nor δA^a is a vector but ∇A^a is, since it is the difference between two vectors in Galilean coordinates transformed to generalized coordinates (a direct algebraic proof may be found in standard references).

The considerations of the last paragraph allow the definition of a true directional derivative in space-time as $\nabla_c A^a \equiv \nabla A^a/dq^c$, that is

$$\nabla_c A^a \equiv \frac{\partial A^a}{\partial q^c} + \Gamma^a_{bc} A^b. \tag{5.139}$$

This derivative is a mixed tensor of rank two since $\nabla_c A^a dq^c = \nabla A^a$ is a vector. We should be able to lower the index 'a' according to $g_{ba}\nabla_c A^a$. A direct calculation using the previous definition together with that of the Christoffel symbol (see Problem) shows that

$$\nabla_c A_b = \frac{\partial A_b}{\partial q^c} - \Gamma^d_{bc} A_d. \tag{5.140}$$

An important conceptual deduction from this argument is the following. The tangent vector to a geodesic is parallel-transported (i.e. held constant) in the same way as any other vector. Hence another way of specifying a constant three-vector is to hold its angle with the tangent three-vector constant during a displacement. This is a generalization of what is always true in inertial space-time in Galilean coordinates, since the spatial geodesics are straight lines.

For the derivative of tensors of higher rank, we apply the parallel-transport correction to each index just as though the tensor were a product of vectors with the corresponding indices. In addition the partial derivative is applied to the whole tensor to give the total change in the tensor along the direction in question. Taking a case of interest to us, the true derivative of the covariant field tensor in generalized coordinates is according to this prescription

$$\nabla_c F_{ab} = \frac{\partial F_{ab}}{\partial q^c} - \Gamma^m_{ac} F_{mb} - \Gamma^m_{bc} F_{am}. \tag{5.141}$$

An important feature of the true change as defined with this geodesic method of parallel transport is that $\nabla_c g_{ab} = 0$. This must also be true since it is true in Galilean coordinates if $\nabla_c g_{ab}$ is a tensor. But since $\nabla_c A^a$ is a vector, it also follows from $\nabla_c A_a = g_{ab} \nabla_c A^b \equiv \nabla_c (g_{ab} A^b)$. Expanding the change in the product according to the Leibnitz product rule gives $\nabla_c g_{ab} = 0$ for arbitrary A^a. Hence $\nabla_c g_{ab}$ is a tensor and the true directional derivative of g_{ab} is zero. It follows also directly from Equation (5.141) as applied to the metric tensor, together with the definition of the Christoffel symbols (see Problem).

Problems

5.14 Show by lowering the index in Equation (5.139) that the true derivative of a covariant component is given by Equation (5.140).

5.15 Show directly from the definition (4.85) and Equation (5.141) that $\nabla_c g_{ab} = 0$.

5.16 Show that Equation (5.113) holds in generalized coordinates.

We may now return to electromagnetism in generalized coordinates. The current four-vector is easily generalized by using the time kept by an observer at rest in the synchronized generalized coordinates, namely $\sqrt{g_{00}} d\tilde{t}$, in place of the same quantity dt for Galilean observers. Thus (restoring c)

$$j^a = \frac{\rho_e}{\sqrt{g_{00}}} c \frac{dq^a}{dt}. \tag{5.142}$$

For an observer at rest we note that $d\tilde{t} = dt$, so we have chosen to use dt in this last equation.

The field tensor in covariant form generalizes to

$$F_{ab} = \nabla_a A_b - \nabla_b A_a \equiv \partial_a A_b - \partial_b A_a, \tag{5.143}$$

where the last form follows from the symmetry of the Christoffel symbols in the lower two indices. This familiarity is somewhat deceptive, however, since now the contravariant field tensor is a quite different quantity $g^{ca} g^{db} F_{ab}$. We discuss this further below.

The electromagnetic equations are generalized from Galilean coordinates by using the true derivative ∇_a in place of the Galilean ∂_a. This gives for the homogeneous pair

$$\nabla_c F_{ab} + \nabla_b F_{ca} + \nabla_a F_{bc} \equiv \partial_c F_a b + \partial_b F_{ca} + \partial_a F_{bc} = 0, \tag{5.144}$$

where the last expression follows because of the antisymmetry of the field tensor and the symmetry in the lower two indices of the Christoffel symbol (see Problem).

The inhomogeneous pair becomes ($c = 1$)

$$\nabla_a F^{ab} = 4\pi j^b = 4\pi \frac{\rho_e}{\sqrt{g_{00}}} \frac{dq^b}{dt}. \tag{5.145}$$

However, standard treatments of differential geometry (see also [5]) show that for an antisymmetric tensor such as the field tensor in contravariant form,

$$\nabla_a F^{ab} = \frac{1}{\sqrt{-g}} \frac{\partial (F^{ab}\sqrt{-g})}{\partial q^a}, \tag{5.146}$$

which shall prove to be very useful.

In one sense there is nothing further to be said, since we are now able to calculate the evolution of the field tensor. The covariant and contravariant components are related by lowering or raising the indices with the metric. However, they are consequently *not* simply related, and we shall postpone this calculation for a few paragraphs. The equation of motion of a charged particle remains Equation (5.94). However, a direct frontal approach normally leads to some algebraic confusion.

Instead of the direct approach, we introduce the names of the components of the field tensor in such a fashion that the field equations written in the synchronized three-space plus coordinate time are closely related to the Galilean forms. This allows an application of electromagnetic theory in generalized coordinates that preserves much hard-won intuition.

Landau and Lifshitz [5] have presented one way of doing this. Let

$$\begin{aligned} F_{0j} &= E_j, \quad F_{ij} = B_{ij} \\ F^{0j} &= \frac{-D^j}{\sqrt{g_{00}}} \quad F^{ij} = \frac{H^{ij}}{\sqrt{g_{00}}}. \end{aligned} \tag{5.147}$$

It transpires that the three-space duals of the antisymmetric three tensors B_{ij} and H^{ij} are both related to the magnetic field, while the three-vectors E_j and D^j are both related to the electric field.

We now insert these quantities into Equation (5.144). Because of the antisymmetry of the field tensor and the cyclic nature of the equation, there are only two independent sets of equations. One set follows by setting the indices $c = 0$, $a = i$ and $b = j$ in the general form, whence using the labels above

$$\frac{\partial B_{ij}}{\partial t} + \frac{\partial E_i}{\partial q^j} - \frac{\partial E_j}{\partial q^i} = 0. \tag{5.148}$$

We define the dual of B_{ij} in the synchronized three-space by

$$B^k = -\frac{1}{2\sqrt{\sigma}} \epsilon^{kij} B_{ij}. \tag{5.149}$$

Just as in generalized four-space, $\epsilon^{kij}/\sqrt{\sigma}$ is the permutation tensor in generalized three-space. Because this space has a positive definite metric σ_{ij} with which indices are lowered, the covariant form $(\epsilon_{kij}\sqrt{\sigma})$ has the same sign. Here ϵ indicates the usual permutation symbol, not the tensor. Normally the dual vector would be defined without the minus sign, but with our metric signature the minus sign is required to obtain the traditional equations.

Let us then multiply Equation (5.148) by $-(1/2)\epsilon^{kij}$. We obtain

$$\frac{\partial(\sqrt{\sigma}B^k)}{\partial t} + \epsilon^{kij}\frac{\partial E_j}{\partial q^i} = 0, \tag{5.150}$$

since the last two terms on the left of Equation (5.148) are the same after multiplication because of the antisymmetry of the permutation symbol. In generalized coodinates the 'curl' operator is actually the dual vector (defined with a positive sign) of $\partial E_j/\partial q^i$, that is $(curl\ \mathbf{E})^k = \epsilon^{kij}(\partial E_j/\partial q^i)/\sqrt{\sigma}$. Note that this can be written in the equivalent expression

$$(curl\ \mathbf{E})^k \equiv \frac{1}{2\sqrt{\sigma}}\left(\frac{\partial E_j}{\partial q^i} - \frac{\partial E_i}{\partial q^j}\right). \tag{5.151}$$

If we write the homogeneous field equations (5.150) in 'three plus one' form as

$$\frac{1}{\sqrt{\sigma}}\frac{\partial(\sqrt{\sigma}B^k)}{\partial t} + \frac{\epsilon^{kij}}{\sqrt{\sigma}}\frac{\partial E_j}{\partial q^i} = 0. \tag{5.152}$$

Introducing the curl notation for the dual in the second term, we arrive at Faraday's law in the (nearly) familiar form (restoring c)

$$curl\ \mathbf{E} = -\frac{1}{c\sqrt{\sigma}}\frac{\partial(\sqrt{\sigma}\mathbf{B})}{\partial t}. \tag{5.153}$$

Stokes' theorem retains its familiar form when the surface element is taken as $dA_k = \sqrt{\sigma}\epsilon_{k\ell m}(dq^\ell dq'^m - dq^m dq'^\ell)/2$. The q and q' refer to the two independent vectors spanning the surface element.

We return to the second independent set of equations that follow from Equation (5.144) by setting $a = i$, $b = j$ and $c = k$. These give directly

$$\frac{\partial B_{ij}}{\partial q^k}\ plus\ two\ cyclic\ terms = 0. \tag{5.154}$$

Multiply this last equation by ϵ^{kij} and note that all three terms are the same because of the permutation symmetry. Then using the definition (5.149), we may write after dividing by $-2\sqrt{\sigma}$ that

$$\frac{1}{\sqrt{\sigma}}\frac{\partial}{\partial q^k}(\sqrt{\sigma}B^k) = 0. \tag{5.155}$$

But the expression on the left of this equation is just what is defined as the divergence in curvilinear coordinates in three space. Hence it becomes

$$div\ \mathbf{B} = 0. \tag{5.156}$$

Gauss' theorem applies when the volume element is taken as $dV = \sqrt{\sigma}\epsilon_{ijk}(dq^i dq'^j dq''^k + dq^k dq'^i dq''^j + dq^j dq'^k dq''^i)/3$, where the three independent vectors span a parallelepiped in space. The surface element is as above.

Equation (5.145) can be treated in a similar way. The zero component is

$$\nabla_a F^{0a} = -4\pi \frac{\rho_e}{\sqrt{g_{00}}}. \tag{5.157}$$

But F^{0a} is a four-vector and the divergence is

$$\nabla_a F^{0a} = \frac{1}{\sqrt{-g}} \frac{\partial(\sqrt{-g}F^{0a})}{\partial q^a}. \tag{5.158}$$

Using Equations (5.147) and (5.137), this gives

$$\frac{1}{\sqrt{\sigma}} \frac{\partial}{\partial q^j}(\sqrt{\sigma}D^j) = 4\pi\rho_e, \tag{5.159}$$

where we remember that $F^{00} = 0$. This becomes in three-vector notation

$$div\ \mathbf{D} = 4\pi\rho_e. \tag{5.160}$$

The spatial components of Equation (5.145) involve the divergence of a four-tensor on the left, so that remembering Equation (5.146) we have

$$\frac{1}{\sqrt{-g}} \frac{\partial}{\partial q^a}(\sqrt{-g}F^{ak}) = \frac{4\pi\rho_e}{\sqrt{g_{00}}} \frac{dq^k}{dt}. \tag{5.161}$$

Equations (5.137) and (5.147) allow this to be written as

$$\frac{1}{\sqrt{\sigma}} \frac{\partial}{\partial q^j}(\sqrt{\sigma}H^{jk}) = 4\pi\rho_e \frac{dq^k}{dt} + \frac{1}{\sqrt{\sigma}} \frac{\partial}{\partial t}(\sqrt{\sigma}D^k). \tag{5.162}$$

However, it may be verified directly by inverting the dual that

$$H^{jk} = -\frac{\epsilon^{jk\ell}}{\sqrt{\sigma}} H_\ell. \tag{5.163}$$

Inserting this into the previous equation, and remembering the definitions of the vector operators, gives the Ampère/Maxwell law (restoring c)

$$curl\ \mathbf{H} = \frac{4\pi\rho_e}{c}\sqrt{g_{00}}\,\mathbf{j} + \frac{1}{\sqrt{\sigma}} \frac{\partial}{\partial t}(\sqrt{\sigma}\mathbf{D}), \tag{5.164}$$

where $\mathbf{j}^k \equiv \frac{\rho_e}{\sqrt{g_{00}}} c(dq^k/dt)$, as per Equation (5.142).

These equations could be used in much the same manner that the Galilean equations of electromagnetism are used in dielectric media. However, we are missing the constitutive equations to relate **E** and **D** and **B** and **H**. For these we must relate the upper and lower index quantities.

It is useful to follow [5] and introduce the three vector g_i that is defined by

$$g_i \equiv -\frac{g_{0i}}{g_{00}}. \tag{5.165}$$

In our calculations we will also want the contravariant form of this vector in the synchronized hyperspace. For this we need the inverse of the metric σ_{ij}. This follows from the condition determining the inverse metric in space-time, namely $g^{ab}g_{bc} = \delta^a_c$. By letting the indices 'a' and 'c' range over temporal and spatial values, these conditions yield (the 'a' and 'c' values are indicated)

$$\begin{aligned} g^{ij}g_{jk} + g^{i0}g_{0k} &= \delta^i_k, \quad (i,k) \\ g^{0j}g_{j0} + g^{00}g_{00} &= 1, \quad (0,0) \\ g^{ij}g_{j0} + g^{i0}g_{00} &= 0. \quad (i,0) \end{aligned} \tag{5.166}$$

We solve for g^{0i} from the third of these equations and substitute into the first equation to find after some rearrangement

$$-g^{ij}\sigma_{jk} = \delta^i_k. \tag{5.167}$$

This shows that the inverse of the spatial metric σ^{ij} equals $-g^{ij}$.

To find our constitutive relations we expand first $F_{0j} = g_{0a}g_{jb}F^{ab}$ to find after some algebra that

$$F_{0j} = -g_{00}\sigma_{jk}F^{0k} + g_{00}g_i\sigma_{jk}F^{ik}. \tag{5.168}$$

Deducing this form requires the definition of σ_{jk} plus the anti-symmetry of F^{ik}.

Now we recall Equation (5.147) and remark that we lower indices on three-space quantities with σ_{jk}, in order to write this last result as $E_j = \sqrt{g_{00}}D_j + \sqrt{g_{00}}g_i\sigma_{jk}H^{ik}$. The second term may be written as $\sqrt{g_{00}}\sigma_{i\ell}g^{\ell}\sigma_{jk}H^{ik}$ or, on recalling Equation (5.163), as $-(\sqrt{g_{00}}\epsilon^{ikm}g^{\ell}H^n\sigma_{i\ell}\sigma_{jk}\sigma_{mn})/\sqrt{\sigma}$. This is equivalent to $-\sqrt{g_{00}}\sqrt{\sigma}\epsilon_{\ell jn}g^{\ell}H^n$. Consequently

$$D_j = \frac{E_j}{\sqrt{g_{00}}} + \sqrt{\sigma}\epsilon_{\ell jn}g^{\ell}H^n. \tag{5.169}$$

Adopting the three-vector notation in this last equation we have finally (g has the components given in Equation (5.165))

$$\mathbf{D} = \frac{\mathbf{E}}{\sqrt{g_{00}}} + \mathbf{H} \wedge \mathbf{g}, \tag{5.170}$$

where the cross or wedge product of two vectors is their dual $\sqrt{\sigma}\epsilon_{n\ell j}H^n g^{\ell}$.

The second constitutive relation follows from $F^{ij} = g^{ia}g^{jb}F_{ab}$ which yields in a direct way $H^{ij}/\sqrt{g_{00}} = B^{ij} - g^{0j}E^i + g^{0i}E^j$. Equation (5.147) must be used, together with the antisymmetry of F_{ab}, in order to obtain this result. We remark that $g^{0i} \equiv -(g^{ik}g_{0k}/$

$g_{00}) = -g^i$ by the third member of Equations (5.166) and the definition of g_i. We may now apply the dual operation in the form $-(1/2)\sqrt{\sigma}\epsilon_{kij}$ and rearrange to obtain $B_k = H_k/\sqrt{g_{00}} + (1/2)\sqrt{\sigma}\epsilon_{kji}(g^j E^i - g^i E^j)$. Adopting the three-vector description in this last expression gives

$$\mathbf{B} = \frac{\mathbf{H}}{\sqrt{g_{00}}} + \mathbf{g} \wedge \mathbf{E}. \tag{5.171}$$

This concludes our formal discussion of electromagnetism in curvilinear coordinates. It is at present much more general than required. Normally the three-space curvilinear metric is constant in time, which simplifies the equations considerably. Moreover, a diagonal metric renders the constitutive relations rather simple, since the cross product terms vanish. We are left with an effective dielectric constant and magnetic permeability each equal to $\sqrt{g_{00}}$.

However, the general equations allow the use of the Clerk-Maxwell equations in a gravitational field. For, as we shall discuss in the next chapter, the modern theory of gravity (due to Einstein) is a metric theory. The metric in this theory is determined not only by a choice of transformations from an inertial frame of reference, but also by the distribution and motion of matter. Once the metric is known, all of the results of this section apply. The metric must be found by solving the Einstein-Hilbert gravitational field equations, which we leave largely to other studies. They add complications to the metric of inertial space-time, in that space-time becomes curved. Moreover the metric may well be a function of coordinate time due to moving matter or passing gravitational waves. This brings us to the entrance to the domain of Gaussian and Riemannian differential geometry.

We conclude this chapter with a small example of the preceding formalism.

Example 5.5

As an example, consider an inertial space in cylindrical coordinates transformed to cylindrical coordinates rotating (with angular speed ω) about the z axis. Then as we have seen (ϕ is measured with respect to rotating axes) the metric becomes

$$ds^2 = dt^2(1 - \omega^2 r^2) - 2\omega r^2 d\phi dt - (dr^2 + r^2 d\phi^2 + dz^2). \tag{5.172}$$

It is readily seen that for this metric $g_{00} = (1 - \omega^2 r^2)$, $\mathbf{g} = (0, g_2, 0)$, and $g_2 = -g_{02}/g_{00} = \omega r^2/(1 - \omega r^2)$. Moreover, the spatial hypersurface metric is diagonal in the form

$$\underline{\underline{\sigma}} = \begin{pmatrix} 1, 0, 0 \\ 0, \dfrac{r^2}{1 - \omega^2 r^2}, 0 \\ 0, 0, 1 \end{pmatrix}. \tag{5.173}$$

Hence $\sqrt{\sigma} = r/\sqrt{(1 - \omega^2 r^2)}$.

Faraday's law (5.153) becomes in explicit form

$$\frac{\epsilon^{kij}}{\sqrt{\sigma}} \frac{\partial E_j}{\partial q^i} = -\frac{1}{c} \frac{\partial B^k}{\partial t}. \tag{5.174}$$

We see that the electric field becomes singularly related to the time-varying magnetic field as $\omega r \to 1$.

Similarly Equation (5.164) in explicit form

$$\frac{1}{\sqrt{\sigma}} \epsilon^{kj\ell} \frac{\partial H_\ell}{\partial q^j} = \frac{4\pi \rho_e}{c} \frac{dq^k}{dt} + \frac{1}{c} \frac{\partial D^k}{\partial t}, \tag{5.175}$$

shows that the magnetic field is singularly related to the current (including the displacement current) as $\omega r \to 1$.

This singular behaviour actually represents the limit to the usefulness of these coordinates. As the rotational speed approaches c, all time derivatives in this system of coordinates tend to zero as seen by an inertial observer (for a finite proper time interval), so the singularities are averted.

Finally, the constitutive relations may be computed from the explicit relations

$$D^j = \frac{E^j}{\sqrt{(1 - \omega^2 r^2)}} + \frac{1}{\sigma} \epsilon^{jn\ell} H_n g_\ell, \tag{5.176}$$

$$B^k = \frac{H^k}{\sqrt{(1 - \omega^2 r^2)}} + \frac{\epsilon^{kji}}{\sqrt{\sigma}} g_j E_i. \tag{5.177}$$

The three-fields are mixed unless the cross products are zero. In that case $\sqrt{g_{00}}$ plays the rôle of a dielectric constant and a magnetic permeability.

References

1. Jackson, J.D. (1999) *Classical Electrodynamics* (3rd edn), John Wiley & Sons Ltd., Chichester.
2. Goldstein, H., Poole, C. and Safko, J. (2002) *Classical Mechanics*, AddisonWesley, San Francisco.
3. Caldwell, A., Lotov, K., Pukhov, A. and Simon, F. (2009) *Nature Physics*, **5**, 363.
4. Rohrlich, F. (1965) *Classical Charged Particles*, AddisonWesley, Reading, MA.
5. Landau, L.D. and Lifshitz, E.M. (1975) *The Classical Theory of Fields*, Pergamon Press, Oxford.

www.wiley.com/go/henriksen

Part II

Relativity With the Gravitational Field

6

Gravitational Structure of Space-Time

Whose genius has the power to utter song, fit for the grandeur of the way things are ...

Lucretius, V, 12: Humphries translation

6.1 Prologue

Previously in this book, we have been able either to ignore the manifold structure of inertial space-time or to treat it as a manifold with a fixed and given Minkowski metric. In the first approach we based the entire treatment on the Lorentzian invariance of the electromagnetic equations plus a positivist view of the resulting structure of space-time. In the second approach, space-time is a Lorentzian manifold that possesses the Minkowski metric structure at least locally. One should remember that space-time is a theoretical construct, while the positivist view is more or less forced upon us by experiment. Nevertheless the positivist results can be expressed in terms of four-vectors and their transformation properties.

Introducing curvilinear coordinates does not change the underlying metric structure, but a general description of four-vectors and spatial distances is required. The description remains simplest in Galilean coordinates, which are always available globally in inertial systems. However, in non-inertial reference systems, such as in a rotating system of coordinates, these are generally available only locally.

These considerations have led us to formulate the metric structure of space-time, and to define constant vector and tensor quantities in a very general way. Without quite realizing it, perhaps, we have been led to construct a physical theory that is invariant in form under coordinate transformations more general than those of Lorentz. This property is often referred to as 'covariance', although it has little to do with the vector component of the same name.

When non-inertial reference frames are chosen, temporal and spatial coordinates can become mixed in non-Lorentzian ways. The metric matrix components may be functions of time as well as of space, as for example when coordinates in uniform acceleration relative to inertial observers are chosen. Nevertheless, global Galilean coordinates always exist, and the underlying space-time is said to be 'flat'. This flatness is the existence of parallel world lines and parallel spatial lines, both of which satisfy the Euclidian axiom regarding parallel lines that never meet. A plane surface is flat in this sense, although the surface of a sphere is not. This difference reflects the 'curvature' of the spherical surface.

We have seen that the physical theory of the electromagnetic force fits this scheme admirably. This is not surprising since the whole apparatus was constructed in order to accommodate its properties. However, there are some unsatisfying conceptual aspects of this theory despite its practical importance.

We have not yet answered Mach satisfactorily in that we have no explanation for the existence of inertial frames. Moreover, Newtonian gravity does not fit readily into Minkowski inertial space-time, since the latter is based on electromagnetic causality. Attempts [1] to construct a special-relativistic theory of gravity fail to allow for the bending of light in a gravitational field. The question raised by Einstein in this connection is in effect 'might we not solve both of these problems together'?

Based on our earlier discussion of the motion of a test particle in generalized coordinates, the gravitational force appears rather insistently as due to a non-inertial metric structure. The motion of a test particle in generalized coordinates (Equation (4.84)) would react to a non-inertial metric through the term containing the Christoffel symbol. An example is the emergence of inertial forces when rotating observers are used. Such forces share with our experience of gravity the property of accelerating all particles independently of their inertial mass. In Newtonian gravity this is due to the equivalence of inertial mass and gravitating mass, but we can dispense with mass altogether if we regard motion under gravity simply as being geodesic motion in the appropriate space-time metric geometry. Mass re-emerges ultimately as the source of the appropriate space-time metric, but it is unique in character. This implies that we adopt as exact the equality of gravitational and inertial mass (which is known to be true to high, but finite, accuracy experimentally). This assumption is known as the 'weak principle of equivalence'.

This description of gravity will require finding a metric of space-time that is more than merely the result of arbitrary transformations from Minkowski space-time. We know that the strength of a gravitational field depends on the quantity and distribution of matter present, and hence the metric structure will have to reflect this dependence. This is an astonishing idea, that the very structure of space-time should be a part of physics! Since matter is frequently in motion and so changing its distribution, the metric of space-time will have to be dynamic in an essential way. The moving matter will establish the dynamic space-time along whose geodesics it moves. Fortunately, as with electromagnetic fields and currents, we can normally separate the matter creating the metric of space-time from the test particle motion in which we are interested. In any case, once the metric is known, motion under gravity is given by Equation (4.84) in such a 'metric theory of gravity'.

Figure 6.1 *Here is a much better inertial frame than the Earth itself. The freely falling space station removes the gravitational field of the Earth (close to the centre of mass). It is not inertial with respect to the distant Universe, but these are small effects. Source: Reproduced by permission of NASA (See Plate 7.)*

The great achievement of Einstein (and formally, Hilbert [2]) was ultimately to write a set of gravitational field equations that allow the metric of space-time to be deduced from the distribution of the matter in it. In general the resulting space-time is 'curved' in the sense that initially parallel geodesics ultimately converge or diverge.

However, what becomes of inertial frames in this picture? Because all matter falls with the same acceleration in a gravitational field (established by experiment to high precision), it is possible to place oneself in a reference frame that accelerates with the local matter. It is only possible locally because of the tidal effects of gravity. These reflect the 'curvature of space-time', since that is responsible for the lack of permanent parallelism of geodesics. The usual example is that of a freely-falling elevator near the surface of the earth. The elevator must be small enough and must fall over a sufficiently small distance (equivalently for a sufficiently short time) that convergence of the geodesics is not observable.

The statement of the 'strong equivalence principle' is, that all physics in such a locally freely-falling frame should be the same as that observed in an inertial frame, that is, the local space-time of a freely falling observer is 'flat' or 'Minkowskian'. Thus locally freely-falling frames become the definition of inertial frames of reference (see Figure 6.1). In this way Mach is partly satisfied, since inertial frames of reference are preferred only in the sense that they 'remove' (by moving geodesically) the local gravitational field. The gravitational field is associated with curvature of space-time, and does reflect the influence of the mass in the wider Universe as Mach would have it. However, a sufficiently isolated space-time should again be 'flat' and globally inertial. That this flatness must reflect the Lorentz invariance of electromagnetic theory (Minkowski metric) may seem mysterious, but we have seen that it stems uniquely from positivism and the assumption of the vacuum speed of light as a maximum signal speed. It is the rôle of the speed of light that remains mysterious.

Strictly speaking this definition of inertial frames only holds exactly at a point in the curved space-time, and so a more precise statement is that inertial space is a 'tangent space' to gravitational space-time. That is, one can construct orthogonal axes at a point in general space-time, each tangent to a geodesic curve through that point. These may be interpreted as Galilean coordinates for the observer at the origin. Although exactly Galilean only at the origin, they hold to some order in a finite region. The useful analogy is that of a tangent plane to any curved surface.

It might be of interest to consider briefly an alternative to the above picture of dynamic space-time, using only the Minkowski nature of inertial frames. We remark that a continuous field of inertial observers also requires a dynamic metric. It does not imply a dynamic space-time, however, but rather is a coordinate property after transforming between freely-falling or inertial frames.

Let the inertial observer at a given point in space-time be O. Moreover, suppose that a continuous field of locally inertial observers are moving with respect to O with velocity $\mathbf{u}(t, \mathbf{r})$. The coordinates of distant events can be found by inverting Equation (2.40) from the distant inertial frame. Every observer uses Galilean coordinates, but the axes may be arbitrarily rotated with respect to those of O. There will necessarily be some uncertainty in the precisely 'punctual' nature of this velocity distribution, so it is better to think of the inertial structure as an array of overlapping small hyper-cubic 'boxes' (each dimension extended along a geodesic) in space-time.

In each box the local inertial observer uses the Minkowski metric to measure the distance between events. How will O measure the distance between closely separated but distant events? A procedure that produces the necessary 'connection' is to use again Equation (2.40), while taking into account the variability of $\mathbf{u}$. With this transformation between inertial frames, O can transform to the coordinates of events in the frame of the local inertial observer. The distance between these events according to the local (say O'' observer) is

$$ds^2 = \eta_{ad} dx''^a dx''^d. \tag{6.1}$$

But $dx''^a = \mathcal{L}^a{}_b dx^b + (\partial \mathcal{L}^a{}_b / \partial x^c) dx^c x^b$ according to Equation (2.40). It is not difficult to show that (see Problem) this becomes

$$\begin{aligned} ds^2 &= \eta_{ad} \left(\mathcal{L}^a{}_b \mathcal{L}^d{}_e + \frac{\partial \mathcal{L}^a{}_c}{\partial x^b} \frac{\partial \mathcal{L}^d{}_f}{\partial x^e} x^c x^f + \mathcal{L}^d{}_e \frac{\partial \mathcal{L}^a{}_f}{\partial x^b} x^f + \mathcal{L}^a{}_b \frac{\partial \mathcal{L}^d{}_f}{\partial x^e} x^f \right) dx^b dx^e \\ &\equiv g_{be} dx^b dx^e. \end{aligned} \tag{6.2}$$

We note that

$$\frac{\partial \mathcal{L}^a{}_f}{\partial x^e} = \frac{\partial \mathcal{L}^a{}_f}{\partial u^k} \frac{\partial u^k}{\partial x^e}, \tag{6.3}$$

according to Equation (2.40).

Hence the metric that would apply everywhere for O, namely g_{be}, is clearly of a general kind wherein every coefficient depends on all four coordinates. It is also a symmetric four-by-four matrix so that it has ten independent components. The choice of four arbitrary functions, corresponding to the four arbitrary coordinate transformations, reduces the number of physical components to six.

The test particle motion for O could in principle now be found from Equation (4.84), *if the field of inertial observers were known*. Such an approach would extend the view that dynamic space-time is only a convenient construct based ultimately on the invariance properties of the electromagnetic equations between inertial observers. There are many special relativistic theories of gravity [1], which are normally rejected experimentally. On the present view they would only apply in a local box, so the general criticism of not allowing a global bending of light would not necessarily apply. For a first approximation, even Newtonian gravity might serve to establish the inertial frames.

It should be emphasized that the preceding discussion is only another way of motivating a metric theory, and it is speculative, untested and inelegant. The Einstein theory regards the general metric of space-time as the fundamental quantity and relates its curvature to the distribution and motion of matter in a very satisfactory way. Understanding this relation will require a brief foray into the realm of Gauss/Riemann geometry, which we append only at the end of this chapter.

One should note as a general remark that at each space-time event it is possible to introduce a local inertial frame that is freely falling, but nevertheless coincident with that event. This is the tangential Lorentz frame of reference. The propagation of a light ray will follow $ds = 0$ in this frame and hence also locally in the background reference frame. This means that one way to constrain the path of a light ray in a general space-time is to require

$$ds^2 = g_{ab}dq^a dq^b = 0. \tag{6.4}$$

Except in simple cases, this does not afford a complete solution for what is called the 'null geodesic'. The complete solution follows from Equation (4.84) when ds is replaced with some other parameter along the path, say $d\lambda$. The resulting equation continues to be the condition for the parallel transport of the tangent vector to the geodesic.

Problem

6.1 Derive the metric imposed on space-time by the inertial field $\mathbf{u}(t, \{x^k\})$ in the form of Equation (6.2).

6.2 The Weak Gravitational Field

A weak gravitational field according to the Einsteinian ideas of the prologue will be contained in a metric of space-time, which one expects to be close to the Minkowski metric. Moreover, the small deviations due to the presence of the gravitational field should result in geodesic motion in the weak-field, low-velocity limit that agrees with Newtonian motion under gravity. Both of these limits are defined in terms of c, since in units with $c = 1$ the spatial velocity $\mathbf{v}$ and the Newtonian potential Φ are small. We should remember that, in these units, a unit time interval corresponds to a very large spatial interval (by a factor c). Hence we can expect the lowest order changes in the Minkowski metric in local space to enter through g_{00}, the coefficient of dt^2.

Minkowski space is not dynamic so it is reasonable to require the space-time metric to be static in the weak-field limit. A static metric must clearly satisfy $\partial g_{ab}/\partial t = 0$, but it must also have $g_{0i} = 0$ for all values of the index i. This is so since a truly static metric must be invariant under a reversal in time direction, which the temporal-spatial cross terms are not. Since every mass moves every other, such a limit applies only for the space-time of isolated masses interacting with test particles.

The weak-field general gravitational metric should thus have the form

$$ds^2 = g_{00}(\{q^k\})dt^2 + g_{ij}(\{q^k\})dq^i dq^j, \tag{6.5}$$

where g_{00} is close to $+1$. The g_{ij} will deviate from -1 to the same order, but because of the smallness of the local spatial intervals these terms can be neglected to lowest order in $1/c^2$ for an isolated mass. For this reason they do not appear in the Newtonian (weak-field, low-velocity) limit [3].

We may calculate the geodesic motion in such a metric using Equation (4.84). For the spatial components in lowest order this equation gives

$$\frac{dv^i}{ds} = -\Gamma^i_{00}(v^0)^2, \tag{6.6}$$

where we have remembered that $v^0 \equiv dt/ds \gg v^i \equiv dq^i/ds$. From the form of the Christoffel symbol (4.86) together with the static form of the metric, we find

$$\Gamma^i_{00} = -\frac{1}{2}g^{ij}\frac{\partial g_{00}}{\partial q^j}. \tag{6.7}$$

Combining these last two equations with $v^0 = dt/ds$ gives

$$\frac{dv^i}{dt} = \frac{1}{2}g^{ij}\frac{\partial g_{00}}{\partial q^j}v^0. \tag{6.8}$$

From the static metric, $g_{ab}v^a v^b = 1 = g_{00}(v^0)^2 - g_{ij}v^i v^j \approx g_{00}(v^0)^2$; that is, to lowest order $v^0 = dt/ds \approx 1/\sqrt{g_{00}}$. This may also be inferred from the next order approximation in the zero component of the geodesic equation (see Problem). Finally, then, in this low-velocity, weak-field limit, the geodesic three-acceleration is

$$\frac{dv^i}{dt} = g^{ij}\frac{\partial\sqrt{g_{00}}}{\partial q^j}. \tag{6.9}$$

Now we must have $g_{00} > 0$ to avoid time reversals and so, to within a units change in the time coordinate, we may without loss of generality set

$$g_{00} = e^{2\Phi(q^i)}. \tag{6.10}$$

In the weak-field limit this is $g_{00} \approx 1 + 2\Phi$ and so Equation (6.9) gives $dv^i/dt = g^{ij}(\partial\Phi/\partial q^j)$. Using orthogonal coordinates and physical components (see for example Equations (1.52) and (1.53) one must use $\sqrt{-g_{\alpha\alpha}}$ for $\sqrt{g_{\alpha\alpha}}$), this is the Newtonian form $dv(\alpha)/dt = -\nabla_{(\alpha)}\Phi$.

Consequently in the weak-field, low-velocity limit the gravitational field can be contained in the metric (restoring c)

$$ds^2 = c^2 dt^2 \left(1 + \frac{2\Phi}{c^2}\right) + g_{ij} dq^i dq^j. \tag{6.11}$$

In this expression the g_{ij} are close to η_{ij} in Galilean coordinates, and hence close to the Euclidian curvilinear form in curvilinear coordinates. An application of the gravitational field equations in standard texts (e.g. [3]) shows that the next order produces a factor $1 - 2\Phi/c^2$ multiplying the η_{ij}. Most importantly, the geodesic motion agrees with motion under Newtonian gravity *if we identify the function* Φ *with the Newtonian gravitational potential*. This will allow us to find some effects of Einsteinian gravity, although clearly not all, since Newtonian orbits do not explain the precession of Mercury's orbit.

Problem

6.2 Use the zeroth component of the geodesic equation to show that $v^0 = 1/\sqrt{g_{00}}$ in lowest order. This will require calculating Γ^0_{00} and Γ^0_{0j}.

As a tangible example of the weak-field, low-velocity limit, consider again the motion of a particle under constant acceleration (Chapter 3). Observer O is at rest in a weak gravitational field and observer O' moves under free-fall along the positive z axis. Over a sufficiently short time starting from O' coinciding with O at rest (for a given acceleration this is effectively $t \ll c/a'$) the velocity will be low and the acceleration may be taken as constant (i.e. a local gravitational field).

Observer O' is inertial according to our basic concepts (see Prologue) and so $ds^2 = c^2 dt'^2$. But according to Equation (3.32) this can be written as $c^2dt^2/(1 + (a't/c)^2)$. Using Equation (3.30) to eliminate t this becomes

$$ds^2 = \frac{c^2 dt^2}{\left(1 + \frac{a'z}{c^2}\right)^2}. \tag{6.12}$$

However, in the low-velocity limit this is also $g_{00}c^2dt^2$ according to the non-inertial observer O, so that $g_{00} \approx 1/(1 + a'z/c)^2$. But if a' is due to gravity, then $a'z = -\Phi$ (zero potential at the origin O) and hence once again $g_{00} = 1 + 2\Phi/c^2$ in this limit.

Although the next phenomenon to be discussed is not restricted to weak constant fields (see the next section), we are able to see its dependence on the Newtonian potential in the present weak-field case. This phenomenon is the dependence on gravitational potential of electromagnetic frequency. This is often referred to as the gravitational 'red shift', but that is the case only if we are receiving waves from a source at a lower gravitational potential than us.

In general the situation is the following. Consider two observers O_1 and O_2, each of whom are at rest in a static space-time. The coordinate time interval is dt and the proper time interval (i.e. interval on the local world line) is $ds = \sqrt{g_{00}}dt$ for each observer. If a light ray is sent from one observer to the other, then the time for a given phase front to traverse the distance between them is (since $ds = 0$ along the path)

$$\Delta t = \int \frac{d\ell}{\sqrt{g_{00}}}, \tag{6.13}$$

where $d\ell^2 \equiv \sigma_{ij} dq^i dq^j$. In a constant metric this does not change in time and so the coordinate period T_o of a complete oscillation will be transported unchanged from one observer to the other. However, this coordinate interval produces the proper period $T = \sqrt{g_{00}}\, T_o$ locally. Hence the proper periods are related by $T_0 = T_1/\sqrt{g_{00}(1)} = T_2/\sqrt{g_{00}(2)}$, or in terms of the wave frequency

$$\frac{\nu_2}{\nu_1} = \frac{\sqrt{g_{00}(1)}}{\sqrt{g_{00}(2)}}. \tag{6.14}$$

This is a general result in a constant gravitational field. One might think, on regarding our argument closely, that it should apply unchanged to stationary gravitational fields since then the g_i are constant. This is indeed true along an open path, but on closed paths the Sagnac effect (e.g. 3.12) will arise. We have seen that synchronization, or 'phase closure' in our present context, cannot in general be effected on a closed path in a stationary (e.g. rotating) field.

The gravitational frequency shift is very important since we know of only two other effects that can change the frequency of a resonant spectral feature. The other two are the linear and transverse Döppler shifts which we have seen arise from different physical effects (see Equation (3.4)) of the finite speed of light. Spectroscopic analysis is vital in exploring the physics of all manner of electromagnetic sources, and the correct transport of frequency is essential to time-keeping and hence to 'micro-navigation' (global positioning systems). This is the complete local census of frequency shifts, but there is one more that is essentially non-local. This is the cosmological red-shift due to the expansion of the Universe, which discussion we leave mainly to others.

However, because the cosmological red-shift is an unavoidable concept in the context of modern physics and astronomy, we pause to discuss it briefly. The latest observations (e.g. WMAP, Wilkinson Microwave Anisotropy Probe) permit describing the Universe as an expanding Euclidian space. That is, it is 'curved' only in the time dimension. The metric description then is rather simply (in Galilean coordinates)

$$ds^2 = c^2 dt^2 - a(t)^2(dx^2 + dy^2 + dz^2). \tag{6.15}$$

The 'scale-factor' $a(t)$ must be found from Einstein's equations applied to the Universe, but this aspect does not concern us here. Although this metric has a simple form, it is neither static nor stationary.

Consider a plane electromagnetic wave propagating towards us from a vast distance along the z axis. Two successive wavefronts are separated by the coordinate distance Δz, which correponds to the proper distance $a(t)\Delta z$ according to Equation (6.15). A local observer will take this proper distance as the wavelength λ. Hence at two different coordinate times t_o and t_e, with $t_o > t_e$ say, the ratio of the wavelengths will be

$$\frac{\lambda_o}{\lambda_e} = \frac{a(t_o)}{a(t_e)}. \tag{6.16}$$

The 'red-shift' of a wave emitted at cosmic coordinate time t_e as observed on Earth at cosmic coordinate time t_o is defined as $Z \equiv \lambda_o/\lambda_e - 1$. Hence finally

$$1 + Z = \frac{a(t_o)}{a(t_e)}. \tag{6.17}$$

There is an easy conversion to frequency, so that this is indeed another type of frequency variability. It relies directly on the global dynamic of space-time. Although it may be approximated locally as a Döppler shift, we see that relative velocity does not arise in its derivation. It is worth remarking that for Mach, the metric (6.15) would be the candidate for the metric far from any local matter. At any fixed cosmic coordinate time, it becomes the Minkowski metric by a units change in spatial measurements. It is in this sense the archetypal inertial frame, but it changes with the cosmic epoch.

We return to more local matters. In the weak-field limit our result (6.14) becomes (restoring c)

$$\frac{\nu_2}{\nu_1} = 1 + \frac{\Phi_1 - \Phi_2}{c^2}. \tag{6.18}$$

This result can be obtained by a relativistic/quantum description of light grafted on to a classical view of gravity. Thus by conserving the energy of a photon moving in a gravitational potential and attributing to it an inertial mass $h\nu/c^2$, we obtain $h\nu + (h\nu/c^2)\Phi = constant$. Writing this at two points and solving for ν_2/ν_1 in the weak-field limit yields Equation (6.18). However, there is no coherent reason to assign an inertial mass to a photon.

An electromagnetic wave, leaving the surface of the sun at position 1 and arriving at an Earth observer at position 2, travels from a potential $\Phi_1 \approx -1.91 \times 10^{11}\mathrm{m}^2/\mathrm{s}^2$ to $\Phi_2 \approx -6.25 \times 10^7\mathrm{m}^2/\mathrm{s}^2$. In each case the zero point of the potential is at spatial infinity. Hence by the result (6.18), the received frequency is shifted to the red according to

$$\frac{\Delta\nu}{\nu_1} \approx -\frac{GM_\odot}{c^2 R_\odot} \approx -2.1 \times 10^{-6}. \tag{6.19}$$

Hence this shift is about two parts in a million. It is not difficult in principle to detect such a shift in solar spectral lines, but the Döppler shift due to stochastic surface velocities tends to conceal it. The magnitude corresponds to about 637 m/sec in the linear Döppler shift, which is, however, already the speed of a modest modern rifle bullet. If the motion is truly turbulent then it contributes mostly to a line broadening, and a shift in a mean line frequency might be measured. Alternatively, lines that form slightly above the main turbulent motion of the photosphere can be measured.

It appears that new techniques are still being developed to reduce the imprecision in the solar red-shift measurement. A relatively modern published value [4] gave 0.9 of the expected value with about 2% uncertainty. This is, however, better than the uncertainty of about 10% that was achieved in the terrestrial Mössbauer measurement [5]. That experiment measured the shift in the frequency of the ^{57}Fe line at 14.4 keV over a difference in height of 22.6 metres. The result was as predicted within errors.

There is, however, a much more familiar test of this effect. It requires a correction to the rate of atomic clocks carried by GPS satellites, in order to synchronize them with clocks on Earth and with one another.

One way to calculate this correction is to take an approximately inertial coordinate system, fixed at the centre of the Earth. We use spherical-polar curvilinear coordinates with the rotation axis of the Earth as polar axis. Then the interval of proper time for a receiver on the Earth (observer 2 say) is $ds_2 = dt\sqrt{1 + 2GM_\oplus/(c^2 R_\oplus)}$. Here the subscript $\oplus$ refers to a terrestrial value. We neglect the rotational velocity of the Earth, basically because objects at rest on the surface of the Earth are *not* in orbit! However, a GPS satellite *is* in orbit at a typical height above the Earth's surface of about 20, 000 km. In general the geodesic motion of each satellite must be carefully tracked by multiple ground stations. For the sake of our example, however, suppose that the satellite (observer 1) is in circular equatorial orbit and the receiver is located on the Earth's equator.

The proper time interval of the satellite will in these coordinates be found from

$$ds_1^2 = \left(1 + \frac{2GM_\oplus}{c^2 R_o} dt^2 - \frac{R_o^2 d\phi^2}{c^2}\right) dt^2, \tag{6.20}$$

where R_o is the orbital radius and $d\phi$ follows the satellite in its orbit. This includes both the gravitational effect and time dilation. In fact in this approximation, $d\phi = \Omega_o dt$, where Ω_o is the Newtonian orbital angular velocity. This allows ds_1 to be rewritten according to

$$ds_1^2 = \left(1 + \frac{2GM_\oplus}{c^2 R_o} - \frac{\Omega_o^2 R_o^2}{c^2}\right) dt^2 \equiv \left(1 + \frac{GM_\oplus}{c^2 R_o}\right) dt^2. \tag{6.21}$$

Equation (6.21) includes the transverse Döppler shift that acts against the gravitational shift. We know that time dilation causes moving clocks to run slow, but this is more than counterbalanced here by the satellite clock being at a higher gravitational potential than the Earth clock. In our typical example the gravitational effect predominates, but it is reduced by a factor of a half due to the time dilation of the satellite clock. The first-order Döppler shift does not affect the rate at which clocks run.

Using the net effect (6.21), it follows from our earlier discussion of the gravitational frequency shift that the ratio of the satellite and terrestrial proper times is

$$\frac{ds_1}{ds_2} - 1 = \frac{\nu_2}{\nu_1} - 1 = -\frac{GM_\oplus}{c^2 R_\oplus}\left(1 - \frac{1}{2}\frac{R_\oplus}{R_\oplus + h}\right), \tag{6.22}$$

where h is the height of the satellite. The term on the right of this last equation is about -6.15×10^{-10} for our configuration. This means that the interval between events (such as emitted and received wave fronts) for the satellite observer is shorter than for the Earth observer. The satellite clock runs fast by about 53 microseconds per day.

It is essential for the operation of the GPS system that the clocks on the satellites be synchronized with clocks on the Earth. Each satellite transmits the time at which it emitted a given wave front, and the difference between that time and the received time is the light distance from the satellite. Several satellites are required to get a fix on a terrestrial receiver position. A timing error of a microsecond is a distance error of 300 metres! There is no way to get the relative clock rates correctly to a fraction of a microsecond without allowing for the gravitational frequency shift.

The differing clock rate is taken into account for each satellite and is checked continually against ground stations with synchronized terrestrial clocks. To avoid the Sagnac effect these clocks are synchronized by slow transport or by correction to a polar clock time.

The naive Newtonian approach of this section serves to suggest yet another gravitational effect on light, namely on its trajectory. A null geodesic will only be a straight line in a locally freely-falling reference frame. An observer O at rest in the local space-time is effectively accelerated with respect to this straight line with minus the local gravitational acceleration $-\mathbf{g}$. If dy is the local displacement in the direction of $\mathbf{g}$ for O, then adopting the Newtonian weak-field limit, $d^2y/dt^2 = g$ and $v_y = gt$. The radius of curvature of the photon trajectory for O is then (using a standard formula from calculus) $r_c \equiv (1 + (gt/c)^2)^{3/2}(c^2/g)$, or simply c^2/g for very short times (localized region in space-time). Hence on traversing a distance $\delta\ell$ for O, the photon appears to deviate from a straight line by the angle

$$\delta\phi = \frac{g\delta\ell}{c^2}. \tag{6.23}$$

For the GPS system of satellites there is an error because of this curvature if the light path is assumed to be a straight line. The above equation suggests as a crude estimate that this effect is only important if roughly centimetre accuracy in GPS position is desired. This is a rough estimate of what is called the Shapiro time delay [6].

In the case of a light ray passing near the limb of the sun with an impact parameter b, we might expect the bending to be produced symmetrically from infinity to $r = b$ and from $r = b$ to infinity. The most important behaviour will be near $r = b$ and we can estimate $\ell \approx 2b$ where $g \approx GM_\odot/b^2$. This suggests a bending angle $\delta\phi \approx 2GM_\odot/(c^2b)$, which is the classical value found essentially by treating a photon as a Newtonian particle. It is, however, wrong by a factor of two (100%) compared with the observed result. This suggests that we are missing something essential in this limit. We return to this question below. The importance is dramatic for astronomy, since we now understand that we observe distant objects in many cases through foreground achromatic lenses that distort and amplify.

6.3 Constant or Stationary Gravitational Field

We begin by summarizing the metric descriptions of the gravitational fields of interest in this section. The constant gravitational field will be described by the metric

$$ds^2 = g_{00}dt^2 + g_{ij}dq^i dq^j, \tag{6.24}$$

where the metric coefficients are independent of t and the g_{ij} are diagonal. The stationary gravitational field is usually expressed as the metric

$$ds^2 = g_{00}dt^2 - \sigma_{ij}dq^i dq^j, \tag{6.25}$$

where σ_{ij} is given by Equation (5.136). The limitation of the diagonalization (time synchronization) to open paths should not however be forgotten.

Occasionally we make use of a spherically symmetric metric (e.g. [3]) in the form

$$ds^2 = e^{\Phi}dt^2 - e^{\psi}dr^2 - r^2(d\theta^2 + \sin^2\theta d\phi^2), \tag{6.26}$$

where the arbitrary functions Φ and ψ are functions only of r and possibly t. If they do depend on t then in these coordinates the metric is evidently not constant, presumably due to moving matter, but it is still spherically symmetric. The arbitrary functions allow for the variation of proper time and proper radial distance in a real gravitational field.

The proper time variation was already encountered in the weak-field approximation of the previous section where it gave rise physically to the gravitational frequency shift. The radial distortion is related to the spatial curvature that is produced in a strong gravitational field. We have not yet encountered curvature, but it is responsible for the correction to the bending of light formula that we estimated in the weak-field approximation. In spherical symmetry there are no angular distortions. The proper distances remain $rd\theta$ and $r\sin\theta d\phi$, although r is not in general the physical radial distance.

The equations of motion are given as the system (4.84), but it is useful often to step back from this general Euler-Lagrange equation to the action itself. We recall that this is

$$\mathcal{S} = -mc\int ds, \tag{6.27}$$

or explicitly

$$\mathcal{S} = -mc\int \sqrt{g_{ab}\dot{q}^a\dot{q}^b}ds, \tag{6.28}$$

Here $\dot{q}$ refers to the derivative with respect to s. It is most important when varying this action to remember that, *after* the variation, the square root factor may be set equal to unity. The Euler-Lagrange equations that result from the variation take the usual form in terms of the degrees of freedom $\{t, q^a\}$ and the effective Lagrangian (the integrand in the action). However, one must recall that d/ds is a total derivative in these equations. Example (6.1) should make these comments clearer.

This procedure allows the Euler-Lagrange equations to be expressed directly in terms of the metric coefficients. In fact, by comparing the results in such cases with the general form of Equation (4.84), one can infer the explicit Christoffel symbols (see Problem and Example). If one is only interested in the variations of the spatial coordinates, then in the action dt can be used as the path parameter rather than ds (see Problem). Indeed, any other coordinate may be used as the path parameter, which allows for different projections of the orbit.

Null geodesics must be treated by using the metric, together with Equation (4.84) and an appropriate path parameter. This parameter may be any one of the coordinates or any other parameter that is continuous on the null geodesic. Such a parameter does not define distance since this is null on the geodesic, and so it is an 'affine' parameter. In affine geometry there is no concept of distance.

Example 6.1

In this example we show how to deduce the Christoffel symbols of the second kind for the general spherically-symmetric metric (6.26). We use the Euler-Lagrange equations that follow by varying the action (6.28). The effective Lagrangian in Equation (6.28) when the spherically symmetric metric is used is

$$L \equiv \sqrt{e^{\Phi}\dot{t}^2 - e^{\psi}\dot{r}^2 - r^2(\dot{\theta}^2 + \sin^2(\theta)\,\dot{\phi}^2)}. \tag{6.29}$$

We begin with the zeroth coordinate t. We put $c = 1$ throughout, but because of the raised index in the definition of the symbols of the second kind, their form is independent of this assumption. From $(d/ds)(\partial L/\partial\dot{t}) - \partial L/\partial t = 0$ there follows

$$\frac{d}{ds}(e^{\Phi}\dot{t}) - \frac{e^{\Phi}\partial_t\Phi\dot{t}^2}{2} + \frac{e^{\psi}\partial_t\psi\dot{r}^2}{2} = 0, \tag{6.30}$$

where we have set $L = 1$ after the differentiations. The first term must be carefully expanded since it is a total derivative as

$$\frac{d}{ds}(e^{\Phi}\dot{t}) = e^{\Phi}\ddot{t} + e^{\Phi}\partial_r\Phi\,\dot{r}\dot{t} + e^{\Phi}\partial_t\Phi\dot{t}^2. \tag{6.31}$$

Hence finally the zeroth geodesic equation is

$$\ddot{t} + \frac{\partial_t\Phi\dot{t}^2}{2} + \dot{t}\dot{r}\,\partial_r\Phi + \frac{e^{(\psi-\Phi)}\partial_t\psi\dot{r}^2}{2} = 0. \tag{6.32}$$

Comparing with Equation (4.84) we may now read off the non-zero Christoffel symbols as (remembering that mixed lower indices appear twice)

$$\Gamma^0_{00} = \frac{\partial_t\Phi}{2}, \quad \Gamma^0_{01} = \frac{\partial_r\Phi}{2},$$
$$\Gamma^0_{11} = \frac{e^{(\psi-\Phi)}\partial_t\psi}{2}. \tag{6.33}$$

For the radial coordinate we must use $(d/ds)(\partial L/\partial\dot{r}) - \partial L/\partial r = 0$, from which there follows (after $L = 1$)

$$-\frac{d}{ds}(e^{\psi}\dot{r}) - \frac{1}{2}\left(e^{\Phi}\partial_r\Phi\,\dot{t}^2 - e^{\psi}\partial_r\psi\,\dot{r}^2 - 2r\left(\dot{\theta}^2 + \sin^2(\theta)\,\dot{\phi}^2\right)\right) = 0. \tag{6.34}$$

Carefully expanding the total derivative once more yields the radial geodesic equation as

$$\ddot{r} + \frac{\partial_r \psi}{2}\dot{r}^2 + \frac{e^{(\Phi-\psi)}\partial_r \Phi}{2}\dot{t}^2 - re^{-\psi}(\dot{\theta}^2 + \sin^2(\theta)\ \dot{\phi}^2) + \partial_t \psi\ \dot{r}\dot{t} = 0. \tag{6.35}$$

Comparing again with the form (4.84) we infer

$$\Gamma^1_{00} = \frac{e^{(\Phi-\psi)}\partial_r \Phi}{2} \qquad \Gamma^1_{11} = \frac{\partial_r \psi}{2},$$

$$\Gamma^1_{22} = -re^{-\psi} \qquad \Gamma^1_{33} = -re^{-\psi}\sin^2(\theta), \tag{6.36}$$

$$\Gamma^1_{01} = \frac{\partial_t \psi}{2}. \tag{6.37}$$

The remaining cases are more straightforward. From the θ Euler-Lagrange equation there follows (after $L = 1$)

$$-\frac{d}{ds}(r^2\dot{\theta}) + \sin(\theta)\cos(\theta)\, r^2\, \dot{\phi}^2 = 0, \tag{6.38}$$

whence on rearranging

$$\ddot{\theta} + \frac{2}{r}\,\dot{r}\,\dot{\theta} - \sin(\theta)\cos(\theta)\,\dot{\phi}^2 = 0. \tag{6.39}$$

From this we read off in the same fashion as previously

$$\Gamma^2_{12} = \frac{1}{r} \qquad \Gamma^2_{33} = -\sin(\theta)\cos(\theta). \tag{6.40}$$

Finally, since there is no dependence on ϕ in the Lagrangian, we have the specific 'angular momentum' as a conserved quantity since

$$\frac{d}{ds}(r^2\sin^2(\theta)\,\dot{\phi}) = 0. \tag{6.41}$$

Expanding the derivative gives

$$\ddot{\phi} + \frac{2}{r}\,\dot{r}\,\dot{\phi} + 2\cot(\theta)\,\dot{\theta}\,\dot{\phi} = 0, \tag{6.42}$$

and hence, on making the comparison with the corresponding component of the geodesic equation,

$$\Gamma^3_{13} = \frac{1}{r} \qquad \Gamma^3_{23} = \cot(\theta). \tag{6.43}$$

The mixed lower index quantities, with the order as given in these results reversed, are equal to the mixed quantities given. All other combinations of indices are zero. These results are useful in dealing with motion under gravity in spherical symmetry. Most frequently they are used in constant form (time derivatives zero), but this is not necessarily the case in continuous matter. It must, however, be so outside isolated point masses, due to a well-known theorem by Birkhoff [7].

Problem

6.3 By using t as the path parameter in Equation (6.28) in place of s, show that one can find the values for the Christoffel symbols given in Equations (6.36), (6.40) and (6.43).

All of the classical techniques that we used in the previous chapter on electromagnetic theory are useful in discussing the motion of a test particle in a given gravitational field. In particular the Hamilton-Jacobi method is important.

We recall that the non-relativistic Hamilton-Jacobi procedure consists of setting the sum of the canonical Hamiltonian and the time derivative of the action equal to zero. The particle momentum in the canonical Hamiltonian is replaced by the spatial gradient of the action, and the numerical value of the Hamiltonian is the energy. In Chapter 5 we saw that for a relativistic charge in an electromagnetic field, both the time derivative of the action and the spatial gradient appear squared. In general for a relativistic particle the energy and the momentum are both contained in the normalization $g_{ab}p^a p^b = m^2c^2$, where p^a is the four-vector momentum mv^a. The relativistic Hamilton-Jacobi equation is thus naturally taken to be (with $p_a \equiv \partial\mathcal{S}/\partial q^a$)

$$g^{ab}\frac{\partial\mathcal{S}}{\partial q^a}\frac{\partial\mathcal{S}}{\partial q^b} - m^2c^2 = 0, \tag{6.44}$$

and we see that the derivatives appear squared as in the charged case.

In Galilean coordinates and in the weak-field limit this equation becomes

$$\left(\frac{1}{c}\frac{\partial\mathcal{S}}{\partial t}\right)^2 - (\nabla\mathcal{S})^2 = m^2c^2. \tag{6.45}$$

This identifies the particle specific energy to within an additive constant and a sign ambiguity as (restore c for clarity)

$$\mathcal{E} - \frac{\partial\mathcal{S}}{\partial t} \equiv cp_0. \tag{6.46}$$

This quantity is conserved along the trajectory of a particle since Equation (6.44) does not contain the time in a constant or stationary metric. This implies that $\partial\mathcal{S}/\partial t$ must be constant.

We choose the positive sign for the energy to agree with the classical limit of a free particle. That is, with $\mathcal{S} = -mc^2\int\sqrt{1 - v^2/c^2}\,dt$ for a free particle, one finds from our definition $\partial_t\mathcal{S} \equiv \mathcal{E} = -mc^2 + mv^2/2$ in the low-velocity Galilean limit. An additive constant is always arbitrary in the definition of $\mathcal{S}$, so we only insist on agreement with the sign of the kinetic energy. We find a better form for the energy in the gravitational context in the next few paragraphs.

Thus $\mathcal{E} = \partial_t\mathcal{S} = cp_0 = c^2 m g_{0a}(dq^a/ds)$, since p_0 is a covariant component of the four-momentum. For a strictly constant field, this becomes $\mathcal{E} = mc^2 g_{00}(dt/ds)$ since the $g_{0j} = 0$. It is conceptually useful to define this energy in a familiar way in terms of a

Lorentz factor and a corresponding three-velocity [3]. To do this in a way that applies to both constant and stationary metric fields, we must consider the quantity $g_{0a}(dq^a/ds)$ more carefully.

For either a constant or stationary metric we have ds from Equation (5.135), which we rewrite slightly as

$$ds^2 = g_{00}d\tilde{t}^2\left(1 - \sigma_{ij}\frac{dq^i}{d\tau}\frac{dq^j}{d\tau}\right). \tag{6.47}$$

Recall that in a constant metric field the synchronized (or orthogonal) coordinate time, $d\tilde{t} \equiv dt - g_j dq^j = dt$, and so the coordinate time has a global significance. For a merely stationary field the synchronization only holds along open curves in general, which does allow, however, for an aperiodic single particle trajectory.

We have taken the opportunity above to define the proper time for an observer at rest in the coordinates $\{q^j\}$ as

$$d\tau = \sqrt{g_{00}}d\tilde{t} \equiv \sqrt{g_{00}}(dt - g_j dq^j). \tag{6.48}$$

It is clear that even in a constant metric, the proper time varies from point to point if g_{00} does. We found an example of this when discussing the weak field limit earlier. This leads to the gravitational frequency shift.

The metric (6.47) together with the definition of proper time for an observer at rest suggests that we define a three-velocity relative to rest observers as

$$v^2 \equiv \sigma_{ij}v^i v^j = \sigma_{ij}\frac{dq^i}{d\tau}\frac{dq^j}{d\tau}, \tag{6.49}$$

so that we may write Equation (6.47) in the familiar form

$$ds^2 = d\tau^2(1 - v^2). \tag{6.50}$$

We can now rewrite the energy as (recall that $g_j \equiv -g_{0j}/g_{00}$)

$$\mathcal{E} = mc^2 g_{0j}\frac{dq^j}{ds} = mc^2\left(g_{00}\frac{dt}{ds} - g_{00}g_j\frac{dq^j}{ds}\right), \tag{6.51}$$

which becomes on using Equation (6.50) and the definition of $d\tau$

$$\mathcal{E} = \gamma mc^2\sqrt{g_{00}}. \tag{6.52}$$

We have set the generalized Lorentz factor equal to

$$\gamma^2 \equiv \frac{1}{1 - \frac{\sigma_{ij}v^i v^j}{c^2}} \equiv \frac{1}{1 - \sigma_{ij}\frac{dq^i}{cd\tau}\frac{dq^j}{cd\tau}}. \tag{6.53}$$

In the constant weak-field, low-velocity limit the energy becomes

$$\mathcal{E} = mc^2 + \frac{m\mathbf{v}^2}{2} + m\Phi, \tag{6.54}$$

as is to be expected. This defines particle energy in our metrics of interest.

One nice feature of the Hamilton-Jacobi procedure for particles is that the motion of light rays or photons can be found in a very similar way. We recall that the motion of a general electromagnetic wave front $S(t,\mathbf{r})$ is written in Galilean coordinates as in Equation (2.10). In Minkowski space this would be $\eta^{ab}(\partial_a S)(\partial_b S)=0$, which becomes in a general metric field

$$g^{ab}\frac{\partial S}{\partial q^a}\frac{\partial S}{\partial q^b}=0. \tag{6.55}$$

This is the null Hamilton-Jacobi equation, but it is often referred to as the 'eikonal' equation ('eikonal' from the Greek word for 'image').

The equality that corresponds to that between the momentum of a particle and the space-time gradient of the action, is the equality between the four wave-vector and the space-time gradient of the phase S. In Minkowski space this vector is $k_a=(\nu/c,-\mathbf{k})$, which is the space-time gradient of the phase $\Phi=\nu t-\mathbf{k}\cdot\mathbf{r}$. In a general metric space-time we must write this gradient as

$$k_a=\frac{\partial S}{\partial q^a}, \tag{6.56}$$

which is the desired relation. Thus $g^{ab}k_ak_b=0$ from the eikonal equation so that the wave vector is null.

Analogously to the energy for a massive particle, we find the coordinate frequency of the light ray from

$$\nu_0=ck_0=\frac{\partial S}{\partial t}, \tag{6.57}$$

but the local proper frequency for a rest observer in the coordinates is found from

$$\nu=\frac{\partial S}{\partial \tau}=\frac{dt}{d\tau}\frac{\partial S}{\partial t}. \tag{6.58}$$

In a constant metric field $d\tau/dt=\sqrt{g_{00}}$. Moreover, in Equation (6.55) there is no dependence on t of the coefficients along the path of the light ray, so the equation requires that $\partial_t S$ (the coordinate frequency) be constant along a light ray. The proper frequency at two points will therefore satisfy Equation (6.14) according to Equation (6.58).

In a stationary metric field the preceding argument applies, but now because of the different definition of orthogonal proper time, we find

$$\nu(1)\sqrt{g_{00}(1)}(1-g_j\dot{q}^j)_{(1)}=\nu(2)\sqrt{g_{00}(2)}(1-g_j\dot{q}^j)_{(2)}, \tag{6.59}$$

which applies on open paths. This mixes the gravitational and Döppler frequency shifts. For example, on a uniformly rotating disc we obtain on an open path ($c=1$)

$$\nu\left(\frac{1-\omega r^2(\omega+\dot{\phi}')}{\sqrt{1-\omega^2r^2}}\right)=constant, \tag{6.60}$$

where as usual ω is the angular velocity and ϕ' is the azimuth on the disc.

When $\dot{\phi}' = 0$ the last relation is $\sqrt{1 - \omega^2 r^2}\, \nu = constant$, between disc observers. Hence the observed frequency on the disc at r increases without limit as $\omega r \to 1$. The frequency relation between an observer at rest on the disc (O') and one at rest in the background inertial frame (O) can be found by taking $\dot{\phi}' = 0$ and $\dot{\phi}' = -\omega$ respectively. This gives $\nu = (1 - \omega^2 r^2)\nu'$, so that the frequency observed by O vanishes as $\omega r \to 1$.

Finally in this section we may ask how a gravitational field interacts with an electromagnetic field acting on a charged massive particle. Effectively this was answered in the previous chapter when electromagnetism was discussed in terms of a general stationary or constant metric. That is, *gravity appears only in the space-time metric*. So long as we use true derivatives and the appropriate metric with which to raise or lower indices, the previous discussion of electromagnetism holds.

6.4 Strong Gravitational Field

In this section we investigate exact solutions of the Einstein equations. These equations determine the space-time metric in the presence of a given distribution of matter. They produce metrics that go well beyond the simple transformation to curvilinear or moving coordinates from a background inertial space-time. We will not in this text discuss the technology of solving these equations, although we will present their form briefly later. However some of their properties are obvious *a priori*.

It is clear that they must be tensor equations of the second rank, since they should allow a free choice of coordinates, and they must allow the $\{g_{ab}\}$ to be found. In order to achieve correspondence with Newtonian theory, they should be second-order partial differential equations. Due to the arbitrary coordinate choice permitted, four of the ten g_{ab} must not be physical. This is because with four arbitrary functions $q'^a(\{q^b\})$ ($a = 0, 1, 2, 3$) available, four of the g_{ab} can be varied at will. It is similar to the choice of gauge in electromagnetism, and just as in that theory, a judicious choice can greatly simplify the problem at hand.

In general we have six unknown functions to find in order to specify the metric field. However, additional assumptions of symmetry such as axial or spherical symmetry reduce the number of these unknowns considerably. With spherical symmetry and a constant space-time, the number of unknowns is reduced to two. This is due to the additional three arbitrary rotations in spherical symmetry, plus an arbitrary origin of time as represented by $t' = t + f(r)$ with f arbitrary.

6.4.1 The Schwarzschild Metric

The simplest case imaginable and yet one of the most important is the space-time metric produced by an isolated spherical mass. This is analogous to Newton's law for the gravitational attraction outside of a spherical mass, or to Coulomb's law for the electric field around a point charge. This exact solution of the gravitational field equations was found by K. Schwarzschild in 1916 [8]. It was extended slightly by Birkhoff in 1923 [7]. He showed that the external metric was inevitably static and hence equal to that of Schwarzschild, no matter what were the internal motions of the spherical mass. The motions must remain spherically symmetric, however, and thus the solution does not apply outside a rotating mass. The remarkable solution in that case is due to Kerr [9], found only in 1963 for a point mass.

The Schwarzschild solution for the metric of space-time around a spherical mass M is usually written as

$$ds^2 = c^2dt^2\left(1 - \frac{r_s}{r}\right) - \frac{1}{\left(1 - \frac{r_s}{r}\right)}dr^2 - r^2(d\theta^2 + \sin^2(\theta)d\phi^2), \qquad (6.61)$$

where the 'Schwarzschild radius' r_s is defined as

$$r_s \equiv \frac{2GM}{c^2}. \qquad (6.62)$$

For calculational purposes it is convenient to measure r in units of r_s and to take $c = 1$ otherwise, to obtain

$$ds^2 = dt^2\left(1 - \frac{1}{r}\right) - \frac{1}{\left(1 - \frac{1}{r}\right)}dr^2 - r^2(d\theta^2 + \sin^2(\theta)d\phi^2). \qquad (6.63)$$

This metric is an example of the general form of a constant gravitational field in spherical symmetry (6.26). As such we know the Christoffel symbols (see the example in the previous section). Therefore we can write the geodesic equations for test particle motion, and the formula for the true derivative of a vector. From the latter formula we may find, for example, the four-acceleration of a test particle as

$$a^b \equiv \mathbf{v}^c\nabla_c\mathbf{v}^b, \qquad (6.64)$$

since ∇_c replaces ∂_c in general coordinates. Moreover we could use the true derivative together with this metric to write the Lorentz equations for the motion of a charged mass in this metric field.

Before exploring particle motion, we should examine some peculiarities of the space-time itself. There are two puzzling limits. As $r \to \infty$ the space-time becomes Minkowski space in spherical polar curvilinear coordinates. It is easy to express this in Galilean coordinates by using $g'^{ij} = (\partial_{q^k}x^i)(\partial_{q^\ell}x^j)g^{k\ell}$, where the $\{q^k\}$ are $\{r, \theta, \phi\}$ and the x^i are $\{x, y, z\}$ (see problem). The metric becomes indeed $ds^2 = dt^2 - (dx^2 + dy^2 + dz^2)$.

The infinite limit is peculiar in that there is no match to the wider Universe, except in so far as the Universe is characterized by Equation (6.15). The source of this preferred frame is due to the whole Universe and so the solution appears 'Machian'. However, the cosmological solution depends on cosmic coordinate time, while the Schwarzschild metric is static. One is missing a cosmological boundary condition, which can however be constructed. It suffices to observe here that wherever the local spherical mass dominates gravitationally, the Schwarzschild metric should be applicable.

The second peculiarity arises as $r \to r_s$, where $g_{00} \to 0$ and $g_{11} \to \infty$. The vanishing of g_{00} implies, according to Equation (6.14), that any signal frequency emitted at r_s will be gravitationally shifted to zero for an observer at $r > r_s$. The divergence of g_{11} implies that 'proper distance' for an observer at rest at r_s becomes infinite for a finite coordinate interval dr. Moreover, on setting $ds = 0$ and considering only radial propagation, the local light speed is $(1/\sqrt{g_{00}})(dr/dt) = c\sqrt{(1 - r_s/r)}$. This vanishes in the limit, so it is impossible to exchange light signals between $r \leq r_s$ and a larger radius r.

This impossibility of 'seeing' beyond r_s from the outside identifies the spherical surface as an 'apparent horizon', which is also the 'event horizon' in this static metric. We found previously, when discussing hyperbolic motion, a similar horizon that separates a rest observer from one with a constant acceleration.

Problems

6.4 Show either from the transformation $x = r\sin\theta\cos\phi$, $y = r\sin\theta\sin\phi$, $z = r\cos\theta$ or its inverse, that the Schwarzschild metric becomes $ds^2 = c^2dt^2 - (dx^2 + dy^2 + dz^2)$ as $r \to \infty$.

6.5 Show that the radial acceleration of an object in the Schwarzschild space-time goes to $c^2/(2r_s)$ as $r \to r_s$. It is convenient to use $c = 1$ and r/r_s as the radial variable in the calculation. Show that it is zero on the radial geodesic.

We have learned that observers construct their spatial 'hypersurfaces' by synchronizing clocks. This clearly cannot be done by light signals at or inside r_s. Even the slow transport of clocks is impossible near $r = r_s$ as the radial acceleration for a constant small proper radial velocity becomes infinite as $r \to r_s$ (see Problem). Consequently our conclusion is that the set of coordinates that are Minkowskian at infinity cannot be extended to or beyond r_s. There are no observers there who can be in contact with external observers.

We need therefore another set of observers who can be in contact over the whole space-time. We follow Lemaître [10] in considering a set of observers who are following radial geodesics (they are freely falling along a radius). These observers may be distributed uniformly over spherical surfaces, each of which is found at a different radius r at a given time t. Together these spherical surfaces plus their synchronized time (see below) will give another reference system for space-time. We must then consider the radial geodesics that are the world lines of these observers.

A convenient treatment of the radial geodesics in the Schwarzschild space-time for our purposes is found from the Hamilton-Jacobi Equation (6.44) which takes the form ($c = 1$, $r \leftarrow r/r_s$)

$$\frac{r}{r-1}\left(\frac{\partial S}{\partial t}\right)^2 - \frac{r-1}{r}\left(\frac{\partial S}{\partial r}\right)^2 = m^2, \tag{6.65}$$

where m may be thought of as either the mass of an individual particle or the mass of a spherical shell. We solve this equation for Hamilton's principal function in the usual way (see Problem) by assuming a separated sum solution in the form $S = \mathcal{E}t + S_r(r)$ ($\mathcal{E}$ is the conserved energy) to find

$$S = \mathcal{E}t \pm \int^r \frac{r}{r-1}\sqrt{\mathcal{E}^2 - m^2 + m^2/r}\, dr. \tag{6.66}$$

We recall from classical mechanics that Hamilton's principal function generates a transformation to new coordinates that are constants. In this context, the constants will be the coordinates that move with the particles, and we designate them by the continuous variable R. We find these constants from $R = \partial_{\mathcal{E}}\mathcal{S}$ or

$$R = t \pm \mathcal{E}\int^r dr \frac{r}{r-1}\frac{1}{\sqrt{\mathcal{E}^2 - m^2 + m^2/r}}. \tag{6.67}$$

For simplicity we cover our space with spherical surfaces that have all fallen in from infinity (starting presumably at different times). Hence for each shell $\mathcal{E} = m$, according to the form of the Schwarzschild metric and Equation (6.52). Moreover we select the positive sign, since this corresponds to in-falling shells as t increases postively. Thus our 'comoving' coordinate (R is constant on each particle or shell) is given by

$$R = t + \int^r \frac{r^{3/2}}{r-1}dr. \tag{6.68}$$

This last expression is a transformation to a new radial coordinate $R(r,t)$. We effect this transformation of the Schwarzschild metric by using $dR = dt + r^{3/2}/(r-1)dr$ in Equation (6.63) to write

$$ds^2 = \left(\frac{r-1}{r}\right)^2 dt^2 + 2\left(\frac{r-1}{r^2}\right)dRdt - \left(\frac{r-1}{r^2}\right)dR^2 - r^2 d\Omega^2, \tag{6.69}$$

where for brevity $d\Omega^2 \equiv (d\theta^2 + \sin^2(\theta)d\phi^2)$. A spatial hypersurface can be established by introducing a synchronized time according to Equation (5.134) as

$$d\tilde{t} = dt + (1/(r-1))dR. \tag{6.70}$$

Eliminating dt from the metric (6.69), one obtains

$$ds^2 = d\tau^2 - \frac{1}{r}dR^2 - r^2 d\Omega^2, \tag{6.71}$$

where the synchronized proper time is

$$d\tau = (r/(r-1))d\tilde{t}. \tag{6.72}$$

Note that there is no difficulty with this synchronized time on closed paths, so that we have a global hypersurface when $d\tau = 0$.

To complete the transformation we must find $r(R,\tau)$. Using Equations (6.70) and (6.72) together with $dR = dt + r^{3/2}/(r-1)dr$ yields

$$d\tau = dt + \frac{r^{1/2}}{r-1}dr, \tag{6.73}$$

or in integral form

$$\tau = t + \int^{r} \frac{r^{1/2}}{(r-1)} dr. \tag{6.74}$$

The logarithmic infinity in the second term is a property of the original coordinates and will not affect the discussion in terms of the new coordinates.

The coordinate t may be eliminated between Equations (6.68) and (6.74). One can write the result in terms of a difference of integrals as

$$R - \tau = \int^{r} dr \frac{r^{3/2} - r^{1/2}}{r-1} = \int^{r} dr r^{1/2} = \frac{2}{3} r^{3/2}, \tag{6.75}$$

so that inverting

$$r = \left(\frac{3}{2}(R-\tau)\right)^{2/3}. \tag{6.76}$$

The Schwarzschild metric is now known entirely in terms of R and τ, since substituting into Equation (6.71) yields

$$ds^2 = d\tau^2 - \frac{dR^2}{\left(\frac{3}{2}(R-\tau)\right)^{2/3}} - \left(\frac{3}{2}(R-\tau)\right)^{4/3} d\Omega^2. \tag{6.77}$$

It is, however, no longer a constant metric and the particle energy $(\partial \mathcal{S}/\partial \tau)$ is not conserved. The slice $R = constant$ is freely falling and the metric looks Minkowskian on this slice provided that the circumferential radius is r. This is implied by the original definition of angles. Thus the Schwarzschild metric is equivalent to prescribing a field of inertial observers, wherever the metric is non-singular.

There is no difficulty with this metric for freely-falling observers until $R = \tau$ which corresponds to $r = 0$, and represents the location of the point mass at the centre of the system. It is a true singularity by assumption of a point mass, just as it would be in the Newtonian theory. The singularity results from an over-idealization of the structure of matter, which must satisfy quantum mechanical rules.

The limit at infinity is somewhat peculiar, as the radial separation of the shells vanishes there. However, given that we have started all of the observers at radial infinity, this is not so unreasonable. It is a feature shared by Kantowski-Sachs metrics [11]. In any case we are certainly limited by our construction of space-time to a region dominated by the central mass.

A massive particle moving radially in this metric (6.77) is not following a geodesic, and the Hamilton-Jacobi equation is difficult to solve in these coordinates in a non-trivial fashion. It is more readily done in the constant form of the metric (6.63) (see below). A particle at rest in the Lemaître metric is falling into the centre at $r = 0$. We have indicated this in Figure 6.2 for the unhappy shells R_1 and R_2.

The radial null geodesics, however, are readily found in these (Lemaître) coordinates by setting $ds = 0 = d\theta = d\phi$. This yields

$$\frac{d\tau}{dR} = \pm \frac{1}{((3/2)(R-\tau))^{1/3}} \equiv \pm \frac{1}{\sqrt{r}}, \tag{6.78}$$

where the sign corresponds to outward- or inward-going light rays.

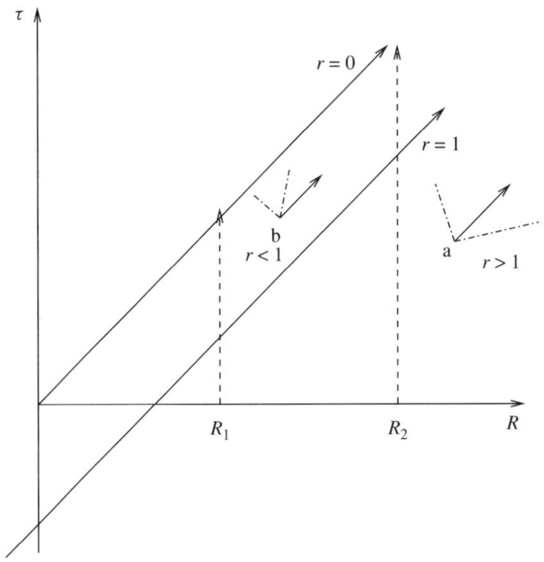

Figure 6.2 *The sketch shows the $\tau - R$ slice of the Lemaître space-time. World lines of constant r are labelled and are straight lines of slope 1 in the figure. At event 'a' where r > 1 the light-cone is shown to contain any line of slope 1. At event 'b' where r < 1, any line of slope 1 lies outside the local light cone. When r < 1, all massive particles and photons will inevitably intersect the line r = 0. Shells of constant R fall into the central singularity in a finite interval of τ as shown*

We are now able to understand the significance of $r = 1$. Each straight line $r =$ *constant* has a world line of slope $d\tau/dR = 1$ in Lemaître coordinates, according to Equation (6.68). When $r > 1$ this is within the light-cone defined by Equation (6.78), which has generators of slope $\pm(1/\sqrt{r})$. Consequently a particle sitting at a fixed $r > 1$ has a time-like world line and is able to remain there, or even travel to larger r given sufficient propulsion. The required acceleration could be calculated from Equation (6.64) if the desired path is known.

At $r = 1$ a particle must be null to remain on such a world-line, which is on the light cone. In fact this can be shown in the Schwarzschild coordinates by studying the velocity of a massive particle traversing a radial geodesic (see Problem).

For $r < 1$ the world line of a particle at fixed r is outside the local light-cone, and is therefore space-like and superluminal. That is, no massive particle and indeed no photon can remain on an $r =$ *constant* world line when $r < 1$. A massive particle must remain within the local light cone, and both generators of the light cone will inevitably intersect the line $r = 0$. No force yet known in nature can prevent particles and photons from falling to the centre once $r < 1$. Figure 6.2 represents these arguments in a $\tau - R$ space-time diagram.

For a solar mass, which is more or less a typical stellar mass, the Schwarzschild radius is $r_s = 2GM_\odot/c^2 \approx 3$ km. We are thus a long way from worrying about the breakdown of Schwarzschild coordinates, when discussing the orbit of a planet or the bending of a ray of light grazing the limb of the Sun ($R_\odot = 695,000$ km). We address these phenomena in the next section.

If the mass of the Sun were slightly less than three km in radius it would be collapsing to the central singularity behind the horizon. Such an object is commonly referred to as a 'black hole', although what exactly is the interior state is both hidden from us and beyond our current physics. Externally, however, the gravitational field is well known through the Schwarzschild metric, provided that there is no angular momentum in the 'hole'. Even the presence of angular momentum can be incorporated in the external metric given by Kerr [9]. It appears now that such collapsed objects play a vital rôle in the structure and activity of astronomical objects. They are inferred to exist from stellar masses to subgalactic masses at present, by their dark gravity.

Problems

6.6 Derive the solution for Hamilton's principal function in the form of Equation (6.66).

6.7 Write the velocity (i.e. dr/ds) of a massive particle falling on a radial geodesic in Schwarzschild coordinates. Note that you need both the radial geodesic equation and the metric to obtain a direct solution. The Hamilton-Jacobi method may be used as in the text to find (implicitly) $r(t)$.

6.8 Show that a particle orbiting on a equilibrium circular path, so that $dr/ds = d^2r/ds^2 = 0$, satisfies the Newtonian condition $(d\phi/dt)^2 = GM/r^3$.

There are many astrophysical applications of this metric, but these are mostly another story. We shall discuss two traditional (and very important) applications in the next section. To conclude this section we give a recent example due to [12] regarding the relative ageing of twins in a gravitational metric.

Suppose that observer A is at rest in Schwarzschild coordinates at radius r. A has to be accelerated in order to achieve this. The proper time of this observer is related to the coordinate time by $ds_A = \sqrt{(r-1)/r}\, dt$. A twin observer B is on a stable circular orbit (inertial or free-fall) at r so that the proper time is related to the coordinate time by $ds_B = \sqrt{(r-1)/r - \dot{\phi}^2 r^2}\, dt$. The appropriate $\dot{\phi} \equiv d\phi/dt$ is given in the last Problem of this section. In any case it is clear that over a given coordinate time interval, observer B will age less than observer A. The novelty is that only observer A is accelerated.

In the purely Minkowski discussion of the twin paradox we identify the asymmetry as being due to the acceleration. Moreover, the acceleration is applied to the twin who is younger when the two reunite. In the arrangement above, however, *the accelerated twin is older at each reunion*. One might argue that it is always the twin who moves relative to some Machian background standard of rest who is the younger, and not the accelerated twin. However, the essential difference is that here, geodesic motion can be on a closed path due to gravity. This cannot be in Minkowski space. Moreover, the acceleration of observer A is also due to gravity, which is not present in Minkowski space. It seems that we should not really compare the two situations.

6.4.2 Orbital Precession and Light Bending in a Schwarzschild Geometry

We do not wish to discuss orbital motion in a strong gravitational field in general, as this is beyond the scope of this text. It is available in many places and it is not beyond the techniques that are now available to the reader. However, the precession of Keplerian orbits and the curvature of a null geodesic passing near a point mass are grand classics. These effects will be discussed as an example of the techniques that we have advocated throughout this text.

We first discuss these problems from the point of view of the Hamilton-Jacobi formulation (cf. [3]). One finds that the solution of Equation (6.44) for planar motion ($\theta = constant$, $\phi - r$ plane) in the Schwarzschild metric of a particle of mass m is

$$\mathcal{S} = \mathcal{E}t + L\phi + \mathcal{S}_r(r), \tag{6.79}$$

where

$$\mathcal{S}_r \equiv \pm \int^r dr \sqrt{\frac{\mathcal{E}^2}{c^2(1-\frac{r_s}{r})^2} - \frac{m^2c^2 + \frac{L^2}{r^2}}{1-\frac{r_s}{r}}}. \tag{6.80}$$

Once again, $r_s = 2GM_\star/c^2$, if $M_\star$ is the central mass. For the Sun we have $M_\star = M_\odot$.

In the case of the solar planetary system, or of a similar system, we expect r_s/r to be small. Hence to find the lowest order relativistic gravitational effects, we wish to expand the solution in terms of r_s/r. It transpires that we must keep terms up to second order in this ratio. This is because it is not clear in the above expression how to compare a term like $(L^2/r^2)(r_s/r)$ with $(\mathcal{E}^2/c^2)(r_s/r)^2$. The trick is first to separate the terms by powers of r. This is facilitated by using a new radial coordinate $R^2 = r^2(1 - r_s/r)$, since this substitution renders the angular momentum term solely in the power R^{-2} just as in the classical limit.

Inverting the definition of R for $r(R)$ gives, after keeping only the first-order terms (higher orders are clearly small in this expression),

$$\frac{r}{R} \approx \frac{r_s}{2R} + 1 \tag{6.81}$$

for $r_s/R \ll 1$. In the same limit $dr = dR$. By keeping *all* terms of second order in r_s/R, one has also

$$\begin{aligned} \frac{1}{1-\frac{r_s/R}{1+r_s/(2R)}} &\approx 1 + \frac{r_s}{R} + \frac{1}{2}\left(\frac{r_s}{R}\right)^2, \\ \frac{1}{\left(1-\frac{r_s/R}{1+r_s/(2R)}\right)^2} &\approx 1 + \frac{2r_s}{R} + 2\left(\frac{r_s}{R}\right)^2. \end{aligned} \tag{6.82}$$

One proceeds by changing the variable to R as indicated, using the preceding expansions, and collecting powers of r_s/R to obtain

$$\mathcal{S}_r = \pm \int^R dR \sqrt{\frac{\mathcal{E}^2}{c^2} - m^2c^2 + \frac{r_s}{R}\left(2\frac{\mathcal{E}^2}{c^2} - m^2c^2\right) + \left(\frac{r_s}{R}\right)^2\left(2\frac{\mathcal{E}^2}{c^2} - \frac{m^2c^2}{2} - \frac{L^2}{r_s^2}\right)}. \tag{6.83}$$

To detect the lowest-order gravitational terms one must now consider the particle energy, and separate the special relativistic effects from the gravitational term. We recall Equation (6.52) in the form

$$\mathcal{E} = \frac{mc^2\sqrt{1 - r_s/r}}{\sqrt{1 - \mathbf{v}^2/c^2}}, \tag{6.84}$$

which in the low-velocity, weak-field limit becomes

$$\mathcal{E} \approx mc^2 + \frac{m\mathbf{v}^2}{2} - \frac{r_s}{2r} \equiv mc^2 + \mathcal{E}_N. \tag{6.85}$$

Here $\mathcal{E}_N$ is the sum of the classical kinetic energy and the Newtonian potential.

It is now possible to express each of the coefficients of the powers of r_s/R in Equation (6.83) by using the square of the energy, namely

$$\frac{\mathcal{E}^2}{c^2} \approx m^2c^2 + 2m\mathcal{E}_N + \frac{\mathcal{E}_N^2}{c^2}. \tag{6.86}$$

Compared with the first term, the second and third terms of this expression are of order $\mathcal{E}/mc^2$ and $\mathcal{E}^2/(m^2c^4)$ respectively. They may be neglected for planetary motion, except in the zeroth order term in Equation (6.83) where the rest mass energy is subtracted out. Substituting into the coefficients in the radial action (6.83), while keeping only lowest-order terms in the rest mass energy, yields

$$\mathcal{S}_r = \pm\int^R \sqrt{2m\mathcal{E}_N - \frac{L^2}{R^2} + \left(m^2c^2\right)\frac{r_s}{R} + \frac{3}{2}(m^2c^2)\left(\frac{r_s}{R}\right)^2}. \tag{6.87}$$

In the limit where $r_s \to 0$, this yields the classical action $\mathcal{S}_r^N$ for Keplerian motion. This limit requires that $u \equiv 1/R$ be periodic in ϕ. The additional terms in powers of r_s/R are small for planetary motion, and we can evaluate them when necessary by substituting the Keplerian value for $1/R$ as a function of ϕ.

However, we are interested in the precession of the orbit, and this follows from the Hamilton-Jacobi transformation equation

$$\beta_\phi = \frac{\partial\mathcal{S}}{\partial L} = \phi + \frac{\partial\mathcal{S}_r}{\partial L}. \tag{6.88}$$

Over some time the change in angle will therefore be

$$\Delta\phi = -\frac{\partial\Delta\mathcal{S}_r}{\partial L}, \tag{6.89}$$

where Δ indicates a finite change in the quantity to which it is applied. But the interesting precession will take place over many orbits when the non-Keplerian effects are small, so that an integral over time will be necessary in the preceding equation. The term in the action (6.87) that is linear in $u \equiv 1/R$ will not contribute to this secular behaviour, since with the Keplerian ansätz it sums to zero. We therefore ignore it in what follows, in favour of the u^2 term. This is why the expansion had to be carried out to second order.

The secular change in the action (6.87) can, after factoring, now be written as

$$\Delta\mathcal{S}_r = \Delta\int^R dR\sqrt{2m\mathcal{E}_N - \frac{L^2}{R^2}}\left(\sqrt{1 + \frac{(3/2)m^2c^2(r_s/R)^2}{2m\mathcal{E}_N - \frac{L^2}{R^2}}}\right). \tag{6.90}$$

By expanding the second square root and remembering the form of $\mathcal{S}_r^N$ from Equation (6.87) with $r_s = 0$, this can be written finally as

$$\Delta\mathcal{S}_r = \Delta\mathcal{S}_r^N - \left(\frac{3}{4}\right)\left(\frac{m^2c^2r_s^2}{L}\right)\frac{\partial\Delta\mathcal{S}_r^N}{\partial L}. \tag{6.91}$$

By Equation (6.89) we obtain after n orbits that the orbital angle will have changed by

$$\Delta\phi = n\left(2\pi + \left(\frac{3\pi}{2}\right)\frac{m^2c^2r_s^2}{L^2}\right). \tag{6.92}$$

Here we have used the Keplerian result that $-\partial\Delta\mathcal{S}_r^N/\partial L = 2\pi n$.

The last equation is the classical (Einstein) calculation of the precession of a planetary orbit in the metric theory of gravity. The excess angle turned after n orbits relative to $2\pi n$ may be measured from an axis of the Keplerian ellipse at some initial time. As such it yields the precession of such an axis. We introduce the specific angular momentum $\ell \equiv L/m$ and the definition of the Schwarzschild radius in terms of a central mass $M_\star$ to write this precession per orbit as

$$\Delta\phi = \frac{6\pi G^2M_\star^2}{c^2\ell^2}. \tag{6.93}$$

In terms of the semi-major axis a and the eccentricity e of the approximate Keplerian orbit, we have $\ell^2 = GM_\star a(1-e^2)$ and so

$$\Delta\phi = \frac{6\pi GM_\star}{c^2a(1-e^2)}. \tag{6.94}$$

The period of the Keplerian orbit P is $P^2 = 4\pi^2a^3/GM_\star$, so that one can write a precession per unit coordinate time averaged over an orbit as

$$\frac{\Delta\phi}{\Delta t} = 3\left(\frac{GM_\star}{a}\right)^{3/2}\frac{1}{c^2a(1-e^2)}. \tag{6.95}$$

This expression would need to be corrected to the proper time of the observer at radius r_O by the factor $\sqrt{1/(1-r_s/r_O)}$. If in addition the observer is moving, then special relativistic corrections to coordinate time would also be required in principle. Both of these corrections are small and are generally ignored.

The traditional application of this result is to the precession of the perihelion of the orbit of Mercury. A careful calculation from Equation (6.95) using current values of the constants and orbital parameters (e.g. [13]) yields

$$\frac{\Delta\phi}{\Delta t} = 2.0850 \times 10^{-4}\ \mathit{radians/century}, \tag{6.96}$$

that is, 43.0 arcsec/century. The century is based on the tropical year.

Remarkably, the expected Newtonian precession of the perihelion of Mercury relative to terrestrial axes is about 5600 arcsec/century! This is due primarily to the precession of the Earth's axis, but about 10% is due to the influence of the other planets. An accurate observation of the precession [6] reports an excess relative to the best Newtonian value of some 42.98 ± 0.04 arcsec/century. This resolves a discrepancy of long standing and is generally taken to be a vindication of the metric theory of gravity.

However, this remains a very small part of a large effect and therefore relies on many separate measurements. Moreover, the Sun is not a point mass and is distorted by its rotation in such a way as to have a gravitational quadrupole moment. In Newtonian gravity such a moment will produce a precession of a planetary orbit. This effect is thought to be negligible at present, but it is important to find other examples of planetary motion where the effect is isolated and large. Such examples arise in binary neutron star systems.

The two-body problem in Newtonian mechanics is solved as a one-body problem in terms of a Keplerian orbit of the reduced mass object (which has the entire angular momentum of the system) about a fixed member of the system at the focus. The result allows the centre of mass motion of each object to be found. The result for the relativistic precession found above applies to the centre of mass orbit, the only difference being that, in the expression for the period of the Keplerian orbit, the total mass of the system appears.

Such two-body systems may be very relativistic, however, as in the case of close binary neutron stars, and thus additional corrections may be required. These systems are of great interest when one or both of the neutron stars is a visible pulsar. The pulsation is produced by the rotation of the neutron star, and it is a natural clock rivalizing laboratory atomic clocks. The timing of these clocks permit many more parameters of the system to be measured, including the mass of the neutron stars.

It is not our place here to discuss these measurements, but we note that in the case of the binary system PSR B1913+16 the orbital precession is 4.22659 degrees per year. The orbital period is only 7.75 hours! This is to be contrasted with the extremely small result found for Mercury, which has a period of about three months. Other such neutron binary systems are known, some of which are even more extreme. Moreover, in some cases both stars are visible pulsars, rather than just the single star as in the PSR B1913+16 system. We can expect these systems to become the most rigorous relativistic laboratories. They have already been used to 'infer' (necessary to the observed evolution but not detected directly) the presence of gravitational waves (e.g. [14]), which is a higher-order prediction of the theory of gravity.

Problem

6.9 Derive the solution of the Hamilton-Jacobi equation in the form given by Equations (6.79) and (6.80).

We now turn to discuss the bending of a light ray by a weak, central gravitational field. We say 'bending' since a weak gravitational field is never able to trap a photon by definition, but only to slightly perturb its path. This bending allows a foreground gravitating mass to act as a 'lens' for background sources. This has become supremely important in cosmology. Such an effect never arises in inertial space-time (Minkowski) wherein a light ray continues always in the same direction with the same global speed c. By contrast, generally the light ray will follow a geodesic in the metric of the gravitational field and, although in a local inertial frame its speed will always be c, its speed will necessarily vary for coordinate observers because of the curved geodesic path. We conclude that, globally, both the direction and the speed of a light ray or photon are variable.

We illustrate the bending or lensing phenomenon by allowing a light ray to pass close to a non-rotating, spherically symmetric mass. Beyond the radius of this mass the Schwarzschild metric will apply, even if the ray passes quite close to its surface. Such a description describes the passage of a light ray from a distant star near the limb of the sun. It was the first test of the predictions of the metric theory.

This problem may be solved in parallel with the preceding discussion of precession by using the eikonal Equation (6.55). We know that the constant coordinate frequency replaces the energy (cf. Equation (6.57)) and we write the azimuthal integral as $\alpha_\phi \equiv (\nu_0/c)b$ for convenience. The significance of the constant b will become apparent. The solution of the eikonal equation can thus be written as

$$S = \nu_0 t + \frac{\nu_o}{c} b\phi + S_r(r). \tag{6.97}$$

Here, $S_r(r)$ has the form $\mathcal{S}_r$ given in Equation (6.80) when we set $m = 0$, $\mathcal{E} \equiv \nu_0$ and $L = b\nu_0/c$.

We may make the same substitution for $r(R)$ and expand again to order $(r_s/R)^2$ to arrive at Equation (6.83) in the form

$$S_r = \pm\frac{\nu_0}{c}\int^R dR\sqrt{1 + \frac{2r_s}{R} + 2\left(\frac{r_s}{R}\right)^2 - \frac{b^2}{R^2}}. \tag{6.98}$$

Setting $r_s = 0$ restores the Minkowskian law of light propagation that we expect to be a straight line.

In Equation (6.98) we may compare the powers of r_s/R with 1, and so discard the squared term. By expanding the integrand to first order in r_s/R after factoring out $1 - b^2/R^2$, we obtain the eikonal to lowest order as

$$S_r = S_r^M \pm \frac{\nu_0}{c} r_s \int^R \frac{dR}{\sqrt{R^2 - b^2}} = S_r^M \pm \frac{\nu_0}{c} r_s cosh^{-1}\left(\frac{R}{b}\right). \tag{6.99}$$

Here S_r^M is the Minkowski radial action written in the form

$$S_r^M = \pm\frac{\nu_0}{c}\int^R \sqrt{1-\frac{b^2}{R^2}}dR. \tag{6.100}$$

Just as in the case of planetary precession, the angular deflection on passing near a spherical mass is given by the Hamilton-Jacobi coordinate transformation in the form

$$\beta_\phi = \phi + \frac{\partial S_r}{\partial \alpha_\phi}, \tag{6.101}$$

from which follows

$$\Delta\phi = -\frac{\Delta\partial S_r}{\partial\alpha_\phi} = -\frac{c}{\nu_0}\frac{\Delta\partial S_r}{\partial b}. \tag{6.102}$$

We can check how this works explicitly in the Minkowski limit by writing $\beta_\phi = \phi + \partial S_r^M/\partial\alpha_\phi$ as

$$\beta_\phi = \phi \mp b\int^R \frac{dR}{\sqrt{R^2-b^2}} \equiv \phi \mp \arccos\frac{b}{R}. \tag{6.103}$$

Inverting this last equation for R yields $R = b/\cos(\phi-\beta_\phi)$, which shows the trajectory to be a straight line with b the point of closest approach. This may be confirmed by a simple sketch. We may take $\beta_\phi = 0$ with no loss of generality since it means only that the axis on which $\phi = 0$, is drawn from the central point mass towards the point of closest approach. Then $\phi = -\pi/2$ at incoming infinity and $\phi = +\pi/2$ at outgoing infinity. Consequently $\Delta\phi \equiv -\partial S_r^m/\partial\alpha_\phi = \pi$ on passing from one infinity to the other.

We may now calculate the bending effect of a gravitating spherical mass by using the result (6.99) in Equation (6.102) and remembering the Minkowski limit to find

$$\Delta\phi = \pi \mp \Delta\left[\frac{d}{db}\left(r_s\left(cosh^{-1}\left(\frac{D}{b}\right)\right)\right)\right] = \pi \pm \Delta\left(\frac{r_s}{\sqrt{D^2/b^2-1}}\frac{D}{b^2}\right), \tag{6.104}$$

where D is an arbitrarily large value of R at which we temporarily consider the photon to be coming in or going out. Letting this value go to infinity, and taking the change of sign into account, gives the gravitational bending effect in excess of π as equal to

$$\Delta\phi = \frac{2r_s}{b} = \frac{4GM_\star}{c^2 b}. \tag{6.105}$$

This bending effect for the Sun amounts to a deflection of 1.75 arcsec. Originally this was observed by regarding a field of background stars during an eclipse of the Sun (see e.g. [15]), but the solar corona makes this a difficult measurement. Direct tests using a radio interferometer [16,17] have confirmed this prediction to 1% accuracy. The related time delay measurement [6] of radar waves reflected from a planet whose line of sight passes near the Sun has achieved similar accuracy.

The nature of the lens formed by a spherical mass is somewhat peculiar in an optical sense. It is achromatic, but the parallel rays of an initially plane wavefront passing by an effectively point mass will all have different impact parameters b. Consequently they will not converge to a focal point but rather along a focal line. In the approximation where all of the bending takes place in the plane perpendicular to the rays that passes through the point mass, the distance along this line where a particular ray converges is $\approx b^2/(2r_s)$. This has the disadvantage of diluting the received flux for any given observer, but the advantage of rendering the distant source visible for an ensemble of observers.

The most spectacular (and currently still developing) cosmological applications of the gravitational bending of light fall under the heading of 'weak lensing' (see Figure 6.3 for an example). In general, of course, the lensing distribution of matter is hardly spherical or point-like, but it turns out that a treatment of the lens as a planar distribution of point sources gives the basic Schwarzschild effect a major rôle. The theoretical development and early applications is treated admirably in the book [18].

In the course of our calculations we have implicitly found the time delay effect first noticed and verified by Shapiro [6]. Let us recall the eikonal solution in the form (6.97). We obtain the time to travel between two different radial points R_I and R_O by differentiating this expression with respect to ν_o and setting the result equal to a constant.

Figure 6.3 *This picture of the cluster of galaxies (CL0024+1654: courtesy of Hubble Space Telescope) demonstrates the phenomenon of gravitational lensing. The mass of the foreground cluster of galaxies is bending and magnifying the light from background galaxies. These appear as lenticular, distorted objects of colours that differ from those of the cluster galaxies. They are intrinsically blue galaxies. Individual galaxies may also act as gravitational lenses. In this way nearby massive objects act as gravitational telescopes for the more distant Universe. Source: Reproduced by permission of NASA, ESA, and H. Ford (Johns Hopkins University) (See Plate 8.)*

In the course of the differentiation one must remember that the constant that factors ϕ is equal to α_o and is actually independent of ν_o (i.e. the constant b is arbitrary). Thus that term does not appear. Moreover we use in the differentiation the form of S_r given in Equation (6.98), but keeping only the lowest order in r_s/R. This procedure gives

$$\beta_o = t \pm \frac{1}{c}\left(\int^R dR\sqrt{1-\frac{b^2}{R^2}} + r_s\int^R \frac{dR}{\sqrt{R^2-b^2}}\right). \qquad (6.106)$$

Performing the integrals gives more explicitly

$$\beta_o = t \pm \frac{b}{c}\left(\arcsin\left(\frac{b}{R}\right) + \sqrt{\frac{R^2}{b^2}-1}\right) \pm \frac{r_s}{c}\left(\ln\frac{R}{b} + \ln\left(1+\sqrt{1-\frac{b^2}{R^2}}\right)\right), \qquad (6.107)$$

where we have taken the positive branch of the inverse hyperbolic cosine in the second logarithm.

We apply Equation (6.107) at each of the events R_I, t_I and R_O, t_O. We choose the positive sign for the inward journey from R_I to b, and the negative sign for the outward journey in order to keep the cordinate time t increasing in each case. Taking the difference to find the total elapsed coordinate time gives

$$t_O - t_I = \frac{b}{c}\left(\arcsin\left(\frac{b}{R_I}\right) + \sqrt{\frac{R_I^2}{b^2}-1} + R_O \rightleftharpoons R_I\right)$$
$$+\frac{r_s}{c}\left(\ln\frac{R_I R_O}{b^2} + \ln\left(1+\sqrt{1-\frac{b^2}{R_I^2}}\right) + \ln\left(1+\sqrt{1-\frac{b^2}{R_O^2}}\right)\right). \qquad (6.108)$$

The notation $R_O \rightleftharpoons R_I$ implies writing again the first two terms in the bracket but with R_I replaced by R_O.

If we were to set $r_s = 0$ in this expression we would obtain the straight line transit time between R_I and R_O when b is the impact parameter with the Schwarzschild mass. Consequently the additional delay time due to the gravitationally curved path is, to lowest order, just the term in r_s, namely

$$t_{delay} = \frac{r_s}{c}\left(\ln\frac{R_I R_O}{b^2} + \ln\left[\left(1+\sqrt{1-\frac{b^2}{R_I^2}}\right)\left(1+\sqrt{1-\frac{b^2}{R_O^2}}\right)\right]\right). \qquad (6.109)$$

Once again negelecting squares of b/R, this may be simplified to

$$t_{delay} \approx \frac{r_s}{c}\ln\frac{4R_I R_O}{b^2}. \qquad (6.110)$$

However, this approximation may not always be valid.

In the example of the GPS system discussed earlier, the radii and the impact parameters are comparable. A crude estimate of the time delay in this case remains r_s/c (specific cases will deviate by the logarithm of small numbers), so that the distance error is essentially r_s for the Earth. This is about 0.9 cm.

The application made by Shapiro and collaborators was to reflect radar signals from the planet Mercury. As Mercury and the Earth traversed their orbits, b as well as R_I and R_O changed. This allowed the effect to be plotted as it varied in time and the predictions to be verified to 0.1%!

Our interest is not in pursuing these applications, despite their ingenuity and importance to astronomical science. We wish only to use these effects as an illustration of various techniques of the metric theory. To this end we consider now different derivations of the same effects using orbital methods that are close to those of Newtonian theory. We present these in the two following examples.

Example 6.2

In this example we consider the planetary precession problem from the point of view of geodesic motion.

Either directly from the effective Lagrangian (6.29) or from the corresponding geodesic equations, we have two integrals of planar geodesic motion as (e.g. $\dot{t} \equiv dt/ds$)

$$c\left(1-\frac{r_s}{r}\right)\dot{t} = constant \equiv \frac{\mathcal{E}}{mc^2},$$
$$cr^2\dot{\phi} = \ell. \tag{6.111}$$

The expression of the constant in terms of the energy follows from Equation (6.51).

Rather than use the radial geodesic directly, we may substitute these integrals into the proper time following the particle, as provided by the Schwarzschild metric, to find

$$\dot{r}^2 + \left(1-\frac{r_s}{r}\right)\frac{\ell^2}{c^2r^2} = \left(\frac{\mathcal{E}}{mc^2}\right)^2 - \left(1-\frac{r_s}{r}\right). \tag{6.112}$$

One may now follow the standard treatment of Keplerian orbits (e.g. [1, 19]) by using the variable $u = 1/r$ and transforming to $d\phi$ from ds through $d\phi = (\ell/(cr^2))ds$ from the integral. This procedure yields the orbit equation

$$\left(\frac{du}{d\phi}\right)^2 + u^2 = \frac{c^2}{\ell^2}\left(\left(\frac{\mathcal{E}}{mc^2}\right)^2 - 1\right) + \frac{c^2r_s}{\ell^2}u + r_s u^3. \tag{6.113}$$

We may differentiate this equation with respect to ϕ to obtain

$$\frac{d^2u}{d\phi^2} + u = \frac{c^2r_s}{2\ell^2} + \frac{3}{2}r_s u^2 \equiv N(u). \tag{6.114}$$

Normally this differentiation will introduce spurious solutions to Equation (6.113) and so we must take care to find the appropriate solution of (6.114) that also solves (6.113). However, we know that we are looking for periodic solutions in both cases. Moreover, with the appropriate choice of amplitude, such a periodic solution will satisfy both equations when $r_s = 0$. In the case of solar planetary orbits the eccentricities are small, and the relativistic perturbation will be small. Thus we expect to be close to a circular orbit for which $u = u_o$ (a constant), and therefore $N(u_o) = u_o$ from the last equation.

This allows us to perturb about a circular solution by writing $u = u_o + \delta u$, where $\delta u(\phi)$ will be small. By substituting into Equation (6.114) and using $N(u_o) = u_o$ we find

$$\frac{d^2\delta u}{d\phi^2} + \delta u = \left(\frac{dN}{du}\right)_o \delta u, \tag{6.115}$$

whence we find

$$\delta u = A\cos(w\phi + B). \tag{6.116}$$

Here A and B are constants that must be chosen to satisfy Equation (6.113) and $w \equiv \sqrt{1 - (\frac{dN}{du})_o}$. Now the radius of the perihelion repeats only after $2\pi/w$ radians, which for small $(dN/du)_o$ becomes $2\pi(1 + (1/2)(dN/du)_o)$. This shows that the precession (excess rotation) per orbit is $\pi(dN/du)_o$. However, from the definition of $N(u)$ in Equation (6.114) one finds $(dN/du)_o = 3r_s u_o$. Moreover $N(u_o) = u_o$ implies $u_o = (c^2 r_s)/(2\ell^2)$ provided that $c^2/(2\ell^2) \gg 3u_o^2$. The latter condition holds (using $\ell^2 = GM_\star a(1 - e^2)$) for small e provided that $r_o/r_s \gg\approx 3(1 - e^2)/2$.

We have thus found the planetary precession angle per orbit for small e to be $\Delta\phi = 3\pi c^2 r_s^2/(2\ell^2)$, that is

$$\Delta\phi = \frac{6\pi GM_\star}{c^2 a(1 - e^2)}. \tag{6.117}$$

This agrees with the result of Equation (6.94), which however is *not* restricted to small eccentricity (although radial orbits with $e = 1$ are excluded). The present calculation really only requires, however, that the expansion of $N(u)$ about a circular orbit converges rapidly. This it does for small r_s/r, and so the perturbative approach is valid even for large eccentricity. This derivation is typical of the calculation of orbits in a Schwarzschild geometry.

Example 6.3

A similar calculation to that of the previous example is available for the bending of light in the Schwarzschild lens.

We must deal with a null geodesic coming from, and returning to, infinity. An affine parameter λ that varies smoothly along the ray or photon trajectory must be used as the independent variable since $ds = 0$. We choose it to be the local proper time so that $d\lambda = \sqrt{(1 - r_s/r)}dt$. Hence from the metric null interval, there holds along the path of the photon in a plane

$$c^2 - \frac{1}{(1 - r_s/r)}\left(\frac{dr}{d\lambda}\right)^2 - r^2\left(\frac{d\phi}{d\lambda}\right)^2 = 0. \tag{6.118}$$

From the geodesic equations we need only the azimuthal integral $r^2(d\phi/d\lambda) = cb$, where the constant b is the impact parameter in zero gravity. Changing the dependent variable to $u = 1/r$ and the independent variable to ϕ using the integral, one finds the null geodesic as

$$\left(\frac{du}{d\phi}\right)^2 + u^2 = r_s u^3 + \frac{1}{b^2}(1 - ur_s). \qquad (6.119)$$

When $r_s = 0$ this equation has the solution $u^M = \cos\phi/b$, which confirms that b is the radius of closest approach. This is the impact parameter in the absence of gravity.

When $r_s u$ is small compared with unity we may neglect it in the second term on the right of Equation (6.119). The first term on the right, although small, contains the relativistic correction. Differentiate this equation again with respect to ϕ and look for a solution of the resulting equation close to u^M. That is, we set $u = \cos\phi/b + \delta u$ assuming that δu is small, and find

$$\frac{d^2\delta u}{d\phi^2} + \delta u = \frac{3r_s}{2b^2}\cos^2\phi. \qquad (6.120)$$

This has the solution $\delta u = (3 - \cos(2\phi))(r_s/(4b^2))$ so that along the null path

$$u = \frac{\cos\phi}{b} + \frac{r_s}{b^2}\frac{(3 - \cos(2\phi))}{4}. \qquad (6.121)$$

Now we recognize the symmetry between the inward-going path and the outward-going path. Let each part of the total path contribute a bending angle ϵ. Then taking $\phi = 0$ at $r = b$, we should have $\phi = -\pi/2 - \epsilon$ at ingoing infinity ($\phi = \pi/2 + \epsilon$ at outgoing infinity, which gives the same result) where $u = 0$. Placing these values in the preceding equation yields

$$0 = -\frac{\sin\epsilon}{b} + \frac{r_s}{b^2}(1 - \epsilon^2/2), \qquad (6.122)$$

so that to first order $\epsilon = r_s/b$. The total deflection in and out is evidently 2ϵ, so that as in the text, $\Delta\phi = 2r_s/b$.

Problem

6.10 Derive Equations (6.113) and (6.119) as used in the examples, following the procedures indicated there.

This concludes our discussion of the gravitational theory as represented by the Schwarzschild metric. We have omitted some essential considerations, such as in what sense is this metric 'curved' beyond the choice of curved coordinate lines? At present it is distinguished mathematically from Minkowski space in curvilinear coordinates, mainly in the sense that it can be transformed to Galilean coordinates locally but not globally.

We shall have to capture the essence of this curvature below before leaving the theory of gravity. The general curved and dynamic metric is, according to Einstein, to be interpreted as the actual Riemannian metric of the space-time manifold. The metric might also, as we have suggested, be interpreted as the prescription for a field of inertial frames. However, the Einstein/Riemann theory has been logically completed whereas the notion of the inertial field has not. If the inertial field $u(t, \mathbf{r})$ is not found directly from mass sources but rather only the metric as in the Einstein theory, then the inertial field interpretation is identical in content.

In the next subsection we record briefly the solution for the metric of space-time outside a rotating mass. This is the Kerr solution [9].

6.4.3 Kerr Metric Outside a Rotating Mass

The Kerr metric is thought to be the unique metric outside a collapsed object (e.g. [20]) that has constant angular momentum. As such it falls into the category of a stationary, axially symmetric metric. It has not yet been possible to connect this metric with the interior of a finite central body, so that it is discussed only in the context of a collapsed object that is usually termed a 'black hole'. The Schwarzschild solution can also describe a (non-rotating) collapsed object, but it can also be matched to an interior metric that describes a finite sphere of matter [21].

The Kerr metric coefficients take the form

$$
\begin{aligned}
g_{00} &= \left(1 - \frac{r_s r}{R^2}\right), \\
g_{11} &= -\frac{R^2}{\Delta}, \\
g_{22} &= -R^2, \qquad (6.123)\\
g_{33} &= -\left(r^2 + \frac{a^2}{c^2} + \frac{r_s r a^2}{c^2 R^2}\sin^2\theta\right)\sin^2\theta, \\
g_{03} &= \frac{r_s r a}{c R^2}\sin^2\theta, \qquad (6.124)
\end{aligned}
$$

where

$$R^2 \equiv r^2 + \frac{a^2}{c^2}\cos^2\theta, \qquad (6.125)$$

and

$$\Delta \equiv r^2 - r_s r + \frac{a^2}{c^2}. \qquad (6.126)$$

The coordinates used are the Boyer-Lindquist coordinates $\{q^a\} = \{ct, r, \theta, \phi\}$ [22], which give the Kerr metric its standard form.

At large r the coefficient g_{00} becomes Schwarzschild in lowest order, so that r_s has its familiar meaning in terms of the mass of the object. The mixed component g_{03} asymptotes to $(ar_s/r)\sin^2\theta$. We may regard this term as due to a rotational perturbation of the Schwarzschild solution. Hence by comparison with the usual transformation to rotating coordinates, we obtain that the angular velocity must satisfy $\omega r^2 \sin^2\theta = (ar_s/r)\sin^2\theta$. That is, more explicitly, $\omega r^2 = (2GM_\star a)/(c^2 r)$.

In general one expects the lowest-order rotational perturbation to be $2GL/(c^2 r)$ where L is the total angular momentum of the body [3]. Consequently a is seen to be just the specific angular momentum of the body. We normally denote this by ℓ, but convention dictates a.

As $r \to \infty$ the metric approaches (slowly) the Minkowski metric in spherical polar coordinates. If the mass goes to zero then at any radius it becomes the Minkowski metric in oblate spheroidal coordinates. The transformations [3]

$$\begin{aligned} x &= \sqrt{(r^2 + a^2)} \sin\theta \cos\phi, \\ y &= \sqrt{(r^2 + a^2)} \sin\theta \sin\phi, \\ z &= r \cos\theta \end{aligned} \tag{6.127}$$

restore the metric to Galilean form in this case. This behaviour demonstrates that there is no effect on the metric of space-time by rotation alone, that is, without the presence of gravitating mass.

The Kerr metric is still under study for its intrinsic properties. A masterful survey of the complexity plus a rare derivation of this metric is to be found in [23]. Nevertheless, some properties are evident. There is an apparent singularity where $g_{00} = 0$ (just as in the Schwarzschild metric), which is explicitly (taking the larger root)

$$r_e(\theta) = \frac{r_s}{2} + \sqrt{\frac{r_s^2}{4} - \frac{a^2}{c^2}\cos^2\theta}. \tag{6.128}$$

In addition there is also an apparent singularity where $\Delta = 0$, for which the larger root is

$$r_h = \frac{r_s}{2} + \sqrt{\left(\frac{r_s^2}{4} - \frac{a^2}{c^2}\right)}. \tag{6.129}$$

This radius defines the event 'horizon' for the central object since light cannot exit from inside this radius (consider the radial null geodesics).

The Boyer-Lindquist coordinate system cannot be extended beyond r_h. This is not true on the surface $r_e(\theta)$, however, except at the poles where it touches the spherical horizon. This three-surface forms the outer boundary of what is called the 'ergosphere' lying between the spherical horizon and this boundary or 'ergosurface'. The name of this region is appropriate because of its peculiar energetic properties. We recall the definition of the energy in the form of Equation (6.51), which becomes here

$$\mathcal{E} = mc^2 \left(g_{00} c \frac{dt}{ds} + g_{03} \frac{d\phi}{ds} \right). \tag{6.130}$$

Inside the ergosurface (but above the horizon), $g_{00} < 0$ and g_{03} remains positive and finite.

To consider the implication of this expression, we note that inside the ergosurface where $g_{00} < 0$ no particle can be at rest in these coordinates. This would require $dr = d\theta = d\phi = 0$ and then because of the sign of g_{00}, we would have $ds^2 < 0$. This means that this trajectory is space-like and is therefore not available to a massive particle. That is, no massive particle can remain at rest inside the ergosurface; at least rotation is inevitable.

We can find a possible trajectory in this region by writing the Kerr metric in the equivalent form

$$ds^2 = \left(g_{00} - \frac{g_{03}^2}{g_{33}}\right)c^2dt^2 + g_{11}dr^2 + g_{22}d\theta^2 + g_{33}\left(d\phi + \frac{g_{03}}{g_{33}}cdt\right)^2, \tag{6.131}$$

where now the coefficient of c^2dt^2 takes the form

$$g_{00} - \frac{g_{03}^2}{g_{33}} \equiv \frac{r^2 + a^2/c^2 - rr_s}{r^2 + a^2/c^2 + (a^2rr_s/R^2)\sin^2\theta}. \tag{6.132}$$

This is clearly positive everywhere outside the horizon since the numerator is positive. Consequently a trajectory $dr = d\theta = 0$ and $d\phi = -cdtg_{03}/g_{33}$ is time-like because only the first term in Equation (6.131) is non-zero. For such a trajectory the energy (6.130) is also positive since $\mathcal{E} = mc^2(g_{00} - g_{03}^2/g_{33})(cdt/ds)$ and $dt/ds > 0$.

The observers following this trajectory are rotating with the ergosphere as seen by distant observers, even though they have no rotation at infinity. The coordinate rotation rate is

$$\frac{d\phi}{dt} = -c\frac{g_{03}}{g_{33}} = \frac{ar_sr}{R^2(r^2 + a^2/c^2 + rr_sa^2/c^2\sin^2\theta)} \tag{6.133}$$

and is an extreme example of the 'dragging of inertial frames'. This effect was the object of the major experiment named 'Gravity Probe B', which has not yet reported as of this writing.

Suppose now that $d\phi/dt$ is some fraction of cg_{03}/g_{33} but that it is negative. Then by Equation (6.130) the particle energy is negative in the ergosphere. This would be a particle that is orbiting retrograde to the black hole rotation. Such a particle will reduce the angular momentum of the Kerr object even while it reduces the energy of the object by adding a negative amount. The rôle of the external energy of the Kerr object is essentially played by the area of the horizon (Equation 6.129 for small a).

It is possible to use this phenomenon to extract energy from a black hole and give it to a particle that escapes to infinity, by leaving part of the mass of the particle in a negative energy orbit (referred to as the 'Penrose Process' [24]). This could be achieved by the spontaneous disintegration of an unstable particle or by pair production assuming sufficiently energetic photons.

Energy may also be extracted by invoking charged particles and electrodynamics in the vicinity of the ergosphere of a Kerr object. Accreting interstellar gas will carry a magnetic field in general and, as we discussed in Chapter 5, we expect this to produce electric fields even if the accretion is steady. Thus Equation (5.170) shows that there will be a **D** equal to $\mathbf{h} \wedge \mathbf{g}$ even in this stationary metric. If it has a divergence then a net charge will be produced and accelerated in principle, although charge neutrality must be maintained overall to maintain a steady state.

If the magnetic field is largely radial and changes sign across the equatorial plane, the necessary neutrality can be arranged. Moreover in that case **D** is directed outwards along the axis both above and below the equator in this arrangement. Of course we must remain on open paths for this formulation to remain valid. We might, for example, integrate the electric field along an open helix to find the available energy. In any case such a mechanism is thought to be responsible for the bipolar jets of relativistic particles issuing from active galactic nuclei (AGN: e.g. [25]).

It is important to note that the Kerr metric changes character dramatically if $a \geq cr_s/2 \geq GM_\star/c$. The horizon disappears in this case (see definition) so that a singularity is revealed. This feature leads to predictability and causality problems in the wider world. For this reason such objects are not thought to exist, although observations suggest that the limit $a_c = GM_\star/c$ may be approached rather closely.

For a solar mass this critical value is about 4.4×10^{15} cm^2/s, but the actual value for the sun is about 1.6×10^{10} cm^2/s. This assumes rigid rotation with a surface equatorial period of 25 days so it is a crude estimate. Nevertheless, it is easy to see by examining the numerical values of the Kerr coefficients that rotation does not produce a substantial deviation from the Schwarzschild metric at the solar radius.

The behaviour of geodesics in the Kerr metric is in general a numerical problem. They are of major importance when calculating the appearance of hot gas in orbit about a collapsed object, which requires both null and time-like geodesics. Recently a numerical code has become available [26] that permits the general calculation of null geodesics. A discussion of time-like geodesics may be found in many places. For geodesics in the equatorial plane one may cite [27] and [28], while Brandon Carter studied the Hamilton-Jacobi approach to the orbits [29].

Stationary and static metrics of the Kerr and Schwarzschild type do not satisfy true cosmological boundary conditions at infinity. They do not know about the rest of the Universe except that there is an inertial frame at infinity even in the absence of mass. They are true analogues of the Coulomb electrostatic field. In fact we know nothing about their origin (just as for charges) without modelling an actual gravitational collapse.

We do know initial conditions under which singularities of a general type will eventually appear, thanks to theorems by Hawking and Penrose (for a summary see [30]), but the detailed evolution is very contingent [3,31]. In view of the continuing discovery by astronomers of objects that appear to have these properties, such problems both of principle and of astrophysics are very current and the literature is immense.

The Schwarzschild and Kerr metrics contain central singularities that translate to infinite energy density of the gravitational field. Normally in a physical theory these metrics would be discarded as possessing unphysical behaviour. It must be the case that a quantum theory of gravity would remove the singularities, but we do not have this theory for strong gravitational fields. Fortunately for astronomers, it seems that the singularities are normally hidden behind a event horizon. This means that they are causally disconnected from the rest of the Universe. It has not been proven that all naturally occurring astrophysical collapsed objects will be so hidden, but the known exceptions are of a rather artificial kind (e.g. [32]). It is generally assumed to be the case, however, which assumption is known as the 'Cosmic Censorship hypothesis'.

This concludes our introduction to the strong gravitational force as incorporated into non-Galilean metrics. We proceed briefly in the next two subsections to indicate how in principle these solutions are found, without going into the technical details which are to be sought elsewhere.

In the next section we consider the relativistic motion of continua, mainly that of pressure-free 'dust' or of ideal fluids. This is a step towards describing the matter source of the non-Galilean metrics.

6.4.4 Relativistic Continua

We recall that for a classical ideal fluid the conservation of momentum takes a three-tensor form, which is in Cartesian components

$$\frac{\partial(\mu v^i)}{\partial t} = -\frac{\partial T^{ij}}{\partial x^j} + f^i_{ext}, \tag{6.134}$$

where the momentum flux or 'stress' tensor is

$$T^{ij} = p\delta^{ij} + \mu v^i v^j. \tag{6.135}$$

Classically μ is the inertial mass density, p is the pressure and $\mathbf{v}(\mathbf{r})$ is the fluid velocity field. In an inertial frame that is moving instantaneously with the fluid, $T^{ij} = -p\delta^{ij}$ and the internal force per unit volume is $-\nabla p$. The external force per unit volume is f^i_{ext}.

The conservation of energy is derived from the momentum conservation equation plus thermodynamic considerations and takes the form

$$\frac{\partial(\mu(e + \mathbf{v}^2/2))}{\partial t} + \nabla \cdot (\mu\mathbf{v}(h + \mathbf{v}^2/2)) = \mu\Theta\frac{ds}{dt} + \mathbf{v} \cdot \mathbf{f}_{ext}, \tag{6.136}$$

where e is the specific internal energy, h is the specific enthalpy, Θ is the thermodynamic temperature, and s is the specific entropy. For an isentropic (adiabatic), isolated ideal fluid in the instantaneously co-moving inertial frame, this says that the time rate of change of the internal energy per unit volume is $-h\mu\nabla \cdot \mathbf{v}$.

In special relativity the instantaneously co-moving frame is the rest frame of a small volume of the fluid. Recall that a single particle in its co-moving frame is characterized only by its inertial mass, which determines how it reacts to an external force. Similarly for a small volume of an ideal fluid we expect to describe the rest frame state and evolution in terms of the inertia, which becomes the *total* rest frame energy density ρ. In addition we must take account of the external force per unit volume, if any. There is (more essentially) an internal force, since a given volume may be acted upon by neighbouring volumes in the continuum due to an internal pressure gradient.

This description of a relativistic fluid should on general principles be contained in a four-tensor formulation of the fluid motion. The formulation must reduce at low velocity and non-relativistic temperature to the classical equations. A natural form for the four-tensor of the matter only, which is now an energy density/momentum flux tensor rather than just a momentum tensor, in the rest frame of a small volume of the fluid is

$$T_o^{ab} = \begin{pmatrix} \rho, 0, 0, 0 \\ 0, p, 0, 0 \\ 0, 0, p, 0 \\ 0, 0, 0, p \end{pmatrix}. \tag{6.137}$$

The subscript o indicates the co-moving frame of the volume element.

This form is 'natural' because only ρ and p are characteristic of the fluid and have the correct dimensions to appear in the tensor. Moreover, the diagonality ensures that there is in the ideal fluid neither internal friction nor an internal energy flux (such as might be due to conduction). Finally, an isotropic pressure is assumed for an ideal fluid.

This cannot be the form that holds in any inertial frame since the four-velocity of the small volume must then appear in the description of its energy and momentum. However, we have only the invariants ρ and p to work with, plus the four-velocity $v^a \equiv dx^a/ds$.[1] One soon discovers that the general tensor form that holds in any inertial frame and that reduces to the rest frame form is

$$T^{ab} = (p+\rho)v^a v^b - p\eta^{ab}. \tag{6.138}$$

This could be found directly by transforming the rest frame quantity (p, ρ treated as invariants) to the frame moving with velocity $-v^a$ by a Lorentz transformation. We note for consistency that $T^{ab}v_a v_b = \rho$ which is indeed invariant, and $T^{bc}a_b a_c = -pa_b a^b$ showing that p is also an invariant ($a^b = dv^b/ds$).

The only possible equation of motion that would be valid in all inertial frames and that could reduce to the non-relativistic evolution is

$$\frac{\partial T^{ab}}{\partial x^b} = f^a, \tag{6.139}$$

where f^a is an external force properly described by the four-vector f^a.

It is in fact delicate to show that the energy equation (zeroth component) in an arbitrary inertial frame reduces to that expected classically at small c and low temperature, because one must separate carefully the rest mass energy from the classical internal energy (see e.g. [33]; pp 104–105; see also a Problem). It is straightforward to show that it reduces to that expected in the rest frame (see Problem). It is also straightforward (see Problem) to show that the spatial components of this last equation reduce to the classical conservation of momentum at small c and small internal energy. One needs only to remember that the total energy density is dominated by μc^2, and that the pressure is also.

We know that the number of properly formulated physical four-forces is strictly limited. If, for example, the fluid is able to carry a current, then one example would be the electromagnetic force in the form $f^a = j_b F^{ab}/c$. Maxwell's equations would then have to be solved simultaneously with the equations of motion.

However, the electromagnetic force does not concern us in this chapter. The force that interests us is the gravitational force. We know that this can be included in a curvilinear metric that is not transformable to Galilean coordinates. But curvilinear coordinates can be easily accommodated by writing the equations of motion using the true derivative as

$$\nabla_b T^{ab} = 0, \tag{6.140}$$

where we neglect non-gravitational external forces. The energy-momentum tensor becomes in turn, after transformation to curvilinear coordinates,

$$T^{ab} = (\rho + p)v^a v^b - pg^{ab}. \tag{6.141}$$

[1] Note that when $c \neq 1$ the coordinate $x^0 \equiv ct$.

Because the energy-momentum tensor is symmetric, Equation (6.140) takes a form [3] that is useful in curvilinear coordinates namely

$$\frac{1}{\sqrt{-g}}\frac{\partial(\sqrt{-g}T^{ab})}{\partial q^b} = 0. \tag{6.142}$$

As usual, g denotes the determinant of the metric.

To include a gravitational field we need only supply a metric that contains the effect of the gravitational field. It does not appear as an external force, but rather as an inevitable property of the system of coordinates, just as do inertial forces in curvilinear coordinates. As in the classical problem, an equation of state that relates p and ρ must be assigned by physical considerations in order to complete the definition of the model.

In general, as we see by taking the Kerr metric as an example, the resulting equations will be rather complicated. They are the subject of much current research. When only weak gravitational fields are involved with non-relativistic temperatures (i.e. $k\Theta/mc^2 \ll 1$), the equations are more manageable (see e.g. [34]).

Problems

6.11 Show that the spatial components of Equation (6.139) reduce to the classical equation of motion at low velocities and non-relativistic temperature.

6.12 Show that in the instantaneous rest frame, the zero component of Equation (6.139) agrees with the classical conservation of energy equation at low velocity and non-relativistic temperature.

6.13 Write the four Equations (6.139) in Galilean coordinates in the absence of an external force.

6.14 Show that if we introduce a density function μ related to the energy density ρ and p by [33],

$$\frac{d\mu}{\mu} = \frac{d\rho}{p+\rho}, \tag{6.143}$$

then the invariant $v_a(\partial T^{ab}/\partial x^b) = 0$ implies

$$\frac{\partial(\mu v^b)}{\partial x^b} = 0. \tag{6.144}$$

where μ is the invariant mass or particle number. You need to recall that $v_a v^a = 1$ and $v_a(dv^a/ds) = 0$. When $p = 0$ the two quantities are the same.

The case of a continuous 'dust' is somewhat easier to deal with, as no equation of state relating p and $\rho \equiv \mu$ (see Problem) need be assigned. This is because by 'dust' we mean cold, collisionless matter, so that the pressure is never a significant factor compared with inertia. A shower of meteors is a good example. Hence one sets $p = 0$. The only non-trivial application is in the presence of gravity (we ignore electromagnetism) when Equation (6.140) applies. Together with particle number conservation (see Problem) in the form

$$\nabla_b(\rho v^b) = 0, \tag{6.145}$$

the problem is completely defined except for boundary conditions. This formalism is appropriate for cosmological models. These are concerned with dark matter particles for which the approximation is excellent, or with the ensemble of galaxies for which the approximation is less good over cosmological time because of collisions.

These topics could be the subject of a book in themselves. However, our main interest is in the formal statement of Equation (6.140). This is the statement of conservation of mass and energy in a continuum that must hold even in a gravitational field. We have written T^{ab} explicitly only for an ideal fluid, but most forms of matter can be described by some such tensor so long as their dynamic evolution is describable by an action (e.g. [3]). The conservation laws will still take the form of Equation (6.140).

In the final section of this chapter we attempt to isolate the internal property that distinguishes curvilinear coordinates in a Minkowski space-time from gravitational metrics. These cannot be transformed to Galilean coordinates globally, and we wish to know why. The secret, it transpires, lies in recognizing the curvature of space-time as a dynamic quantity coupled to the local energy density of the matter.

6.4.5 The Curvature of Space-Time

There is some property of the metrics taken to describe a gravitational field that prevents them from being globally equivalent to Minkowski space. We know that the gravitational metrics are equivalent to a field of local inertial frames with an appropriate velocity function, but this does not permit us to determine the nature of this velocity field in the presence of matter. We turn rather to consider the metric of space-time as the fundamental physical reality, and seek intrinsic geometric distinctions between metric manifolds. In the process we are interpreting the metrics as containing locally the geometry of space-time itself.

Since at least the studies of Gauss and Riemann, Lobachevsky and Bolyai, we have been aware of geometries in which the geodesics do not satisfy the parallel-line axiom of Euclid (fifth postulate). A simple example is the two-sphere for which the geodesics are the great circles on the surface. Thus meridian lines are geodesics, and in a patch very near the equator (which they meet perpendicularly) they are very nearly parallel in the Euclidian sense.[2] However, these meridians meet at the pole so that globally they do not satisfy Euclid's axiom.

The spherical surface is an example of a space with uniform positive 'curvature' in geometers' terms, and it is the curvature of the space that makes it disobey the parallel line axiom. Otherwise put, the fifth postulate of Euclid defines Euclidian space to be 'flat'. A bowl with, say, a hyperbolic profile is an example of a surface with uniform negative curvature. Surfaces with mixed positive and negative curvature such as (at least locally) mountain cols or horse saddles also exist.

[2] Indeed by a spatial manifold we mean that such a small patch satisfies the Pythagorean theorem and is Euclidian.

This geometric distinction between tangent (local) flat patches and global curvature is also the relation between local inertial frames that are Minkowskian, and the global gravitational manifolds. In the gravitational manifolds, Minkowski space holds only in a local patch. It is the 'tangent space' at each event in space-time. Algebraically this means that we can reduce the metric of the manifold locally to η_{ab} and reduce its first derivatives to zero, but not the second derivatives.

Indeed, we have deduced the vanishing of the first derivative previously in terms of tensor derivatives, but it is easy to show directly by performing the differentiation given the metric form of the Christoffel symbols (see Problem). To repeat, the true first derivative of the metric in a Minkowskian manifold (i.e. rather than a Pythagorean metric patch in a spatial manifold, we have a Minkowski metric patch for a space-time manifold) satisfies

$$\nabla_a g_{bc} \equiv 0. \tag{6.146}$$

In local Galilean coordinates this is just $\partial g_{bc}/\partial x^a = 0$. We used the coordinate-independent relation to deduce the form of the Christoffel symbols in terms of the metric, so demonstrating this directly in the Problem really just reverses the argument.

The distinction between space-time and space is of no consequence for manifolds of arbitrary geometry. Once the metric is known[3] the formalism applies. Thus it is natural to look for the distinction between local and global space-time in terms of the curvature of the manifold. We have not discussed how the curvature of a given metric manifold is determined, but happily the problem was solved by Gauss and Riemann.

Consider Figure 6.4. On the left we sketch a geodesic triangle on the surface made from the great circle that is the equator and two meridians. We can parallel-transport a vector around this triangle by keeping constant the angle it makes with the tangent to the geodesic. This is because the tangent vector is parallel-transported along the geodesic curve by definition of the shortest path. Parallel transporting the vector (which we recall keeps it constant) is equivalent to keeping the angle with the geodesic tangent constant.

The sketch indicates that a vector transported around the geodesic triangle in this fashion, starting at postion 1 and ending with the return at 4, *is not equal to its initial self*. The equality would certainly hold after traversing a rectangle on the Euclidian plane. It is the curvature of the surface that is reponsible for this difference. The question that was posed by Gauss and Riemann, and subsequently solved by them, is how can we recognize the presence of curvature in a space using only intrinsic quantities and independently of the coordinates used? The intrinsic part is essential since for higher-dimensional spaces we are unable to view them embedded in three-space, as we do with the spherical two-surface. The coordinate-independent part requires us to seek a tensor criterion for intrinsic curvature.

The right hand side of Figure 6.4 is meant to indicate a small patch on the $a - c$ curved surface in a general space. The bounding sections are coordinate lines measured along geodesic curves. This makes the coordinates locally flat and makes parallel transport

[3] In more general spaces the 'connection' that allows parallel transport of vectors may be defined non-metrically. This requires an independent definition of the Christoffel symbols and is not used in the standard Einstein theory.

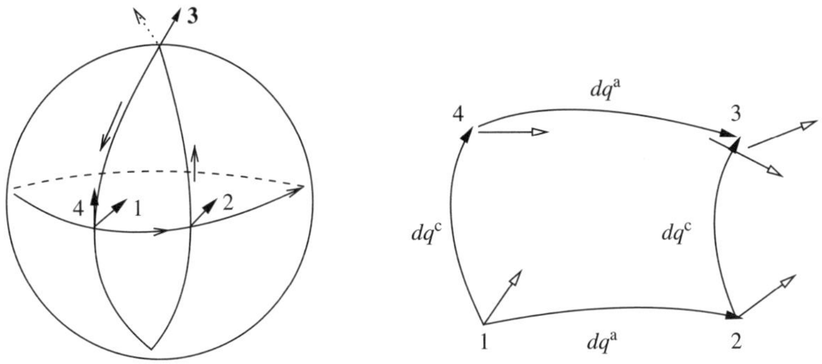

Figure 6.4 *The sketch on the left shows a global geodesic triangle (123) on the two-sphere. A vector that is transported around the triangle by keeping the angle with the local geodesic constant should be parallel-transported and remain constant. However, because of the curvature of the surface that is reflected in the curvature of the geodesics, the vector returned at position 4 is not the same as the original vector. The sketch on the right shows a locally flat small patch made from coordinates chosen to lie along geodesic curves. The change in a vector parallel-transported around the loop due to the curvature is the difference between the vectors transported on the two possible paths. In the text this difference is shown to depend algebraically on the second derivatives of the metric through the Riemann curvature tensor*

easy. A vector is shown parallel-transported from point 1 to point 3 along the two different possible paths (123 and 143). They do not yield the same resulting vector. This must be due to the appearance of curvature, which we expect to be hidden in the second derivatives of the metric. These second derivatives will appear in the second true derivatives of an arbitrary vector as we proceed to demonstrate.

Transporting a vector A_b from 1 to 2 will produce a new vector component given by $A_b + \nabla_a A_b dq^a$. Transporting this vector in turn from 2 to 3 yields $A_b + \nabla_a A_b dq^a + \nabla_c(A_b + \nabla_a A_b dq^a)dq^c$. Proceeding first from 1 to 4 and then from 4 to 3 yields similarly $A_b + \nabla_c A_b dq^c + \nabla_a(A_b + \nabla_c A_b dq^c)dq^a$. By subtracting these expressions one finds that the difference in the vector transported along the two paths is

$$\delta A_b = (\nabla_c(\nabla_a A_b) - \nabla_a(\nabla_c A_b))\, dq^a dq^c. \tag{6.147}$$

We have used the smallness and flatness of the patch to treat it as a rectangle to first order, and the coordinate increments are both fixed and arbitrary, although small.

We recall that the necessary tensor derivative in the last expression is formally

$$\nabla_c(\nabla_a A_b) = \frac{\partial(\nabla_a A_b)}{\partial q^c} - \Gamma^e_{ca}\nabla_e A_b - \Gamma^e_{cb}\nabla_a A_e, \tag{6.148}$$

which follows by applying the covariant rule

$$\nabla_a A_b = \frac{\partial A_b}{\partial q^a} - \Gamma^d_{ab} A_d \tag{6.149}$$

to each index. Hence explicitly

$$\delta A_b = dq^a dq^c \left\{ \frac{\partial^2 A_b}{\partial q^c \partial q^a} - \frac{\partial \Gamma^d_{ab}}{\partial q^c} A_d - \Gamma^d_{ab} \frac{\partial A_d}{\partial q^c} \right. \tag{6.150}$$

$$\left. - \Gamma^e_{ca} \left(\frac{\partial A_b}{\partial q^e} - \Gamma^d_{eb} A_d \right) - \Gamma^e_{cb} \left(\frac{\partial A_e}{\partial q^a} - \Gamma^d_{ae} A_d \right) - (a \rightleftharpoons c) \right\}. \tag{6.151}$$

The notation $a \rightleftharpoons c$ indicates the same terms with a and c interchanged. On taking the indicated difference one finds that all terms with coordinate derivatives of A_b cancel. This difference is also what one would find after completing the circuit in the anticlockwise sense.

One can write the net difference in the vector around such a small circuit in a coordinate-independent tensor form that follows from the last equation as

$$\delta A_b = R^d_{bac} A_d dq^a dq^c. \tag{6.152}$$

This introduces the Riemann curvature tensor whose form follows from the above argument as

$$R^d_{bac} \equiv \frac{\partial \Gamma^d_{cb}}{\partial q^a} - \frac{\partial \Gamma^d_{ab}}{\partial q^c} + \Gamma^e_{cb}\Gamma^d_{ae} - \Gamma^e_{ab}\Gamma^d_{ce}. \tag{6.153}$$

It follows from the definition of the Christoffel symbols that the Riemann curvature is basically proportional to the second derivatives of the metric. Indeed, in the local Galilean patch these are the only non-zero quantities.

We know the Riemann quantity to be a tensor because δA_b, A_b and dq^a, dq^c are all vectors. Hence R^d_{abc} must transform as the fourth-order tensor as indicated by the indices. Since it is a tensor, it may be combined with the metric tensor to form different tensors by raising or lowering indices. These are given the same symbol and distinguished only by the position and/or number of the indices. A clear demonstration of the tensor character follows by combining Equations (6.147) with Equation (6.152) to write

$$R^d_{bac} A_d \equiv \left(\nabla_c (\nabla_a A_b) - \nabla_a (\nabla_c A_b) \right), \tag{6.154}$$

since everything on the right is a tensor. This also shows neatly that the non-zero Riemann curvature is equivalent to the non-commutation of the true tensor derivatives.

It is remarkable that the Riemann curvature is a tensor since the Christoffel symbols are not. They cannot transform linearly under a coordinate transformation, being zero in Galilean coordinates and non-zero in curvilinear coordinates. This difference in behaviour is due to the dependence of the Riemann curvature on the second derivatives of the metric.

From our derivation of the curvature tensor we see that the four indices of the Riemann curvature tensor correspond to two for the choice of subsurface, one for the choice of vector component, and a dummy index that allows it to sum over all components of a vector. That is, the change in a vector component transported anti-clockwise around a small loop depends on a sum over the areas of all coordinated subsurfaces and over all vector components. The sum over all of the coordinated areas corresponds to the possible arbitrary orientation of an arbitrary physical surface with respect to the coordinate axes. The sum over all of the components of the vector is due to the parallel transport part of the true derivative. Parallel transport gives the change in a vector component to depend on all of the others, because the total vector is constant under the transport.

We have derived the change around a small loop for the covariant component of a vector. However, since we must have $\delta(A_b A^b) = 0$ around the circuit (it is a number), it follows that

$$A_b \delta A^b + A^b R^d_{bac} A_d dq^a dq^c = 0, \tag{6.155}$$

whence for arbitrary A_b

$$\delta A^b = -R^b_{dac} A^d dq^a dq^c. \tag{6.156}$$

The Riemann tensor has many symmetries that are most easily proven by writing R_{dbac} in local Galilean coordinates (e.g. [3]). The symmetries that concern us chiefly are: (i) that it is invariant under two cyclic permutations of the indices; and (ii) that it is antisymmetric under an interchange of indices in each of the first and second pair.

Two additional quantities that are derived from the Riemann curvature tensor are the Ricci tensor (convention as in [3])

$$R_{bc} \equiv g^{da} R_{dbac} \tag{6.157}$$

and the scalar curvature

$$R \equiv g^{ab} R_{ab}. \tag{6.158}$$

The scalar curvature is a gross measure of the curvature of a space. In two dimensions it reduces properly to the Gaussian curvature $R = 2/(r_1 r_2)$ where r_1 and r_2 are the principal radii of curvature at a point on the surface [3] with the appropriate signature.

The Ricci tensor is a coarser measure of the curvature after summation with the metric. It is clearly a symmetric tensor because the interchange of 'b' and 'c' is equivalent to two cyclic permutations in the Riemann tensor, and 'a' and 'd' may be interchanged in the metric coefficient. Therefore there are only ten independent components rather than the $n^2(n^2 - 1)/12$ components of the Riemann tensor (after symmetries) in an n dimensional space (e.g. [35]). Much like T_{ab} for stresses, the Ricci tensor is sensitive to the curvature in a hyperplane that is formed from a vector lying in a surface and a vector normal to the surface.

Problems

6.15 Show directly by expanding $\nabla_a g_{bc}$ that it is identically zero. The Christoffel symbols of the second kind must be known in terms of the metric.

6.16 Show that for a two-dimensional space there is only one (i.e. $n = 2$ in the previous formula for the number, but this does not explain which one) independent component of the Riemann curvature tensor, which may be taken as R_{1212}. List the components that are non-zero. How are they related?

6.17 Using the symmetries of the Riemann tensor, and the results of the previous Problem, show that in two dimensions the curvature scalar is $R = 2(R_{1212})/(g_{11}g_{22} - g_{12}^2)$. To show that this is equal to $2/(r_1 r_2)$ requires an expansion of local geodesic coordinates to at least second order in the coordinates.

If we are looking for a tensor measure of the curvature of space-time to determine the six non-trivial metric coefficients associated with gravity in space-time, we are evidently closer with the average represented by the Ricci tensor than with the full Riemann tensor. The formula quoted above for the independent components gives 20 independent components of the Riemann curvature in four-dimensional space-time, 6 in three-space, 2 in two-space and none for a curve (one-space; i.e. a line is bent in a two-surface).

However, there is still an apparent excess of four independent components in the Ricci tensor. These are removed due to the existence of certain differential identities called the Bianchi identities (e.g. [3]). These identities are properties of the full Riemann tensor, but they may be contracted with the metric tensor to give a constraint on the Ricci tensor as

$$\nabla^b R_{bc} = \frac{1}{2}\frac{\partial R}{\partial q^c}. \tag{6.159}$$

These four conditions reduce the number of independent components of the Ricci tensor to six.

We are now poised to acknowledge the stroke of genius due to Einstein. The Ricci tensor has only six independent components. It is a measure of the curvature of space-time and it depends only on the metric and its first and second derivatives. The curvature of space-time shares with gravity the property of distinguishing between the local tangent space (flat in geometry and inertial in gravitational metrics) and global properties (curvature in geometry and curved geodesics in gravity). In a vacuum, where the only sources of the gravitational field are on the boundaries, Einstein makes the hypothesis that such gravitational metrics are a solution of

$$R_{bc} = 0. \tag{6.160}$$

This implies that $R = 0$ in a vacuum.

These are the field equations of the gravitational field in a matter-free region. They are the analogues of the Maxwell equations in a vacuum. From these six independent equations the metric coefficients of a non-flat space-time should follow. However, the Bianchi constraints are not contained in this statement. For this reason another equivalent form may be assumed, namely

$$R_{bc} - \frac{R}{2}g_{bc} = 0 \equiv G_{bc}. \tag{6.161}$$

By contracting this equation with g^{bc} one obtains again that $R = 0$, and by applying ∇^b to the equation one sees that the Bianchi identities hold. The tensor on the left is commonly called the Einstein tensor G_{bc}, and because of the Bianchi identities $\nabla_b G_{bc} = 0$.

Such a vacuum theory is of extreme beauty and purity. The strong field solutions of Schwarzschild and Kerr that we discussed earlier follow from these equations. There are no parameters in such a theory of gravity beyond the mass, spin, charge or magnetic flux that may be placed at the centre of the system or on a boundary.

However, the world is filled with matter and all of it gravitates by experiment. Hence the final step is to couple a tensor description of the matter to the tensor description of the geometry. Unfortunately there are very few descriptions of matter that are as elegant and precise as is the differential geometry. A fairly obvious description that can be coupled to the Einstein tensor is the energy-momentum tensor of matter. This is because this quantity is designed to conserve energy and momentum of the matter by the Equation (6.140). This is automatically preserved because of the Bianchi identities by a coupling of the form

$$G_{bc} = \kappa T_{bc}, \tag{6.162}$$

where κ is a numerical 'coupling constant'. These are essentially the Einstein (sometimes 'the gravitational') field equations. The coupling constant in cgs units is established by requiring low-velocity and weak-field correspondence with Newtonian gravity. This fixes it to be ($x^0 = ct$)

$$\kappa = \frac{8\pi G}{c^4} \approx 2.1 \times 10^{-48} cm/erg. \tag{6.163}$$

The coupling is exceedingly weak, from which comes the difficulty of detecting disturbances in space-time produced by evolving matter. Such disturbances include waves in space-time and the dragging of inertial frames. Astronomical masses in motion are required.

The most precise form of classical 'matter' that we know of is the energy-momentum tensor of the electromagnetic field for which [36]

$$T_{bc} = -\frac{F^a_b F_{ca}}{4\pi} + \frac{g_{bc} F^{de} F_{de}}{16\pi}. \tag{6.164}$$

David Hilbert actually derived Equation (6.162) from a variational principle with this matter source, slightly before Einstein. However, he did not make the generalization to all matter that Einstein foresaw from the beginning.

A theory with this matter source automatically contains Maxwell's equations in curved space-time. Such solutions exist, but have not yet found major application for various reasons. One principal reason is the enormous electromagnetic fields that would be required to be comparable to the rest-mass energy of matter. A magnetic field of 10^{11} gauss is required to be roughly equal to the rest-mass energy of one gram in one cubic centimetre. We remark that even this theory is not a 'unified theory' of gravity and electromagnetism, since Clerk-Maxwell's equations remain in origin quite separate from space-time. Einstein tried to incorporate Maxwell's equations into the structure of space-time but this failed [2].

The algebraic gymnastics required to unfold these equations in applications are formidable if attempted manually. This has led to the development of computer codes that can perform this algebra with more or less transparent instructions. One remarkable example runs under the mathematics system called MAPLE. This module is called GRTensor, has been very well tested and is freely available [37].

Most practical applications of Einstein's equations use the ideal fluid form of matter that we discussed in the previous section. Cosmological applications use a mixture of 'dust' and relativistic matter such as radiation. The dramatic difference from Newtonian cosmology is that there are solutions expanding in coordinate time (the Friedmann solutions [27]). This means that not only matter but space-time itself is exploding away from an origin event (commonly called the 'big bang'). Another physically motivated type of matter is that of a mixture of non-interacting particles with a distribution of velocity. This is the realm of relativistic kinetic theory.

It is important to have *a priori* physical restrictions on the energy-momentum tensor. Otherwise any metric at all that produces G_{bc} would generate a corresponding matter tensor. This would render the theory completely without content except for vacuum solutions. Fortunately this is not the case. However, ever more realistic descriptions of matter are necessary and should be sought.

So here at last we reach the conclusion of our summary of relativistic theory. The applications and implications of this theory are legion, and many more remain to be discovered. It has so far passed every experimental test, but remains aloof from the quantum world except in so far as expectation values of the matter tensor may be used. That approach is similar in spirit to the use of the electromagnetic tensor, and does not represent a quantum theory of gravity.

The picture of dynamic space-time that we have been led to in the last few sections is philosophically exceedingly radical. Were it not for the weakness of the coupling constant, our clocks, our metabolic rates, our sizes and the distances between us would all be in constant flux.

To a certain extent the theory is also inconsistent when used for local matter. It produces Minkowski space-time as a kind of aether at a large distance from matter so that it is not Machian. What it should asymptote to is ultimately a question of cosmology, together with a smooth embedding of local solutions into the Universal solution. That the averaging of all these local solutions gives the global Friedmann solution has not been decided.

Moreover, we have postulated space-time as a dynamic object, constantly flowing and rippling as a continuous manifold of events. It is coupled to a similar description of matter. If matter is ultimately atomic in nature, then space-time between the atoms might be expected to show a certain 'granular structure'. However, an individual atom is not coupled significantly to space-time. Hence space-time locally becomes a 'stage' on which physics is played out. This may not be the case if particles are in fact the quanta of fields whose ground state might contribute structure to space-time.

In fact, in connection with this last remark, since we have $\nabla_a g_{bc} = 0$, we may also add the metric to the Einstein tensor multiplied by some coupling constant without ruining the conservation laws. Then one has the field equations as (the negative sign is by convention; the term imparts a negative curvature when $\Lambda > 0$)

$$G_{bc} - \Lambda g_{bc} = \kappa T_{bc}, \tag{6.165}$$

where the constant Λ is usually called the cosmological constant. It is only significant on a cosmological scale since an astronomical upper limit is $|\Lambda| <\approx 10^{-55}$ cm^{-2}. It is now thought to be positive.

Taken to the other side of the equation, this cosmological term may be interpreted as a quantum vacuum, with $\rho_V = c^4\Lambda/(8\pi G)$ ergs cm^{-3} and $p_V = -\rho$ [38]. This gives the energy/momentum tensor of the vacuum (treated as an ideal fluid) to be $T_V^{ab} = -p_V g^{ab}$.

The startling discovery brought to us by supernovae distance studies [39,40]) and corroborated by WMAP, is that Λ is positive and accounts for more than 70% of the energy density of the Universe! The enduring problem is that a 'natural' deduction of the value of Λ from fundamental physics is based on the expected quantum vacuum at the origin of the Universe. The estimate is more than 114 orders of magnitude larger than the measured value. This has been described as the biggest error in history.

We must leave these fascinating topics to another place. However, the notion of dynamic space-time is not much less radical than absolute (fixed) space-time. For this reason I have tried until the last few sections to present the theory as a theory of measurement.

We should remember that it is the fact that gravity acts like an inertial force in accelerating all objects equally that has led us here. This allows us to remove it by an appropriate choice of observer in local 'free fall'. This fact plus the notion of treating all possible observers and coordinates equally (principle of relativity), leads to Riemannian metrics starting from the Minkowski local patch. The Minkowski structure is, in turn, fundamentally the Lorentz invariance of Maxwell's equations.

The curved Riemannian geometry arises even in comparing measurements between a field of freely falling (inertial) observers. Were there a way of deriving this velocity field from the distribution of matter, it might be possible to avoid dynamic space-time. However, there is not at present such a way. Adopting the metric of space-time as fundamental and identifying gravity with geometric curvature through the field equations has triumphed over all challenges, and exudes an awesome inner beauty.

La Pièce est jouée.

References

1. Robertson, H.P. and Noonan, T.W. (1968) *Relativity and Cosmology*, W.B. Saunders & Co., Philadelphia.
2. Ohanian, H.C. (2008) *Einstein's Mistakes*, W.W. Norton & Co., New York.
3. Landau, L.D. and Lifshitz, E.M. (1975) *The Classical Theory of Fields*, Pergamon Press, Oxford.
4. LoPresto, J., Schrader, C. and Pierce, A.K. (1991) *ApJ*, **376**, 757.
5. Pound, R.V. and Rebka, G.A. (1960) *Physical Review Letters*, **4**, 337.
6. Shapiro, I. (1989) *General Relativity and Gravitation*, Cambridge University Press, Cambridge.
7. Birkhoff, G.D. (1923) *Relativity and Modern Physics*, Harvard University Press, Cambridge, MA.

8. Schwarzschild, K. (1916) *Sitzbereichung. Deut. Akad. Wiss. Berlin, K1, Math. Phys. Tech.*, **189**.
9. Kerr, R.P. (1963) *Phys. Rev. Letters*, **11**, 237.
10. LeMaître, G.E. (1925) *Journal Math. Phys.*, **4**, 188.
11. Kantowski, R. and Sachs, R.K. (1966) *Journal Math. Phys.*, **7**, 443.
12. Abramowicz, M.A. and Bajtlik, S. (2009) arXiv:0905.2428v1, May 14.
13. Irwin, J.A. (2007) *Astrophysics: Decoding the Cosmos*, John Wiley & Sons, Ltd., Chichester.
14. Taylor, J.H. (1994) *Reviews of Modern Physics*, **66**, 711.
15. Eddington, A.S. (1920) *Space, Time and Gravitation*, Cambridge University Press, Cambridge.
16. Fomalont, E.B. and Sramek, R.A. (1975) *Astrophys. J.*, **199**, 749.
17. Fomalont, E.B. and Sramek, R.A. (1976) *Phys. Rev. Letters*, **36**, 1475.
18. Schneider, P., Ehlers, J. and Falco, E.E. (1992) *Gravitational Lenses*, SpringerVerlag, New York.
19. Stephani, H. (1982) *General Relativity*, Cambridge University Press, Cambridge.
20. Carter, B. (1971) *Phys. Rev. Letters*, **26**, 331.
21. Rindler, W. (2006) *Relativity*, Oxford University Press, New York: problem 14.7.
22. Boyer, R.H. and Lindquist, R.D. (1967) *Journal Math. Phys.*, **8**, 265.
23. Carter, B. (1973) In *Black Holes*, Les Houches SummerSchool, 1972 (edited by B.S. DeWitt and C. DeWitt), Gordon and Breach, New York.
24. Penrose, R. (1969) *Nuov. Cim. Ser. I*, **1**, 252.
25. Blandford, R.D. and Znajek, R.L. (1977) *Mon. Not. R. Astr. Soc.*, **179**, 433.
26. Dexter, J. and Algol, E. (2009) *Astrophys. Journal*, **696**, 1616.
27. Hartle, J.B. (2003) *Gravity*, Addison Wesley (Pearson), San Francisco.
28. Bardeen, J.M. (1973) In *Black Holes*, Les Houches SummerSchool, 1972 (edited by B.S. DeWitt and C. DeWitt), Gordon and Breach, New York.
29. Carter, B. (1968) *Phys. Rev.*, **174**, 1559.
30. Hawking, S.W. and Ellis, G.F.R. (1973) *The Large Scale Structure of SpaceTime*, Cambridge University Press, Cambridge.
31. Oppenheimer, J.R. and Snyder, H. (1939) *Phys. Rev.*, **56**, 455.
32. Foglizzo, T. and Henriksen, R.N. (1993) *Phys. Rev. D.*, **48**, 4645.
33. Fock, V. (1964) *The Theory of Space, Time and Gravitation*, Pergamon Press, MacMillan, New York, pp. 104–105.
34. Goldreich, P. and Julian, W.H. (1970) *Astrophys. J.*, **160**, 971.
35. Weinberg, S. (1972) *Gravitation and Cosmology*, John Wiley & Sons Ltd., New York.
36. Jackson, J.D. (1999) *Classical Electrodynamics*, John Wiley and Sons Ltd., New York.
37. Lake, K.W. and collaborators. *GRTensor is distinct from packages distributed with MAPLE and must be obtained independently. GRTensorII is distributed freely at* http://grtensor.org.
38. Zel'dovich, Ya.B. and Novikov, I.D. (1996) *Stars and Relativity*, Dover, New York.
39. Riess, A.G., *et al.* (1998) *Astronomical J.*, **116**, 1009.
40. Perlmutter, S., *et al.* (1999) *Astrophys. J.*, **517**, 565.

Index

Note: page numbers in *italics* refer to figures; page numbers followed by the letter 'n' refer to footnotes.

Practical Relativity: From First Principles to the Theory of Gravity Richard N. Henriksen